ZEITSCHRIFT FÜR GEOMORPHOLOGIE

Annals of Geomorphology – Annales de Géomorphologie **Neue Folge**

A journal recognized by the International Association of Geomorphologists (IAG)

Supplementband 115

Proceedings of the Fourth
International Conference on Geomorphology
Bologna 1997
Volume II

Magnitude and frequency in Geomorphology

Edited by MICHAEL CROZIER and ROLAND MÄUSBACHER

With 87 figures and 20 tables

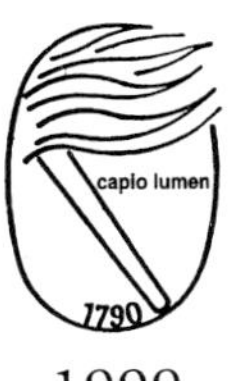

1999

GEBRÜDER BORNTRAEGER · BERLIN · STUTTGART

Die Deutsche Bibliothek – CIP-Einheitsaufnahme

International Conference on Geomorphology <4, 1997 Bologna>:
Proceedings of the Fourth International Conference on
Geomorphology : Bologna 1997. – Berlin ; Stuttgart : Borntraeger
(Annals of geomorphology : Supplementband : ...)
Vol. 2. Magnitude and frequency in Geomorphology. – 1999

Magnitude and frequency in Geomorphology : with 20 tables / ed.
by Michael Crozier and Roland Mäusbacher. – Berlin ; Stuttgart :
Borntraeger, 1999
(Proceedings of the Fourth International Conference on
Geomorphology : Vol. 2) (Annals of geomorphology :
Supplementband ; 115)
ISBN 3-443-21115-1

Zeitschrift für Geomorphologie / Supplementband
Zeitschrift für Geomorphologie = Annals of geomorphology.
Supplementband. – Berlin ; Stuttgart : Borntraeger
Früher Schriftenreihe
Reihe Supplementband zu: Zeitschrift für Geomorphologie
ISSN 0044-2798

115. International Conference on Geomorphology <4, 1997, Bologna>:
Proceedings of the Fourth International Conference on
Geomorphology
Vol. 2, Magnitude and frequency in Geomorphology. - 1999

ISBN 3-443-21115-1 / ISSN 0044-2798

♾ Printed on permanent paper confirming to ISO 9706-1994

Contents

Foreword

This volume was initiated by the Working Group on Magnitude and Frequency of Geomorphic Processes set up by the International Association of Geomorphologists. The contributions to this volume have been invited from the specialists within the geomorphic community and from contributors to the Symposium on Magnitude and Frequency in Geomorphology held as a part of the Fourth International Conference on Geomorphology, Bologna, 28 August – 3 September, 1997.

The concepts of magnitude and frequency acknowledge an episodicity and variability within the operation of geomorphic processes. As such, they have occupied an important focus in both the development of theory of landform evolution and in the more applied aspects of process science. In the early 1960's work on magnitude and frequency was partly directed at redressing the views of gradualism in landform development which, in turn, could be traced as a reaction to the more catastrophic arguments that were still evident at the end of the 19th Century. In the last forty years, an increasing data base has encouraged statistical evaluation of process behaviour which has led to the emergence of the concepts such as 'geomorphic work' and 'dominant landforming events'. As this work has been extended to different spatial and temporal scales, inevitably qualifications had to be made to earlier concepts. The overall approach, however, has stimulated thinking and research. The acquired knowledge on processes behaviour has not only promoted a better appreciation of how geomorphic systems operate but has also enhanced the stature of geomorphology as a relevant science. Understanding episodic behaviour process-response systems, and their implicit thresholds is essential to successful engineering design, resource management, hazard assessment and prediction of environmental change.

The papers in this volume represent a wide range of research subjects on aspects of magnitude and frequency. The concept has been used as a framework not only to characterise the relationships between the forcing processes and system response but also to analyse the effect of temporal and spatial scale on the frequency and magnitude of different geomorphic processes. While the papers in this volume evaluate the importance of magnitude and frequency concepts they also demonstrate how they have contributed to geomorphology. We believe that this volume should make and important contribution to scholarhip and research in geomorphology.

MICHAEL CROZIER
ROLAND MÄUSBACHER

Z. Geomorph. N.F.	Suppl.-Bd. 115	1–18	Berlin · Stuttgart	Juli 1999

The magnitude-frequency concept in fluvial geomorphology: a component of a degenerating research programme?

Keith Richards, Cambridge, UK

with 6 figures

Summary. This paper presents a critique of that magnitude-frequency concept which defines an "effective", or "dominant" event in terms of long-term sediment transport, and then associates this with channel morphology. It argues that the concept was associated with a research programme which emphasised engineering time scales and equilibrium forms, but which is now degenerating in geomorphology. The paper demonstrates that the concept has been protected by a series of ad hoc auxiliary statements and hypotheses, of which a critical one is that the effectiveness of an event is dependent on the initial morphological boundary conditions. This implies that the effectiveness of an event depends on morphology, as much as *vice versa.* A more appropriate geomorphological research programme is suggested, which emphasises the continuing and spatially-distributed feedback between form and process. This is applicable to a wide range of geomorphological problems, in both unconsolidated materials and hard rock, and is considered to lend itself to rigorous examination of observational evidence and critical, rational and scientific evaluation of alternative interpretations.

Zusammenfassung. Diese Arbeit präsentiert eine Kritik an dem Magnituden-Häufigkeits-Konzept, das ein 'effektives' oder 'dominantes' Ereignis hinsichtlich Langzeitsedimenttransport definiert und es dann mit der Gerinnemorphologie assoziiert. Es wird argumentiert, daß das Konzept mit einem Forschungsprogramm verbunden war, welches Ingenieurzeitskalen und Gleichgewichtsformen betonte, das aber heute in der Geomorphologie an Bedeutung verliert. Die Arbeit zeigt, daß dieses Konzept von einer Reihe ad hoc Hilfsaussagen und -hypothesen gestützt wird, wovon eine entscheidende besagt, daß die Effektivität eines Ereignisses von den anfänglichen morphologischen Randbedingungen abhängt. Damit wird impliziert, daß die Effektivität eines Ereignisses von der Morphologie abhängt und umgekehrt. Es wird ein besser angepaßtes geomorphologisches Forschungsprogramm vorgeschlagen, in dem ein kontinuierliches und räumlich verteiltes feedback zwischen den Formen und Prozessen betont wird. Dies ist auf eine große Anzahl von geomorphologischen Problemen anwendbar, sowohl im Locker- wie auch im Festgestein. Darüber hinaus wird davon ausgegangen, daß es für die gründliche Untersuchung von Beobachtungsmaterial und eine kritische, vernünftige und wissenschaftliche Bewertung alternativer Interpretationsversuche geeignet ist.

1 *Origins of the magnitude-frequency interpretation of landforms*

Wolman & Miller's (1960) classic paper on the magnitude and frequency of effective events aimed to examine "..the relative importance of extremes or catastrophic events and more ordinary events with regard to their geomorphic effectiveness expressed in terms of material moved and modification of surface form" (p. 54). At the time it was published, certain pre-conditions had already established a context within which this could provide an appropriate framework for the interpretation of landforms. These included, in particular, (a) the introduction of approaches in fluvial geomorphology that had close parallels with engineering analyses, and (b)

0044-2798/99/0115-0001 $ 4.50

the ascendancy of an equilibrium-based paradigm. The former was exemplified by the development of the hydraulic geometry concept by LEOPOLD & MADDOCK (1953), in a paper which concluded with an emphasis on the similarity between this concept and the regime behaviour of self-formed, sediment-transporting canals. The latter had already been embodied in much empirical research in fluvial geomorphology, particularly that published in the "Professional Papers" of the United States Geological Survey in the 1950s, and was formalised by HACK (1960), in the same year as WOLMAN & MILLER outlined the magnitude and frequency framework. Indeed, on p.69, their discussion of the effective wind and wave conditions required to create beach morphology was based on equilibrium relations for beach slope, wave steepness and wind speed, and the required wind speed was back-calculated from the equilibrium slope and then shown to be a commonly-occurring event.

The consequence of this intellectual framework, which might reasonably be interpreted as a 'research programme' in the terms defined by LAKATOS (1970), was the interpretation of alluvial river morphology in regime terms, with aggregate morphometric properties of channels (width, mean depth) being related to a 'dominant' or 'effective' discharge. This was an event which was considered to be formative on the grounds that it was representative of the overall range of flows imposed by catchment hydrology (ACKERS & CHARLTON 1970), and was also associated with the transportation of a significant proportion of the sediment yield of the river. DURY (1973) was able to encapsulate this model most successfully, by equating the bankfull discharge with the most probable annual flood, and associating the usual quantitative measures of channel geometry (the downstream hydraulic geometry) with the bankfull discharge. ANDREWS (1980), amongst others, has subsequently been able to demonstrate that bankfull discharge equates with the dominant event in terms of the transportation of total load. Consistent with this empirical evidence, and WOLMAN & MILLER's original intention quoted above, the form of "magnitude-frequency concept" subjected to critical evaluation in this paper is that which involves two kinds of statements; one which identifies an event which is dominant in terms of sediment transport, and the other which associates this event with attributes of the channel morphology.

Two critical implications of this model may be highlighted. The first, internal to the model, is that it is purely statistical in several senses. The dominant discharge is not a 'real' event, but is representative of a probability distribution; a river may never experience a discharge that is particularly close to this measure of central tendency. Furthermore, because the concern is with events at the tail of the probability distribution of all flows, the shape of the distribution in this region is poorly defined. Thus it is likely to be difficult to choose between alternative extreme value distributions, and possible that the distribution will be far from the smooth form illustrated schematically in Fig. 1a. In addition, the dominant discharge is only statistically related to the morphology with which it is associated. WOLMAN & MILLER (1960, p. 66), for example, discuss the statistical association amongst discharges of a given frequency, channel width and meander wavelength, with recourse to phrases such as "..the constancy of the relation.." and "..the very good correlation". CARLSTON's (1965) analysis of the relationship between meander wavelength and several different flow indexes is perhaps the most explicit appeal to strength of correlation as an indicator of the effective discharge. The second, 'external' criticism, is that the magnitude-frequency concept is itself devoid of geomorphological content until it is parameterised for a specific geomorphological purpose. It is a concept that may be applied equally to the distribution of 10Hz turbulent velocity fluctuations in a study of micro-scale gravel bedforms, to the relationship between bankfull discharge and channel

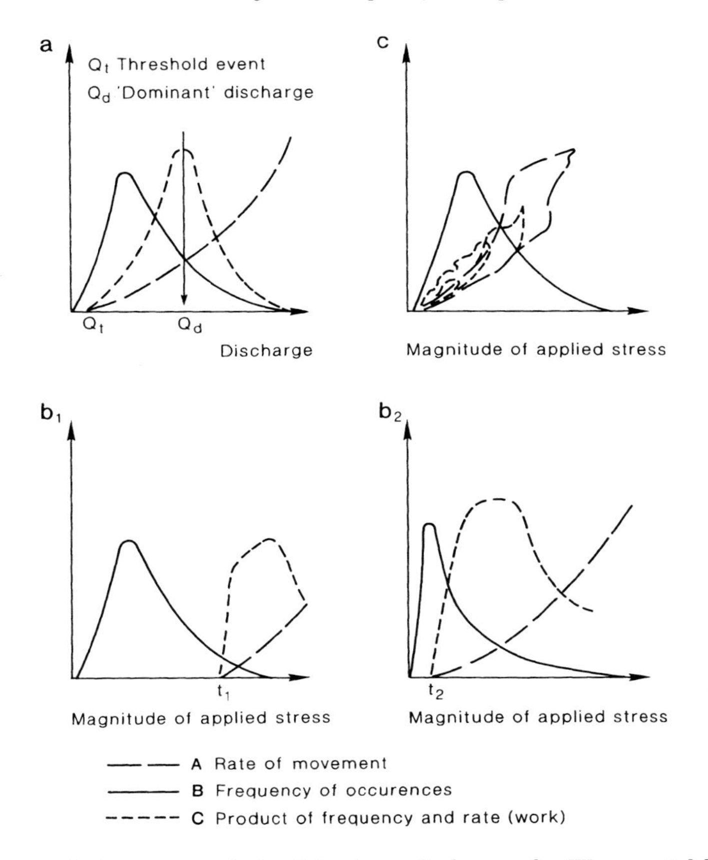

Fig. 1.(a) The magnitude-frequency analysis of dominant discharge, after WOLMAN & MILLER (1960); (b) The effect of variations in the threshold flow and in the distribution of events on the identification of the dominant event (after BAKER 1977); (c) The additional complexity introduced by differential transport hysteresis in successive events.

size and shape, or to the erosional intensity of successive glaciations in relation to valley development. However, it can just as equally be used in economics to measure the dominant income group in terms of contribution to total tax revenue. Significantly, the period in which this research programme and its content became established was one in which such highly-generalised conceptual models were often proposed, and were eventually found empirically wanting. The random model of network topology is another example of this, which although of considerable value in some respects, ultimately proved to be of limited value as a geomorphological tool because of its universal applicability to branching systems.

2 *Developments: auxiliary hypotheses required by the magnitude-frequency concept*

Since 1960 the core of the magnitude-frequency concept has had to be supported by a number of subordinate hypotheses and auxiliary assumptions, some of which were clearly seen as necessay by Wolman & Miller themselves. The most obvious of these relate to the construction of the simple diagram that defines the concept (Fig. 1a). In deriving this diagram, Wolman & Miller (p. 55) reduced a functional relationship of the form $q_s=k(\tau-\tau_c)^n$ to the form $q_s=x^n$, so that the generalised basis of the diagram only holds for an applied stress (x) which is greater than the threshold shear stress (τ_c) required to mobilise sediments. However, the assumed log-normal event distribution may no longer apply given this transformation. Together, these issues suggest that it is necessary to consider more explicitly both the form of the functional relationship relating sediment transport to discharge when there is a threshold of motion, and the shape of the probability distribution of discharge events, both transporting and non-transporting. Baker (1977) has illustrated these effects by defining the consequences, for the return period of the dominant discharge, of variation in the level of the threshold discharge in response to sedimentological conditions, and of variation in the shape of the discharge distribution. This analysis shows that the effective discharge may have a wide range of return periods (Fig. 1b), and that in some circumstances, no clear dominant event exists. Where discharge events are generated by more than one set of hydrological processes, and their probability distribution is a compound one (Waylen 1985) reflecting, for example, both rainfall and snowmelt flood peaks, the statistical description of effectiveness is further confused. The significance of the peak flow magnitude as an index of the transportional effectiveness of an individual event is also influenced by the roles of event duration and sequence. The duration of flow above a given level is not directly related to the peak magnitude of a flood event, but rather combines with it in order to determine the *potential* sediment transport effectiveness of the event; a peaky flood event with a peak discharge above the threshold of motion has a greater transport capacity than a more attentuated hydrograph with a similar total flow.

The significance of these auxiliary assumptions about the threshold of sediment motion and the distribution of events for geomorphological interpretation is nicely illustrated by the debate about valley meandering. One consequence of the equilibrium paradigm, and its conclusion that form is adjusted to an average event, is the assumption that this association can be inverted to yield palaeo-event magnitudes from the morphometry of palaeo-forms. Thus, where meander bends are observed at the valley scale, a prior bankfull discharge supposedly formative of these features may be estimated (Dury 1965). As Tinkler (1971) showed, however, this ignores the fact that the threshold of erosion is a much larger and less frequent event for the bedrock walls of a valley than for the banks of an alluvial channel, so the return period of the effective event in this case is much greater. Tinkler's argument concludes that the valley geometry is eventually fossilized by the development of a valley fill that disperses the power of more extreme events across the flood plain surface. Significantly, this implies that morphological development alters the effective event magnitude and frequency (see below).

The magnitude-frequency interpretation also fails to accommodate the *sequence* with which events occur, which is a critical control of the effectiveness of any individual event (Beven 1981). This effect may be represented by an auto-regressive model, in which a channel property such as width at time t (w_t) is related to width at a previous time (w_{t-1}). Yu & Wolman (1985) in fact define a "memory" parameter λ, for non-zero values of which the channel will always tend to be wider than expected for its mean flood discharge because its "memory" maintains its width, after any large event which initiates widening. In their model, recovery occurs after a widening event, and a simulated sequence of flood-dependent widenings and subsequent exponential "recoveries" is as shown in Fig. 2. The mean of this time series of widths is a purely statistical index that bears little relationship to any formative discharge event. This is suggested by the following "thought experiment", which illustrates the conceptual weakness of a model which associates channel dimensions with a representative discharge. Consider a template-cut channel in a sand-bed flume. A low baseflow is passed down the flume, and randomly-chosen hydrographs are superimposed on this at random times. The sequence of these events produces a time series of measured widths similar to that in Fig. 2a, and this series evolves through time as the events occur. The width at any time, and the mean width over a preceding period, both depend on the history of events; the mean width is therefore essentially an *emergent* property of the developing time series. The physically-important variables are the maximum widths attained after extreme events, and the rate constant for the "recovery" process. No physical process actually relates the mean width for a particular time period to a statistical attribute of the discharge series over that same period (although there may be a statistical relationship between them). Even the concept of a "recovery period" is open to question, as it implies recovery to an expected width, such as the equilibrium width. This is, of course, an anthropo-morphism. The river has no knowledge of any condition to which it should restore itself, it merely responds to the sequence of events it experiences, and the flow and sediment passing through the pre-existing morphology cause erosion and deposition patterns that change its size and shape. This behaviour, as YU & WOLMAN (1985) note "...leads to a serious question about river equilibrium, a concept deeply rooted in the work of many geomorphologists. Given variable discharges, many of which change the channel form when they occur, it is impossible to associate the channel geometry with a single discharge of a certain frequency of occurrence, because the role of discharge depends heavily on the current channel form which is an end result of antecedent flows" (p. 508).

The physical role of the sequence of events in fact often reflects an inter-dependence of discharge and sediment supply events, particularly associated with their temporal non-coincidence. For example, HARVEY et al. (1979) have shown how the supply of sediment from valley-side gulley systems to basal alluvial fans is associated with relatively frequent rain storms (10–20 per annum), but that the river trimming the toes of the fans incorporates that sediment into its bed material and adjusts its morphology only during hydrographs caused by less frequent rainstorms. NEWSON (1980) also indicates that two comparable storms may have quite different

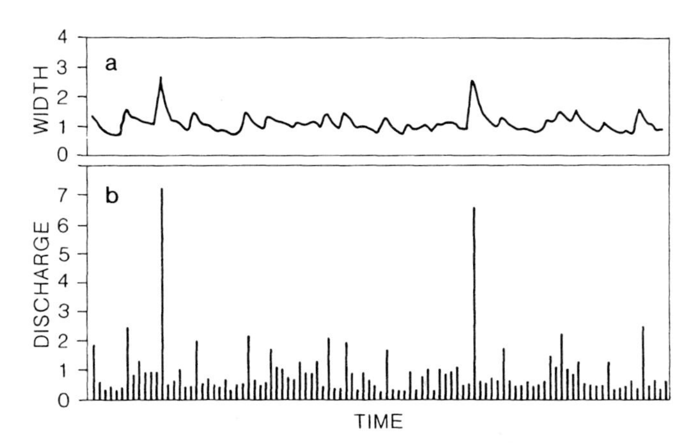

Fig. 2. (a) The simulated series of channel widths created by the simulated sequence of discharge events shown in (b), after YU & WOLMAN (1987).

impacts on channel process and form, because the first generates slope failures that supply sediment which the second event incorporates into the channel morphology. The *actual* effectiveness of a flow event in terms of work done in transporting sediment is thus dependent as much on sediment supply, both absolute and relative, as on transport capacity (Fig. 1c), and the morphological and sedimentological consequences of a storm are therefore dependent on the specific effects of preceding storms. LANE et al. (1996) also conclude, in a study of the dynamics of channel change in a proglacial braided stream, that phases of erosion and deposition can occur independently of changes in discharge, and that "..the dynamics of river channel change are sensitively dependent upon a changing combination of discharge and upstream sediment supply, whose relative timing is critical" (p. 12).

In Fig. 1a, the relative effectiveness of an event of given discharge (or applied stress) is measured by its sediment load, which is estimated from the rating relationship (curve a). This is subject to certain weaknesses. Firstly, as FERGUSON (1986) has pointed out, when load (L) is estimated from the product of a concentration and a discharge, with the former derived from a power-function rating relationship, there is an under-estimation because the concentration is a biased estimate of the arithmetic mean of the log-normal conditional distribution. A correction to the load (L_c) is required, of the form

$$L_c = L\, exp(2.651\, s^2) \tag{1}$$

where s^2 is the standard error of estimate of the logarithmic regression. The true load is consistently under-estimated by the rating curve method, the estimated load being a smaller percentage of the true load the larger is the scatter about the regression. A second complication is that, as sediment supply varies both between successive events and within events (between rising and falling limbs), the functional relationship between sediment load and discharge varies. Thus, rather than assuming a common power-function rating relationship, the rating curve parameters may vary between events. These two sources of error imply that Fig. 1a needs to be revised to allow for variations in curve *a* as function of event magnitude (Fig. 1c), or more precisely, that for each event it should be estimated by the continuous product of concentration and discharge, rather than from rating curves.

One obvious further limitation of the magnitude-frequency concept is that it associates a representative event with a particular process, and with a simple geometric attribute of channel morphology. However, channels are more complex phenomena than this allows; conventional hydraulic geometry variables such as width and depth are insufficient to represent the three-dimensional topographic form of a river channel, and this form in its various details is a function of a variety of processes. These different processes and their resultant forms are then associated with quite different magnitudes and frequencies of event. The dominant or effective event in relation to suspended sediment transport is a more common flow than that which is dominant in the context of bedload transport (because of the higher threshold flow in the latter case). Modification of bed morphology might well be associated with a flow of a different frequency from bank erosion, and therefore cross-section properties such as width. PICKUP & WARNER (1976) suggest that a discharge more frequent than the modal annual flood dominates bedload transport and the modification of bed topography, but more extreme events than this control the erosion of cohesive banks and the formation of the cross-section. Scroll-bar and point-bar deposition, and therefore channel pattern development, may well be controlled by even higher magnitude events. Thus various aspects of channel form are related to different processes that are most effective during different levels of event, and a simple magnitude-frequency interpretation ceases to be applicable. WOLMAN & MILLER (1960) recognised this when they stated that "..the effective 'constructional' discharge is less frequent than the effective erosional discharge" (p. 65), after discussing the example of bank erosion by sub-bankfull flows, followed eventually by point bar deposition by bankfull flows. Since the width of a meandering channel reflects both processes, it cannot be indexed to a "dominant" or "effective" event of a single return period.

These issues tend to provide support for an alternative model, in which channel morphology is a reflection of all events occurring within a certain time period, with more distant events having less expression in the morphology than more recent events of comparable magnitude (PICKUP & RIEGER 1979). In this model, every competent event has some influence on channel form, and the channel form y at time t is a weighted sum of all events up to and including time t, according to:

$$y(t) = \int h(u)\, Q(t-u)\, \mathrm{d}u \qquad (2)$$

where $h(u)$ is an impulse response measuring the effect of discharge on channel form over various periods (u). However, as noted above, the channel form is not simply $y(t)$, but $y(t) = (w,d,v,s,n,d_{max}, \ldots\ldots)_t$. Thus, even this model is an over-simplification of the true relationship between channel form and morphologically-effective events, because each attribute will have its own impulse response and limiting lag time. The model implies, however, that a regional hydraulic geometry relationship of widths to dominant discharges can only be an approximation if variations in discharge are small, and if the sensitivity of y to variations in Q is small.

Perhaps, however, the two most damaging auxiliary assumptions are those that specifically concern the role of morphology. The initial assumption of the magnitude-frequency concept is that effectiveness in terms of long-term sediment transport defines an event which is also morphologically significant. However, WOLMAN & GERSON (1978) recognised that an extreme event whose net effect on long-term material yield was minimal could nevertheless have a highly visible morphological effect for a significant period of time because the recovery following the event is so gradual. This is particularly true of some hillslope processes; slopes may retain the morphological consequences of late-glacial landsliding for 10,000 years, although the long-term transport associated with the failure event, averaged over that period, is negligible. As noted above, there is even a question of what is meant by "recovery", as this concept implies that there is an equilibrium form to which a disturbed system returns. However, the key implication of the direct morphological impact of certain events is that it invalidates the fundamental connection in the magnitude-frequency concept between effectiveness (dominance) in relation to sediment transport, and the responsibility for creating morphology.

The other sense in which morphological factors undermine the magnitude-frequency concept is that they are not simply the product of processes, but actually exert control over processes. Channel forms in fact control the role of discharge. This is even apparent in the interpretation of the origins of river floodplains, which were classically analysed within the magnitude-frequency paradigm. LEOPOLD & WOLMAN (1957), for example, explicitly used the magnitude-frequency hypothesis to relate the floodplain level to inundation by an event which was considered formative of the channel and floodplain association, and which had a return period approximately equivalent to that of the mean annual flood. By contrast, ASSELMAN & MIDDELKOOP (1998) suggest that a highly non-linear relationship exists between river discharge and the efficiency of the floodplain to trap sediment, and explicitly state that "the effective discharge for floodplain sedimentation depends on flooplain characteristics" (p. 607). More precisely, it is not the discharge that is the physically-significant variable determining the work done by an event: rather, it is the shear stress, or the stream power. Significantly, these variables actually depend on the channel morphology and sedimentology. The shear stress is dependent on the channel gradient, and on the width and depth of the channel (since a wide channel will have a shallow flow depth for a given discharge). The grain shear stress component available to drive the transport process depends of the sedimentological structure of the bed. Thus, the physically-significant index of channel change for a given discharge itself depends on the shape of precisely that phenomenon whose shape is supposedly explained by the magnitude-frequency concept. As LANE et al. (1996) argue, "..what probably controls the nature of within-channel dynamics is a combination of externally-imposed discharge and sediment supply conditions, in the context of existing, internal, channel morphology. The latter is important because it controls the within-reach spatial patterns of channel competence, capacity and sediment transport" (p. 13).

This lengthy series of qualifications, additions and modifications to the original magnitude-frequency concept has all the characteristics of a degenerating research programme. As defined by LAKATOS (1970), a research programme has a core, a belt of auxiliary hypotheses, and a number of techniques for solving problems. It is said to be "progressive" if the auxiliary hypotheses can be modified to take account of new data, in a way which leads to the prediction of novel facts, but "degenerating" if this does not occur. A degenerating research programme involves its advocates adding untestable auxiliary assumptions, or modifying assumptions in ways that merely account for already known facts – this represents the *ad hoc* addition of auxiliary statements in order to save the core of the programme. A classical example arose when Kepler dropped the assumption that planets had circular orbits in favour of elliptical (COUVALIS 1997). This enabled him to make new predictions, implying a theoretically progressive addition, and as these predictions were later confirmed, they were also empirically progressive. In contrast, Ptolemaic rivals merely added untestable hypotheses to their core to account for discrepancies between their theory and observations. It is, however, worthy of note that a research programme may appear degenerating for a period, but may return later to a progressive status when its auxiliary hypotheses are capable of being tested; the return of Halley's comet was predicted long before the prediction could be tested. It is important in the context of the present discussion to note that it is not the magnitude-frequency concept that is the degenerating research programme. The degenerating programme is the equilibrium framework with which the concept is associated, and for which it provides the basis for an argument about form-process adjustment, through the combination of statements about a dominant event and its association with overall morphometry.

3 *Initial conditions: morphology as a necessary condition for event effectiveness*

The key limitation of the magnitude-frequency concept to emerge in the above discussion is that it focuses on processes as causes of forms, to the exclusion of a consideration of landforms as controls of process. This issue is explored in further detail in this section.

The development of ideas about intrinsic thresholds in landform development established the notion that the effectiveness of an external event, such as a rainstorm or flood, depends on the state of the system, particularly the proximity to a threshold. Whether a relatively extreme flood is capable of converting a river from a meandering to a braided planform depends on the proximity of the prevailing condition to the threshold channel slope-bankfull discharge threshold (LEOPOLD & WOLMAN 1957). Similarly, the valley fill in a semi-arid valley bottom may gradually steepen over time as sediments accumulate, until it becomes susceptible to incision by the runoff generated by a heavy rainstorm (PATTON & SCHUMM 1982). At an early stage in this morphological evolution, even an extreme event of very low frequency and high magnitude may be ineffective; later, however, steepening of the valley surface results in the incision being triggered by a relatively high frequency, low magnitude event (Fig. 3a). This is also apparent in a time series of Factor of Safety values for a hillslope. Over time, weathering or basal undercutting may cause a slow decline in the Factor of Safety. Increased pore water pressures in a heavy rainstorm may cause failure eventually, but the effect of a rainfall depends on the prior preparation of the slope by the secular process (Fig. 3b). Clearly the implication is that, in certain respects, morphology is a necessary condition determining the effectiveness of events, together with aspects of the nature of the materials which underlie the surface form.

This is more formally reflected in the analysis of chaotic behaviour in dynamical systems, which is particularly dependent on system sensitivity to initial conditions (PHILLIPS 1992). Small

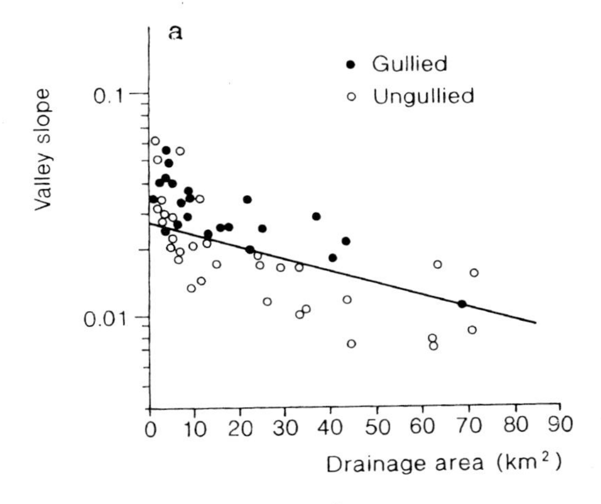

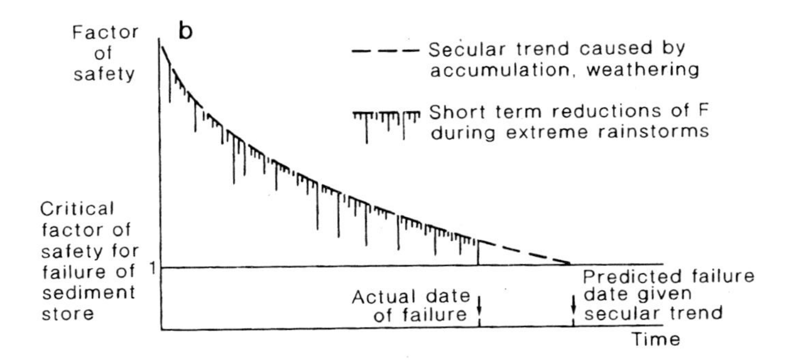

Fig. 3. (a) Gully formation as a function of slope and basin area (after PATTON & SCHUMM 1975): gullying occurs in storm events only if the slope exceeds the threshold condition; (b) Slope failure occurs during an event which is effective because of a prior decline in the Factor of Safety associated with a long-term preparatory process such as weathering.

differences in the initial morphology can result in significantly diverging, or bifurcating, behaviour in the potential paths taken by the system's evolution. This is reflected in a catastrophe theory model such as that illustrated in Fig. 4a. This hypothetical model of channel pattern is based on the second-order cusp catastrophe, which defines a surface of equilibrium patterns, represented quantitatively by the system-state variable of "total sinuosity" (sinuosity defined as the total length per unit reach length of all major channels, which allows a common scale of measurement for multi-thread and single-thread patterns). This sinuosity reflects particular combinations of two control variables measuring a "force" (stream power) and a "resistance" (bank material strength). As the control variables change over time, so the channel pattern varies, but because the surface contains a fold, the path taken by the system across the surface may depend sensitively on its starting point, and there may be a bifurcation in which similar changes in the control variables result in diverging behaviour in the system state, de-

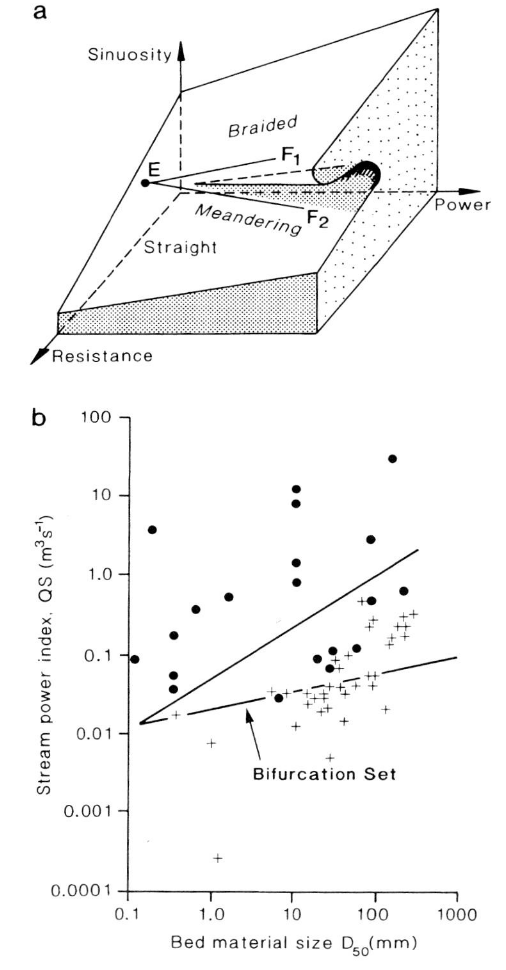

Fig. 4.(a) A qualitative model of non-linear change in channel pattern, based on the cusp catastrophe model; (b) The bifurcation region in a cusp catastrophe model in which bed material size and stream power are the system state variables, and channel pattern the response variable; a region in the bed material-power plane is occupied by both single and multi-thread rivers.

pending on whether it migrates towards the upper sheet of the folded surface (a high total sinuosity, braided state), or remains on the lower sheet (a low total sinuosity, single-thread meandering state). The data tabulated by LEOPOLD & WOLMAN (1957) indicate a possible projection of the edges of the fold onto a "power-resistance" plane in Fig. 4b, with single-thread and multi-thread channels occurring in the zone between these limits, but single-thread channels otherwise plotting in the low power-high resistance region and multi-thread channels in the high power-low resistance region. As PHILLIPS (1992) notes, "..to the extent that equilibrium exists at the system level, it is likely to exist only at restricted spatial and /or temporal scales" (p. 199). The implication of this statement, and the model suggested in Fig. 4, is that global equilibria are improbable, but that locally and temporarily they may exist.

These issues are further illustrated by FURBISH'S (1991) discussion of the evolution of meander patterns, which emphasises the way in which bends impose distinct velocity distributions on their downstream neighbours, and therefore define local and unique initial conditions for further development. The implication is that many different modes of bend development can occur even in bends of identical initial shape, depending on how they relate to adjacent bends. He concludes that "sensitivity of the meandering process to initial conditions thereby contributes to the diversity of bend forms.." with the implication that "..no single geometrical form serves as an asymptotically stable, evolutionary state.." (p. 1587). His appeal to self organization has been taken up by STØLUM (1996), who shows that the sinuosity of a meandering reach is an emergent property which, over time, reflects the interplay between the reach-scale processes of bank erosion that cause increasing sinuosity and the local-scale process of bend cut-off which reduces it. In large, unconfined and freely-meandering rivers, these processes result in a mean sinuosity which is close to 3.14 (that is,). This provides an indication that certain (emergent) morphological properties, and certain of the processes that cause their emergence (in this case, meander cut-off), are largely independent of the regime of effective flow events; and that a close link exists between sensitivity to initial conditions and the self-organizing behaviour of non-linear systems.

Generally, however, these considerations all suggest that a fluvial system may be said to display sensitivity to the initial conditions represented by their morphology and sedimentology at a particular moment in their history. These systems are therefore in a continuous state of change in response to the fluxes of water and sediment imposed upon them, and their sediment transport capacity, which depends on these initial conditions. This has been encapsulated in the diagram in Fig. 5a, initially by ASHWORTH & FERGUSON (1986) following an investigation of flow patterns in a proglacial braided stream similar to those shown in Fig. 5b. The latter diagram indicates that bed velocities reflect the prevailing bed morphology encountered by a flow, and this may be expected to result in a spatial pattern of bedload transport which in turn creates a spatially-distributed response in terms of local aggradation and bed degradation. These bed changes then alter the morphology, and this new form is encountered by the next inflow of water and sediment, and the system continues to evolve. In short, the river morphology and process is recursive, and the form at time t influences the way in which the system develops to time $t+1$. This implies that the basis for understanding channel change is no longer in terms of switches from one equilibrium state to another, but rather in terms of a continual evolution because of the spatially-distributed feedback between form and process. Again, changes wrought by an event of a particular magnitude and frequency will be differentially effective depending on the prevailing morphological conditions. As LANE et al. (1996) argue, under these conditions "..the relationship of measures such as the dominant discharge to river channel change is

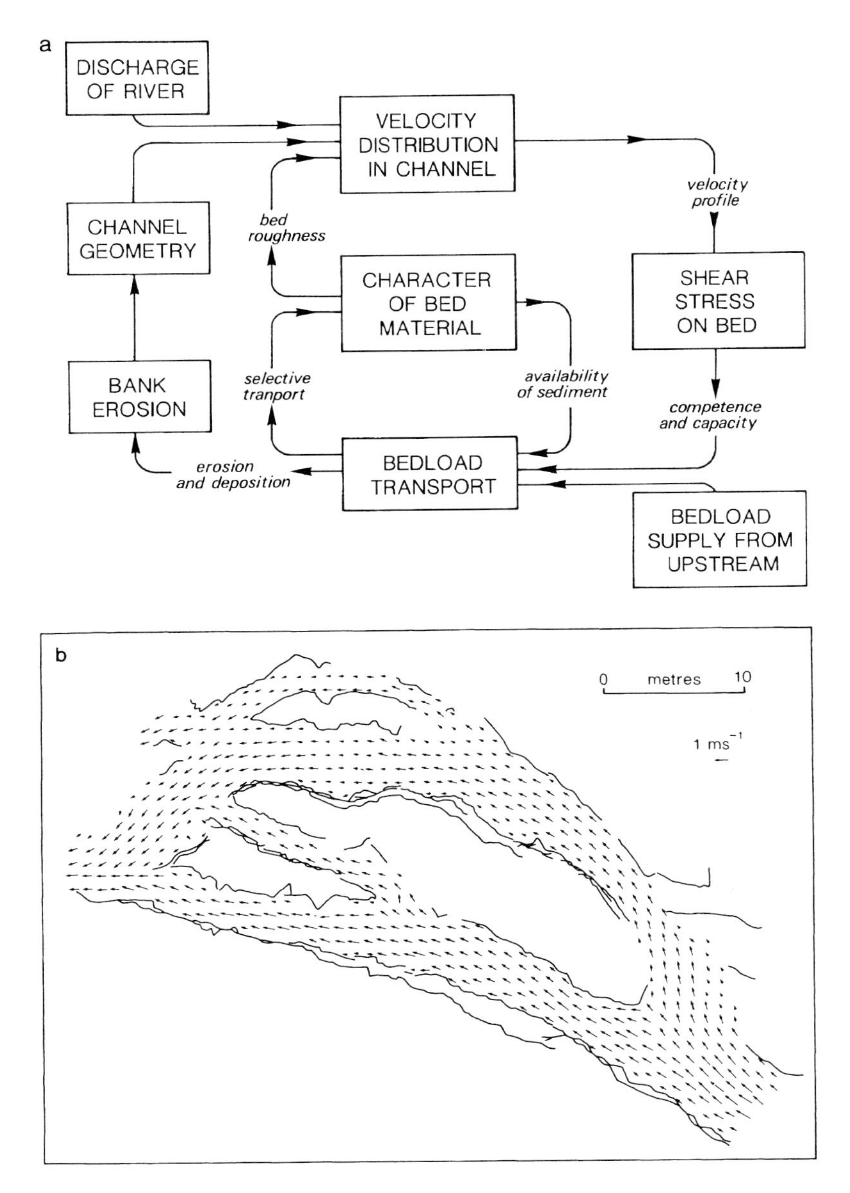

Fig. 5.(a) Form-flow-transport feedback relationships, modified from ASHWORTH & FERGUSON (1986); (b) Depth-averaged flow vectors in a natural gravel-bed braided river (the meltwater stream of the Haut Glacier d'Arolla, Switzerland).

at best tenuous" (p. 14). The best demonstration of this paradigm shift in the interpretation of the relationship between channel form and process is the change in attitude to studies of bedload transport. It is now clear that methods of sampling and measuring bedload transport are fraught with error and uncertainty, and that as a result it is rarely straightforward to account for channel morphology using empirical bedload transport data. However, an alternative is to estimate bedload transport rates by monitoring the change in bed morphology (CARSON & GRIFFITHS 1989; LANE et al. 1995). This implies a methodological as well as a theoretical reversal in the frameworks for relating form and process.

4 *The nature of a truly 'geomorphological' research programme*

There have been criticisms of the emphasis in modern geomorphology on the significance of "process"; DOUGLAS (1982) has indeed referred to the "unfulfilled promise" of process studies. However, the failure of process-oriented geomorphology to account for the origins of landforms reflects not so much the emphasis on the study of process, but rather, the engineering-equilibrium paradigm in which the results of process investigations have been employed. Geomorphic systems involve a complex and spatially-distributed feedback between form and process, but the tendency has been to assume that processes are causal, and forms are consequences. This has been re-inforced by the assumption that the direction of causality is itself conditioned by the choice of time and space scale (SCHUMM & LICHTY 1965). "Form" has thus been neglected, although this is unacceptable in a true feedback, as it defines the initial condition at each stage in its evolution, and therefore controls the trajectory of morphological change.

One obvious question that might be raised in opposition to the view that the form-process feedback constitutes the central objective of a truly 'geomorphological' research programme might be that this is not relevant outside fluvial geomorphology, with its concern for the dynamic behaviour of relatively small-scale, "labile" systems over short time-scales. However, many examples demonstrate that this framework is necessary to provide a proper account of landform development over a range of time scales, in a variety of materials and in very different kinds of system.

Estuarine evolution. Many river estuaries have evolved over a Holocene time scale as sea level has risen over the last 6,000 years or so of the post-glacial period. The effect of changing sea level may be supplemented by the effects of changing sediment supply from the terrestrial catchment, and of coastal subsidence. Being a predominantly depositional environment, these changes may be preserved in the sedimentology and stratigraphy of the estuary infill, which therefore also records information on the changing morphology of the estuary. As these morphological changes occur, they interact with those of sediment supply and sea-level to alter the tidal fluxes of water and sediment, and therefore the spatial patterns of sedimentation and the evolution of morphology. Thus, PATTON & HORNE (1992) have reconstructed the development of the Connecticut River estuary using a methodology that combines contemporary process studies with Quaternary stratigraphy, and conclude that "..the rates of ... processes can be the result of external forces acting on the estuary, such as eustatic sea level rise which can change the rate of submergence, or changes in the upland drainage basin that affect the sediment supply to the estuary. Or, these process rates may change because of phenomena internal to the estuarine system such as coastal uplift *or changes in estuarine circulation brought about by changes in estuarine morphology*" (p. 392) [my italics]. Today, "..the Connecticut River ... has a low sediment trapping efficiency that results from the coincidence of its highly seasonal hydrologic regime and its morphology inherited and modified from an earlier tidal river" (p. 415).

Tower karst development. If this example fails to convince because of the soft-sediment character of the landform in question, consider WILLIAMS's (1987) analysis of tower karst. WILLIAMS concludes that inheritance of form is a key determinant of the mode of tower karst development: "Amongst [explanations proposed for the evolution of tower karst] are two extreme possibilities:

1. that tower karst morphology evolves directly and independently of any previous morphology, given favourable lithological, relief and climatic circumstances; and
2. that the geometry of tower karst depends explicitly on the topographic characteristics inherited directly from a previous phase, such as cockpit karst relief, but becomes increasingly independent of this inheritance as erosion proceeds.

Present evidence points to both occurring but to the second being the most commonplace" (p. 454–5). The detailed argument here has some parallels with that for the development of inselbergs, in that exposure of a rock knoll encourages runoff and reduces the rate of weathering, except at its base where sub-surface weathering creates a notch and controls slope retreat. The example is a valuable demonstration that chemical processes are also dependent on inherited morphology, and that inheritance plays a key role in an account of the evolution of tower karst. Inheritance is in fact a neglected aspect of much geomorphological explanation. However, inheritance in Fig. 5a clearly refers to a continually-changing, dynamic initial condition, rather than a temporally-remote and relatively static condition. This difference can be summarised as the difference between $y_t = f(y_{t0})$, and $y_t = f(y_{t-1}, y_{t-1}, y_{t-2} .. y_{t-n})$. In soft, unconsolidated materials the rapidity of change is such as to necessitate a short time step, while in hard rock the time step may be much longer, the time scale over which forms persist will be much longer, and the nature of the "event" required to effect significant morphological change will be very different – it is more likely to be a stage in the Quaternary climatic record, than a single hydrological event.

Mass movement and slope development. Fig. 3b suggests that the timing of mass movement on a hillslope reflects the combined effects of long-term morphological and geotechnical conditioning processes and short-term events in which rain storms reduce the Factor of Safety by increasing pore water pressures. A particular magnitude and frequency of storm thus has very different effects depending on the gradient and height of a slope experiencing progressive undercutting, or the mechanical strength of the slope material as weathering changes the stability over a period. This implies a similar threshold-related behaviour to that implied for gully erosion in Fig. 3a, and discussed above. BRUNSDEN (1974) has discussed the nature of a "cycle" of landsliding in which small-scale mudslide and talus creep processes progressively degrade the toe-slope deposits of a larger coastal rotational slide until the back-slope is oversteepened again, and a further deep-seated failure occurs after a period of 40–60 years. Thus, an interaction of mass movement events of high and low frequency occurs, the former preparing the morphology of the slope until it favours the next low frequency event. There is therefore a feedback amongst slope form and a range of slope processes, with form and process changing roles in terms of cause and effect. CHANDLER & BRUNSDEN (1995) employed archival photography and analytical photogrammetry to reconstruct the three-dimensional slope form of a coastal landslide complex over a 50-year time scale, and speculated about the existence of a steady state morphology. However, as Fig. 6 implies, this is a *statistical* steady state in which the time-averaged angle of about 18.5 has no geotechnical significance (but is an emergent property), and disguises a cycle of steepening and failure. The conclusion is that "..one view of geomorphological change is that landform change takes place when a state of process equilibrium and morphological stability is perturbed by an impulse of change of sufficient character to over-

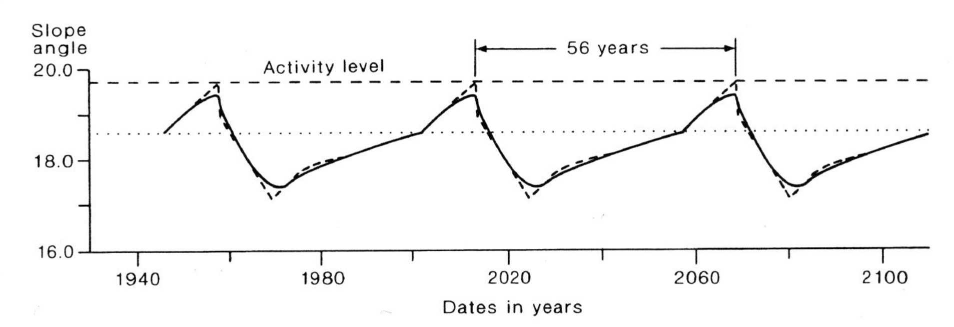

Fig. 6. A speculative model of the evolution of the slopes of a coastal mudslide complex (Black Ven, Dorset, England) over about 100 years (after CHANDLER & BRUNSDEN 1995).

come the tolerance of the system. The impulses may be 'preparatory'.. and 'triggering'. In the case of Black Ven there is evidence that the system was prepared for a new phase of mudsliding by the erosion and steepening of the cliffs" (p. 271).

Styles of glaciation. An excellent example of the relevance of form-process feedback to the interpretation of large-scale, long-term landform development is that of the changing style of glaciation as landforms develop through a series of Quaternary glaciations. This is illustrated well by the example of the Lahul Himalaya, studied by OWEN et al. (1997). The oldest, Chandra stage glaciation in this region involves eroded low-gradient benches and drumlins at heights now over 4300m; this was glaciation of a landscape of broad valleys with minimal fluvial incision, and with extensive ice cover which flowed across interfluve areas. The later Batal (c. 40ka B.P.) and Kulti (c. 18ka B.P.) glaciations involved more-or-less extensive valley glaciation. As OWEN et al. put it, "..glaciation style may reflect terrain conditions, which are altered through time by erosional and depositional events during earlier glaciations"; thus, "..the style of glaciation in the Lahul Himalaya was probably a consequence of increased aridity during late Quaternary times, probably due to a weakening of the Indian monsoon, and .. the glaciers were topographically constrained. ..progressive incision of the landscape would have created narrower and deeper valleys down which the glaciers would have flowed. The glaciers would have been confined within the valleys and could not have spread out to produce extensive ice-caps or ice-sheets" (p. 85). The events whose magnitude and frequency are here relevant are global climatic changes which generate glaciations in the Himalaya, not single flood events. The methodology remains one in which understanding of contemporary processes is coupled with meticulous field investigation of sedimentology, stratigraphy, and morphology in order to unravel the inter-dependency of morphological and climate-related process changes over time.

5 *Conclusion*

The implication of these examples is that geomorphologists working on different systems in a wide range of environments can, in fact, adopt a similar approach; one which recognises that the geomorphological significance of an event depends on the circumstances encountered by that event – because the effect of the event is conditioned by pre-existing morphology. The magnitude-frequency concept requires the assumption that an average, equilibrium state depends on a representative event. This is undermined by the fact of continual evolution in all geomorphic systems, at least on some time and space scales. The system's change in response to an event depends on its history, and on the morphology which forms the initial condition encountered by the event. This is the truly geomorphological paradigm to which we might now usefully subscribe as our research programme, and which would justify the removal of the question-mark from the title of PHILLIPS's (1992) paper, "The end of equilibrium?"

It might be argued that to conclude that "form" is a key element of explanation (rather than simply its end product) is nothing novel. Fig. 5a, for example, is a familiar-looking systems diagram. However, quantification of many of the systems diagrams presented, for example, in CHORLEY & KENNEDY (1971), would commonly have been achieved by using system-averaged parameters, such as whole-system morphometric variables and output process variables. For the kind of feedback system considered here, what is required for each system component is a *map* of the spatially-distributed nature of the property it represents. The *internal* organisation of form and process is what is summarised by the system diagram, rather than the relationships amongst aggregate system properties. The other important change is in viewing the system as recursive, with the boundary condition which determines the spatial pattern of processes itself continually being changed by those processes. Inheritance is not, therefore, from time $t=0$, but from the immediately preceding time steps, and is a continually evolving inheritance.

It might also be argued that the examples quoted in the previous section are essentially local or regional studies, in which it is difficult to identify either a strong conceptual framework or any general geomorphological implications. The response to this may be two-fold. Firstly, in order to interpret these evolving form-process relationships, it is necessary to apply a detailed understanding of the mechanics of the relevant processes, and of the ways in which they are dependent on particular boundary conditions. Thus, while the application of this knowledge may be to a local problem, there remains the need to research the mechanics of process, and the techniques for revealing the changing boundary conditions. These are general scientific objectives. Secondly, studies such as PATTON & HORNE's, or that of LEWIS and his co-workers, are *properly* scientific in that they present detailed observational evidence and their subsequent interpretations, and leave themselves open to criticism, rival hypothesis-testing, and alternative views. They may be proved wrong, both in detail and in general. The magnitude-frequency model, in which a statistically-dominant event is statistically associated with a form, is by comparison sufficiently lacking in detail and precision as to be relatively untestable. One might conclude, in fact, that it is the critique of details of interpretation in local or regional geomorphological studies that leads to advances in the understanding of the generally-occurring mechanisms and their morphological consequences. Such a critique is possible within a research programme which considers how events may lead to observable changes, but not within one in which events are simply associated with a hypothetical equilibrium.

References

ACKERS, P. & F.G. CHARLTON (1970): Meander geometry arising from varying flows. - Journ. Hydrol. **11:** 230–252.

ANDREWS, E.D. (1980): Effective and bankfull discharges of streams in the Yampa River Basin, Colorado and Wyoming. - Journ. Hydrol. **46:** 311–330.

ASHWORTH, P.E. & R.L. FERGUSON (1986): Interrelationships of channel processes, changes and sediments in a proglacial braided stream. - Geograf. Ann. **68A**: 361–371.

ASSELMAN, N.E.M. & H. MIDDELKOOP, H. (1998): Temporal variability of contemporary floodplain sedimentation in the Rhine-Meuse delta. – Earth Surf. Proc. Landforms **23:** 595–609.
BAKER, V.R. (1977): Stream channel response to floods with examples from central Texas. – Bull., Geol. Soc. Amer. **88:** 1057–1071.
BEVEN, K. (1981): The effect of ordering on the geomorphic effectiveness of hydrologic events. – Internat. Assoc. Hydrol. Scis., Publ. **132:** 510–526.
BRUNSDEN, D. (1974): The degradation of a coastal slope, Dorset, England. – Institute of British Geographers, Spec. Publ. **7:** 79–98.
CARLSTON, C.W. (1965): The relation of free meander geometry to stream discharge and its geomorphic implications. – Amer. Journ. Sci. **263:** 864–885.
CARSON, M.A & G. GRIFFITHS (1989): Gravel transport in the braided Waimakariri River: mechanisms, measurements and predictions. – Journ. Hydrol. **109:** 201–220.
CHANDLER, J.H. & D. BRUNSDEN (1995): Steady state behaviour of the Black Ven mudslide: the application of archival analytical photogrammetry to studies of landform change. – Earth Surf. Proc. Landforms **20:** 255–275.
CHORLEY, R.J. & B.A. KENEDY (1971): Physical geography – a systems approach. – Prentice Hall, London, 370.
COUVALIS, G. (1997): The Philosophy of Science. – Sage, London, 206.
DOUGLAS, I. (1982): The unfulfilled promise: earth surface processes as a key to landform evolution. – Earth Surf. Proc. Landforms **7:** 101.
DURY, G.H. (1965): Theoretical implications of underfit streams. – U.S. Geol. Surv., Prof. Pap.: **452-C:** 57.
DURY, G.H. (1973): Magnitude-frequency analysis and channel morphometry. – In: MORISAWA, M. (ed.): Fluvial Geomorphology. – SUNY Binghamton, Publications in Geomorphology, 91–121.
FERGUSON, R.I. (1986): River loads underestimated by rating curves. – Water Resources Res.h **22:** 74–76.
FURBISH, D.J. (1991): Spatial autoregressive structure in meander evolution. – Geol. Soc. Amer., Bull. **103:** 1576–1589.
HACK, J.T. (1960): Interpretation of erosional topography in humid temperate regions. – Amer. Journ. Sci. **258-A:** 80–97.
HARVEY, A.M., D.H. HITCHCOCK & D.J. HUGHES (1979): Event frequency and morphological adjustment of fluvial systems. – In: RHODES, D.D. & G.P. WILLIAMS (eds.): Adjustment of the Fluvial System. – George Allen & Unwin, 139–167.
LAKATOS, I. (1970): Falsification and the methodology of scientific research programmes. – In: LAKATOS, I. & A. MUSGRAVE: Criticism and the growth of knowledge. – Cambridge University Press, Cambridge, 91–196.
LANE, S.N., K.S. RICHARDS & J.H. CHANDLER (1995): Morphological estimation of the time-integrated bedload transport rate. – Water Resources Res. **31:** 761–772.
LANE, S.N., K.S. RICHARDS & J.H. CHANDLER (1996): Discharge and sediment supply controls on erosion and deposition in a dynamic alluvial channel. – Geomorphology **15:** 1–15.
LEOPOLD, L.B. & T. MADDOCK (1953): The hydraulic geometry of stream channels and some physiographic implications. – U.S. Geol. Surv. Prof. Pap. **252:** 57.
NEWSON, M. (1980): The gomorphological effectiveness of floods – a contribution stimulated by two recent events in mid-Wales. – Earth Surf. Proc. **5:** 1–16.
OWEN, L.A., R.M. BAILEY, E.J. RHODES,W.A. MITCHELL & P. COXON (1997): Style and timing of glaciation in the Lahul Himalaya, northern India: a framework for reconstructing late Quaternary palaeoclimatic change in the western Himalayas. – Journ. Quatern. Sci. **12:** 83–109.
PATTON, P.C. & G.S. HORNE (1992): Response of the Connecticut River estuary to late Holocene sea level rise. – Geomorphology **5:** 391–417.
PHILLIPS, J.D. (1992): The end of equilibrium? – Geomorphology **5:** 195–201.
PICKUP, G. & W.A. RIEGER (1979): A conceptual model of the relationship between channel characteristics and discharge. – Earth Surf. Proc. **4:** 37–42.
PICKUP, G. & R.F. WARNER (1976): Effects of hydrological regime on magnitude and frequency of dominant discharge. – Journ. Hydrol. **29:** 51–75.
SCHUMM, S.A. & R.W. LICHTY (1965): Time, space and causality in geomorphology. – Amer. Journ. Sci. **263:** 110–119.
STØLUM, H.-H. (1996): River meandering as a self-organization process. – Science **271:** 1710–1713.

TINKLER, K.J. (1971): Active valley meanders in south-central Texas and their wider implications. – Geol. Soc. Amer. Bull. **82:** 1873–1899.
WAYLEN, P.R. (1985): Stochastic flood analysis in a region of mixed generating processes. – Transactions, Institute of British Geographers **10:** 95–108.
WILLIAMS, P.W. (1987): Geomorphic inheritance and the development of tower karst. – Earth Surf. Proc. Landforms **12:** 453–465.
WOLMAN, M.G. & GERSON, R. (1978): Relative scales of time and the effectiveness of climate in watershed geomorphology. – Earth Surf.Proc. **3:** 189–208.
WOLMAN, M.G. & L.B. LEOPOLD (1957): River flood plains: some observations on their formation. – U.S. Geol. Surv., Prof. Pap. **282-C:** 87–107.
WOLMAN, M.G. & J.P. MILLER (1960): Magnitude and frequency of forces in geomorphic processes. – Journ. Geol. **68:** 54–74.
YU, B. & M.G. WOLMAN (1987): Some dynamic aspects of river geometry. – Water Resources Res. **23:** 501–509.

Address of the author: Prof. KEITH S. RICHARDS, Department of Geography, University of Cambridge, Downing Place, Cambridge CB2 3EN, U.K.

Z. Geomorph. N.F.	Suppl.-Bd. 115	19–33	Berlin · Stuttgart	Juli 1999

Space and time scales in geomorphology

Leszek Starkel, Kraków

with 11 figures

Summary. In studying the origin and evolution of landscape, we use various scales, both in space and time. Most landforms result from interaction of various factors. The author suggests distinguishing the following effective time scales of landform evolution: formative extreme events and secular processes, clusterings of events, phases of higher frequency of extreme events, cyclic changes of longer duration. The assumption about positive correlation between size and age of the forms is frequently not valid.

The existing relief incorporates forms of various origin and age, which have undergone continuous adaptation to new conditions. The author underlines three principal regularities in relief evolution: way to destiny (planation), continuous adjustment to new conditions, coexistence of diverse forms controlled both by tectonic and climatic factors as well as by lithology.

Zusammenfassung. Bei der Untersuchung des Ursprungs und der Entwicklung von Landschaften behelfen wir uns sowohl im Raum, als auch in der Zeit verschiedener Skalen. Die meisten Landschaftsformen resultieren aus der Interaktion verschiedener Faktoren. Der Autor schlägt die folgende Unterscheidung der für die Landschaftsgenese relevanten Zeitskalen vor: formende Extremereignisse und säkulare Prozesse, Häufungen von Ereignissen, Phasen mit höherer Frequenz von Extremereignissen und langfristige zyklische Änderungen. Die Annahme einer positiven Korrelation zwischen Größe und Alter von Formen ist häufig nicht zutreffend.

Das existierende Relief beinhaltet Formen unterschiedlichen Ursprungs und Alters, die kontinuierlicher Anpassung an neue Randbedingungen unterliegen. Der Autor betont drei prinzipielle Regelmäßigkeiten in der Reliefgenese: der Weg zum Schicksal (Einebnung), kontinuierliche Anpassung an neue Randbedingungen und die Koexistenz von diversen Formen, die von tektonischen und klimatischen Faktoren, aber auch von der Lithologie kontrolliert werden.

1 *Introduction*

A long time after G.K. Gilbert (1877) presented the concept of dynamic equilibrium, W.M. Davis (1899) formulated the landscape's way to destiny (being disturbed by tectonic or climatic changes). Leopold et al. (1964) described the aim of dynamic geomorphology and Dury (1975) returned to a catastrophic theory of events; Denys Brunsden (1990) put forward ten spectacular commandments of geomorphology. The last two of them, relating to about sensitivity to change and the ability to resist increasing impulses, formed, indeed, a 'sack full' of various relations. "Let's open this sack" – In this paper I explore some of Brunsden's ideas, keeping in mind, that observations are inherently intertwined with theoretical presuppositions (Rhoads & Thorn 1996).

The complicated architecture of the earth's surface is a joint product of various forces, which mobilise different substances. But relief itself is not a substance, only a geometric shape of the Earth's surface. This product of various transfers of matter, in the mean time, forms a geometric base for all on going transportation of material in the environment.

0044–2798/99/0115–0019 $ 3.75

The elements of landscape we may see, touch, measure-belong to those features of the environment, which are changing very slowly through time.

To explain their origin and evolution the geomorphologist examines substratum (structures building the relief) as well as processes of construction and destruction. Other scientists, geologists and geophysicists, study the same context, but from different perspective (cf. TWIDALE 1996).

With reference to the various space and time scales of a landscape's form we should consider that different types of scale should be commensurable (KIRKBY 1990, CHURCH 1996). Going from a smaller to a larger spatial scale, when explaining landforms, we must replace stochastic models with deterministic theories and dynamic models. When studying large samples we use extensive methods, on the contrary when studying small samples – intensive methods. One of the most difficult methodological questions is generalisation or transfer of methods and results from small objects to large ones (RICHARDS 1996).

2 *Spatial scales*

Using BRUNSDEN's (1996) terminology we may distinguish mega-, macro-, meso-, micro-, nano- and pico-forms. Here, one common feature is their coexistence in space, although the length of their life is different. This means that another common feature of forms of different orders

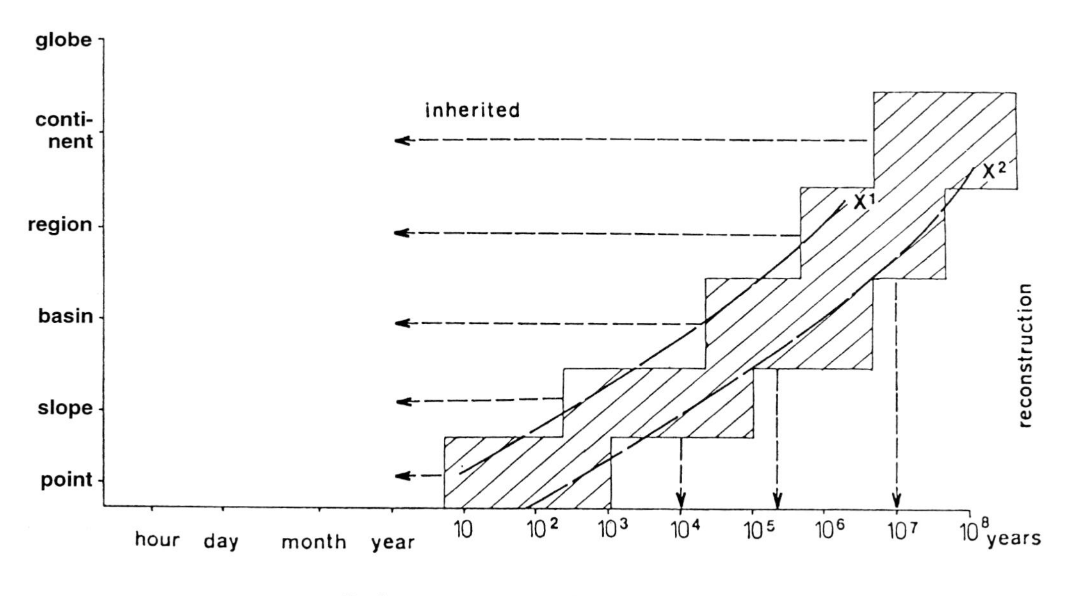

Fig. 1. The age of forms of different size in the gradually developing landscape (the lined area). Two curves indicated the combination of different forms over the less (x_1) and more (x_2) resistant rocks. Horizontal arrows indicate various inherited components incorporated in existing landscape. Vertical arrows indicate the ways of reconstruction of past landscape (based on STARKEL 1992).

is their equal exposure to external factors and simultaneous transformation. A frequent assumption made by geomorphologists is that the larger forms are more stable and the pico- and nano-forms are ephemeral. Therefore, the greater forms should be older in origin than the smaller ones (Fig. 1).

Among landscape elements we can distinguish various forms of genesis, created by different independent factors either of tectonic or climatic origin. In the second group are forms created by gravitational forces (by water, ice, mass movements), difference in pressure (by wind) or difference in density (by freezing, leaching, liquefaction).

The majority of forms was initiated or developed by interaction of various factors e.g. fluvial processes and mass movements, karstic processes and piping etc. This simultaneous, alternate or consecutive passing of various thresholds in a given site creates polygenetic forms.

One of the spatial differences in form evolution is a multifold action of processes (Fig. 2). Different processes may act parallel, accelerating the development of a linear pattern of the forms. This is the case of the Scotch mountains, where the direction of fluvial processes and ice cap expansion were parallel. Two different factors may also act in opposite directions, for instant, the Polish lowland has been drained toward the north and invaded several time by the Scandinavian ice sheet. If two different factors act in transversal directions, then, for example, the dune fields may be formed from sands blown perpendicular, from the braided river channels.

Forms of similar origin may have different areal extent, starting from localised features controlled by lithological differences or by local heavy downpours, up to great forms or form complexes, mainly related to morphostructures or morphoclimatic regions.

There are various types of spatial pattern of forms, controlled by endogenic or by exogenic processes (Fig. 3). Endogenic processes produce active tectonic structures, which, depending on differentiated uplift rate and inherited relief, show the largest depth and density of dissection either at their margins (cf. the Western Ghats bordering the Deccan Plateau) or, in contrast, in their central part (central Himalaya). In the first case the interior of the plateau still preserves the inherited mature landscape.

A different, convergent spatial pattern is represented by a fluvial system (SCHUMM 1977). It is characterised by a directed transfer of water and matter along water courses. The slope and channel subsystems are interrelated, but their time of response and time of passing thresholds may differ. Both subsystems are either sensitive to various forming events (slopes – to downpours, channels – to continuous rains – Fig. 4) or they may react synchronously, during catastrophic events (cf. STARKEL 1976).

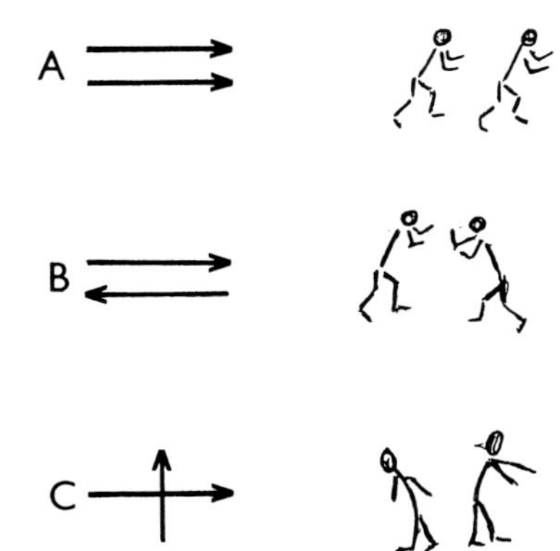

Fig. 2. Directions of the action of processes in space.

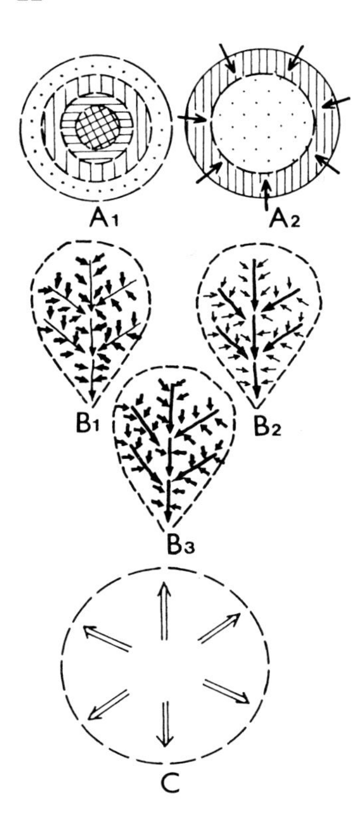

Fig. 3. Spatial distribution of forms and intensity of processes. A. Highest density and relief energy (A_1 – central elevated part. A_2 – margin of uplifted block). B. convergent fluvial system with slope and channel subsystems reacting to various threshold events (B_1 – to heavy downpour, B_2 – to continuous rain, B_3 – to catastrophic event, combined downpour and continuous rain), C – divergent system of the ice sheet.

A completely opposite, divergent system is exposed after the decay of an ice sheet. The radial pattern of subglacial channels, striation, eskers and drumlins mark this divergent pattern.

3 *Time scales*

Relief is a product of long evolution, it is a result of coexistence of forms of various age. Many authors distinguish various time scales (minutes, hours, days, years, decades, millennia, millions of years) and relate to them the forms of various sizes, from ripples to the mountain ranges and lowlands (THORNES & BRUNSDEN 1977, BARSCH 1990).

Recently, BRUNSDEN (1996), after discussion with YATSU (1993) and others, has modified his concept of events (BRUNSDEN 1990) and identified instantaneous events (natural hazards), short term events (of order of centuries), long-term events (glacial stages) and geological events (ice ages, orogenies). Each of these "events" should be characterised by a process intensity higher than the mean, and preceded and followed by a steady phase. During every effective event the threshold of the form stability is passed (WOLMAN & GERSON 1978). During longer phases or cycles some thresholds are also passed. But can we measure the limits of such long events using existing methods, both in time and in space?

In my opinion there is no reason to change the meaning of event and extend it over longer time units. Such longer "events" incorporate events of lower order (shorter duration). Every

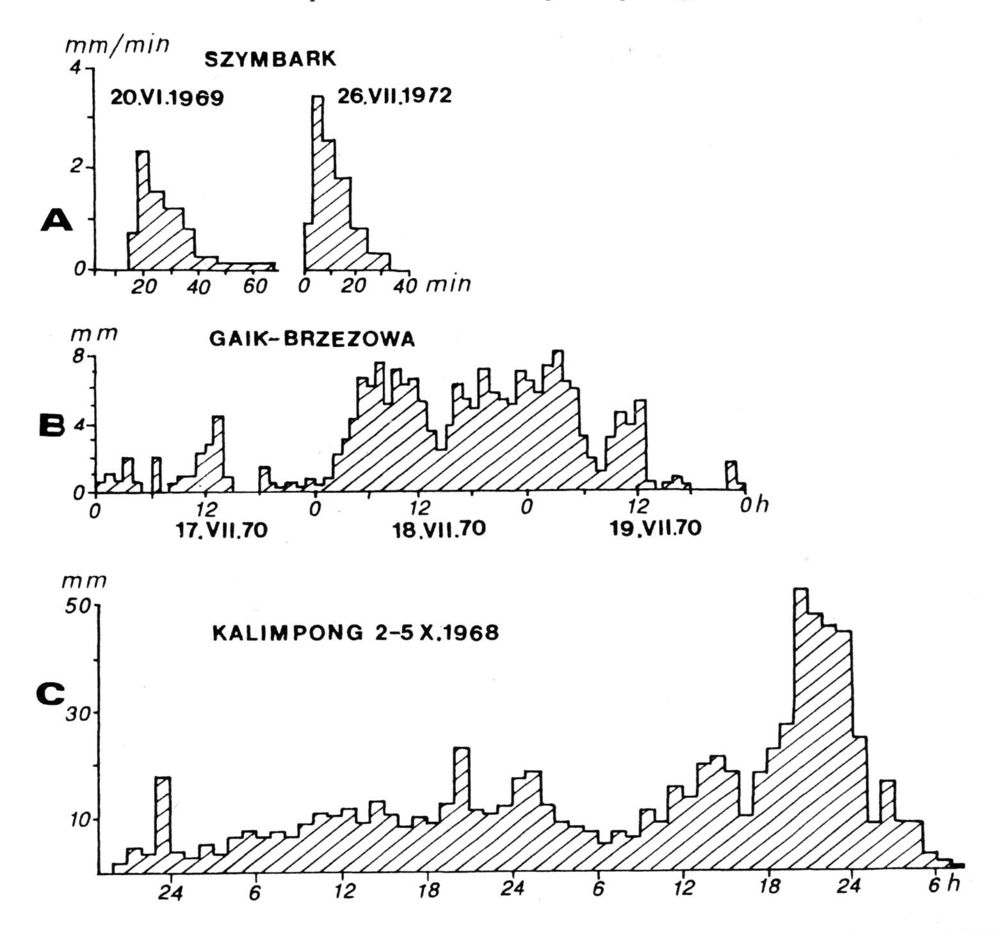

Fig. 4. Types of extreme rainfalls reflected in the threshold of the slope and channel subsystems (cf. Fig. 3B), A. heavy downpours in the Polish Carpathians, B. continuous rain in the Polish Carpathians, C. continuous rain combined with downpour in the Sikkim Himalaya.

longer time period is composed of shorter events of various intensity and duration. The last ones expressed in landform evolution (called formative, effective or extreme) may be concentrated in various time scales which should be named with terms commonly used in geochronology.

As an alternative I present the following hierarchy of time scales, expressed in terms of relief evolution (Fig. 5):

A. Singular extreme events represent the lowest level and exist against a back-ground of secular processes. The extreme event may vary in duration from seconds to days or months and cause an effective or even a creative change of form. These events are generally incorporated in the existing climatic regime (e.g. floods, landslides) or have a character of geocatastrophe or cataclysmic event (Baker 1988) with recurrence intervals from decades to millennia.

In contrast the secular processes operating over a duration of days to years and centuries are represented by processes of lower intensity, continuing during the inter-event or steady

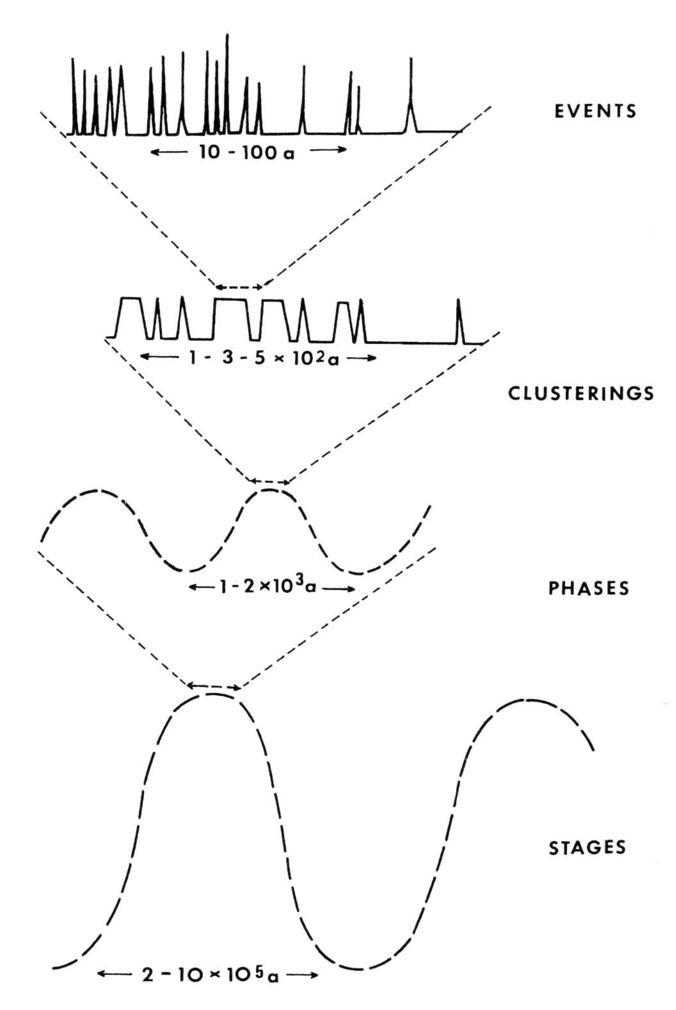

Fig. 5. Hierarchy of time units with various intensity of processes: events, their clusterings, phases and stages. In last case the curve indicates the long-term fluctuations in temperature or precipitation reflected in type, intensity and frequency of events.

period. But even during such periods there are many events associated with active slope wash, solifluction, leaching, deflation etc., when the intensity of processes is still higher than the mean value, measured as the total product of diurnal or annual activity.

B. The clustering of events can be assigned duration from years to decades and was described from the present century as well from the past (GROVE 1979, STARKEL 1984, 1996). KIRKBY (1987) calls this the "Hurst effect". Heavy rains, floods or tornadoes, repeating in the consecutive years, may cause a substantial change in transformation of the forms, before the full relaxation follows.

C. Phases with high frequency of the extreme events of duration from centuries to millennia. These may cause substantial disturbances in the geomorphic systems. Such phases have been well recognised during the Holocene in several advances of mountain glaciers (GROVE 1979, KARLEN 1991), parallel reactivating of solifluction processes and other mass movements (GAMPER 1993, KOTARBA 1995, STARKEL 1985), and in transformation of alluvial plains with channel avulsions (STARKEL 1991).

D. Cyclic changes (during the Quaternary) of duration from tens to hundreds of millennia, which include stages of various climatic or various tectonic activity and are manifested in alternate, different process complexes and their intensity, with numerous lower range phases, clusterings and events. These alternate stages are reflected in such features as sequences of terraces and piedmont surfaces (SOERGEL 1921, PENCK 1924, STARKEL 1986b), in alternate periods of intensive denudation and in soil regolith formation. Due to superposition of cyclic climatic changes over tectonically stable areas, usually in a former periglacial zone, we observe that mainly the sediments and forms of the last cold stage are preserved (DYLIK 1967, STARKEL 1987a). According to various concepts, the highest rate of transformation in the fluvial system takes place during the transitional phases from glacial to interglacial (JAHN 1956, KNOX 1976).

E. The above described cyclic changes may also form a clustering of a higher order, among them glacial epochs (Quaternary) or orogenic phases millions of years long, responsible for the total transformation of landscapes in a regional, continental or global scale (KING 1953). In a general sense for example, the Alpine orogeny was the most important "event" in the evolution of the Mediterranean landscapes.

4 *Forms trough time*

In discussing the evolution of the forms in time, we should distinguish three parameters of their age: time of the onset of their creation , time of their formation and expansion, and time of adaptation (transformation).

Time of creation is an initial stage when, due to climatic or tectonic events (or their clustering) or even due to the passing an intrinsic threshold (cf. SCHUMM 1977), a new form comes to existence.

Time of formation is a total period of a build-up and includes several consecutive events with relaxation phases in-between. With older forms (e.g. developing during Pliocene, mid-Pleistocene etc.) this may include a longer time unit, when this form reaches its stage of maturity.

Time of transformation of old landscapes or adjustment is especially long in the case of river valleys and slopes, which even during the mid-upper Quaternary have been adapting several times to alternating climatic conditions. Each change to a periglacial climate provoked reactivation (rejuvenation) of slope forms, while each change to temperate forest climate caused their stabilisation and soil formation (or dissection).

In the case of tectonic change we observe an uplift followed by rejuvenation (dissection) of existing relief or fossilisation (aggradation) connected with the subsidence.

Relations between age and size of the forms are not simple. The traditional assumption and belief that the greater forms are older and smaller forms – younger (group **A**, Fig. 6) is not valid under all circumstances. There are several deviations from this assumption and these are illustrated in Fig. 7 us follows:

B. Larger forms are younger than the inventory of smaller forms. This is a case, when an old erosional platform has been dismembered by active tectonic faults (cf. in California, Meghalaya Plateau).

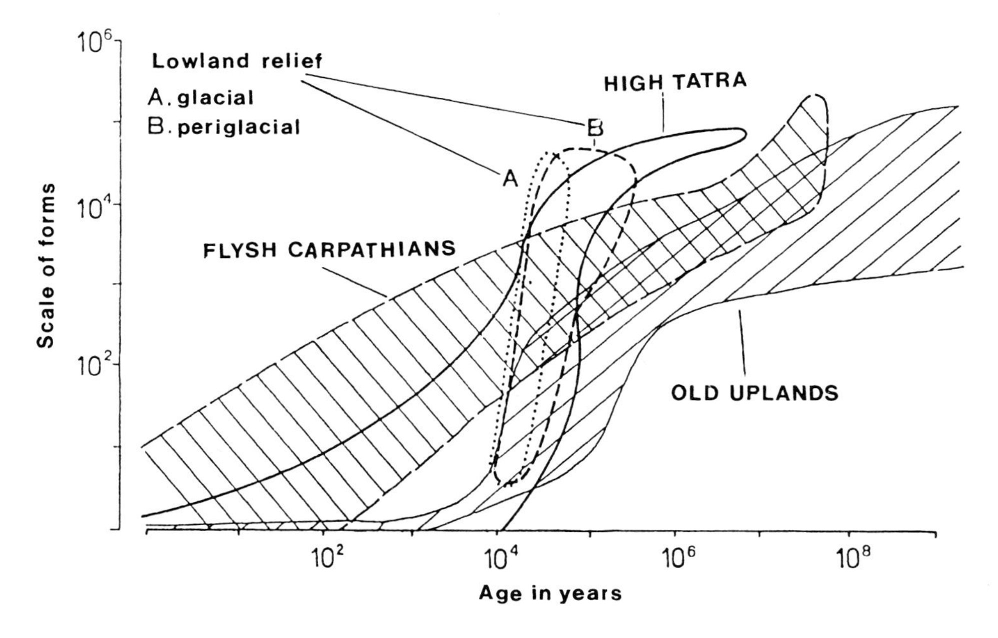

Fig. 6. Size of forms of different age in various landscapes of Poland (based on STARKEL 1992).

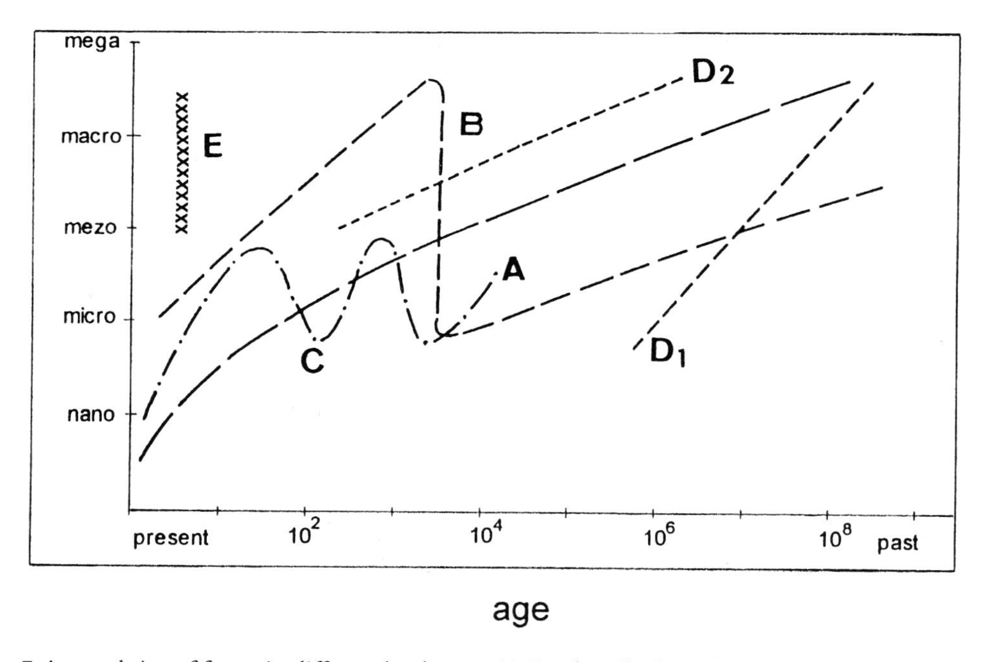

Fig. 7. Age and size of forms in different landscapes (A-E – described in text).

C. A whole landscape is composed of young features, which are dominated by meso- and microforms. This happened during continuous intensive uplift of the Neogene-Quaternary unconsolidated deposits at the margin of Alpine orogenic chains (Southern Italy, Romanian Subcarpathians – cf. STARKEL 1978).

D. A structure controlled relief may be either totally inherited from a distant era (the Australian interior), and the forms of various order are generally old (D_1), or may result from the dissection of structures and the resistant finer beds start to be manifested much later (D_2). Resistance of bedrock controls the rate of transformation. Therefore, depending on variations in lithology and tectonics the relations between the form size and its age may be different and even oscillate from place to place.

E. A landscape generated by a large, cataclysmic event may be composed of young forms – great and small ones, of similar age.

Old and young forms exist together and develop in parallel (STARKEL 1987a). This coexistence is well visible especially in the landscapes which undergo rejuvenation.

There, we find side by side the inherited upper valley reaches which in higher mountains are or were occupied by glaciers (KLIMASZEWSKI 1960). Another example is the Meghalaya Plateau in India where a high uplift rate and resistant limestone beds facilitated the preservation of shallow and mature valley heads (STARKEL 1978).

But there are some specific exceptions from this diversified picture (Fig. 8). Such landscapes as the Western Siwaliks reveal structural elements from their first formation, when a rapid rise of folded sedimentary beds exposed the more resistant beds from the very beginning (STARKEL 1978). Such landscapes may be called "mature from birth".

An opposite situation exists on the fully unconsolidated, Neogene-Quaternary, clays and sands of the eastern corner of the Caucasus Mts where the denudation rate of gentle slopes is high. We may call this landscape as being "young till late maturity".

Fig. 8. A cartoon presentation of relations between size.and stage of evolution of existing landforms (a – older forms greater, b – younger greater, c – whole relief young, d – relief mature from birthday, e – relief young till late maturity).

5 *Continuous adaptation*

From the previous discussion on relief evolution through time we conclude that the continuous adjustment of inherited forms to new conditions is an essential part of all geomorphological considerations (STARKEL 1987a). In the present-day landscape various generations and groups of forms coexist:

older (inherited) forms, adapted
old (inherited) forms, dissected
old (inherited) forms, fossilised
new (superimposed) generation of forms

BÜDEL (1977), discussing the morphogenetic zonation of the earth, differentiates between the clima-genetic and clima-dynamic approaches. BRUNSDEN (1990), in the last two commandments of geomorphology, distinguishes a great spatial variety in sensitivity to change reflecting relations between resisting and disturbing forces and, then formulates a rule that "the landscape's ability to resist impulses of change tends to increase with time".

This rate of adaptation of each landscape is related to climatic and tectonic changes. It also depends on rock resistance and on the relations between the effectiveness of extreme and secular processes.

The climatic changes influence the transformation of different systems at various scales. The slopes of the former periglacial zone have undergone a relatively small transformation during the Holocene (DYLIK 1967, STARKEL 1986a). In contrast, the river channels reacted to every change in water discharge and sediment load (STARKEL 1983, 1990). Therefore, the alternation of meandering and braided channel pattern is particularly well expressed in the middle valley reaches.

The rate of uplift and subsidence is directly reflected in the incision rate and later in the denudation rate (SELBY 1974, STARKEL 1987b, YOSHIKAWA 1985).

Rejuvenation (dissection) of every highland or plateau expands mainly from its margin (base level) while the upper valley segments remain still not dissected and preserve inherited relief features (Fig. 9). But opposed to this, rejuvenation may also start from the upper section. The higher gradient of valley heads and concentration of drainage make formation of a badland network possible (Fig. 9).

The non dissected or non fossilised older forms, nevertheless, are being incorporated into the existing landscape system in various ways. The most diverse incorporation is on the slopes

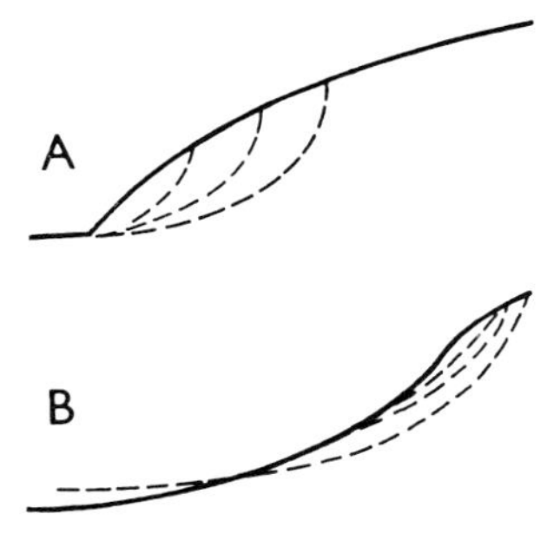

Fig. 9. Expansion of rejuvenation: A. from margin (upper course not dissected), B. from upper section (only incision in steeper valley head).

and in the valley floors. Slope fragments of various origin act as one slope system. This happened to the sides of the former glacial troughs or to the cryopediments over the Mongolian Plateau (Fig. 10, KOWALKOWSKI & STARKEL 1984). The new valley network created after the ice sheet decay is composed of the segments of various origin (KOUTANIEMI & RACHOCKI 1981, KOUTANIEMI 1991). Many of the former wide streamways take the form of underfit river valleys (DURY 1964).

A supplementary factor which disrupted the continuity in the slope or river channel development and even caused a full dispersion of processes over the system is the anthropogenic one. The terraced slopes with cart-roads, regulated and cascaded river channels – create new space patterns and new thresholds for various processes (GREGORY & WALLING 1987).

In the chain of the relief evolution and continuous adaptation, the incorporation of products of the events, of their clustering and of the phases takes place. Depending on the frequency, areal extent and size of transformation the secular (slow and frequent) processes may also produce new shapes of slopes, as well as new forms as cryoplanation terraces, dune fields, karstic caves etc. The extreme (rapid and rare) events are restricted in space but cause substantial changes. One cataclysmic event may destroy the whole construction, until now the enclave of senility. Among them a special role played by extremes which change in time the type and intensity of processes. Several debris flows in the valley floors start and end as hyperconcentrated flows (COSTA 1988). Similar are large landslides; the mechanism of their movement changes both in space and time (FROEHLICH et al. 1992).

In the slowly transforming landscapes the role of extreme events is different. Besides a distinct rejuvenation, after long relaxation phase new features may finally become incorporated in the inherited relief.

On longer time scales, the phases with frequent events creating the separate alluvial fills become finally incorporated in the Holocene floodplains (STARKEL 1983) that might be easily detected from the Last Cold Stage alluvial fills, interfingering with periglacial colluvia and deluvia (Fig. 11).

In this continuous adjustment of the relief an important role is played by symbiosis and convergence of various processes and forms. The symbiosis is expresses by the polygenetic form, created by the co-operation of various processes, acting sometimes in different directions and in different seasons of the year. The convergence of relief features of different origin is relatively rare. A good example is the braided pattern of the river channels, which (1) in some dry valleys of the arid zone may be inherited from the distant past, (2) may reflect the sediment

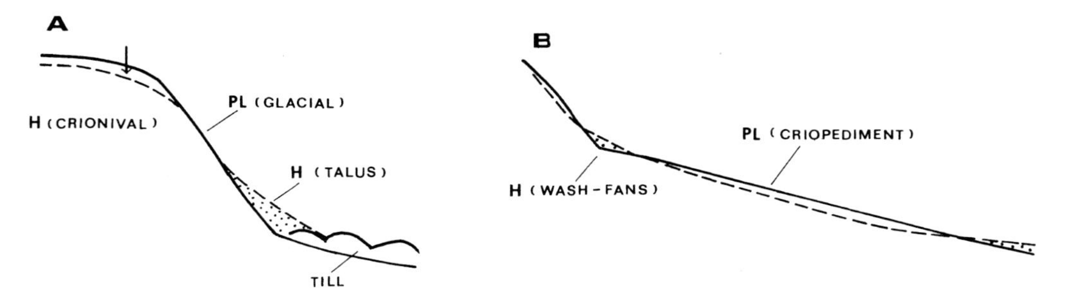

Fig. 10. Adaptation of inherited Pleistocene slopes during the Holocene (after KOWALKOWSKI & STARKEL 1984). A. High mountain glacial slope in the Khangai Mts. (Mongolia), B. Cryopediment transformed at the foothills of the Khangai Mts.

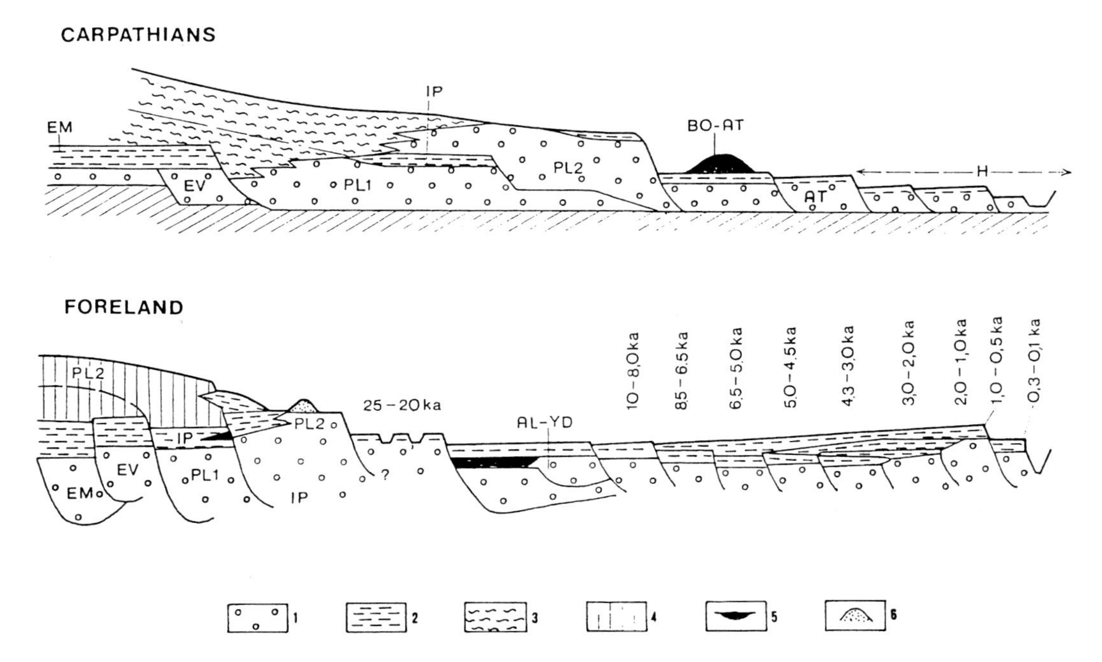

Fig. 11. Synthetic model of evolution of the valley floors in the Polish Carpathians and Subcarpathian Basin showing several Pleniglacial and Holocene fills (after STARKEL 1995). 1. channel facies, 2. overbank facies, 3. slope sediments, 4. loess, 5. organic deposits, 6. dunes, EM - Eemian, EV - early Vistulian, PL - pleniglacial, IP - interpleniglacial, AL - Alleröd, YD - Younger Dryas, BO - Boreal, AT - Atlantic.

load regime in active gravel-bed channels, but also may characterise (3) a section of high infiltration and sinking of water or (4) a section with formation of seasonal icings (in Mongolia). The combination of various factors is also not excluded.

6 *Searching for regularities in the relief evolution*

There are several physical rules describing circulation of matter and regulating for example the evolution of the earth's surface. In reconstruction of the evolutionary chain we simplify and reduce all factors to endogenic and exogenic. For explanation we produce tentative hypotheses and search for evidence (BAKER 1996). The controlling factors are ordered into leading genetic factors (distinguishing events, their clustering, phases, stages ...) and in space (after size, in latitudinal and vertical zones). We evaluate the role of tectonic activity and diversified resistance of bedrock. We try to put all the information into a mathematical model of the land surface development (LAWRENCE 1996). One of the latest attempts to order systems mentioned in the introduction, are the spectacular commandments of geomorphology (BRUNSDEN 1990).

I find that the whole theory of the landscape evolution could be concentrated in the following three principal regularities: A drive at planation, B. adjustment or adaptation, C. diversity and coexistence. These are characterised as:

A. The hypothesis of planation (peneplain). It is a geomorphological concept of a destiny to which all forms modelled by denudational processes tend. The final stage, defined by W.M. DAVIS (1899) and H. BAULIG (1952), may be reached only in the exceptional cases and is balanced by the concept of perpetual dynamic equilibrium of G.K. GILBERT.

This pathway to destiny is mainly controlled by the rock resistance and stability of the old morphostructures. Therefore, the planated landscapes exist especially in stable cratonic regions of the former Gondwana continent.

B. The continuous adjustment (or adaptation) of relief to new conditions is created by endo- and exogenic processes changing their intensity and direction with time. In almost every landscape there are old roots (inherited from the past) and superimposed younger forms (products of events). Going back to the past our knowledge about the mechanism and sequences of changes in time is more and more hypothetical (cf. DORN 1996, STARKEL 1991). Most of universal forms like slopes are the joint product of alternate formative and stable phases or of perennial adaptation during thousands and millions of years. Therefore, only as the exception (from the rule) may be preserved the "tote Landschaften" (*sensu* BÜDEL 1977) mainly related to the penultimate morphogenesis (e.g. Last Cold Stage).

C. Diversity and coexistence. The great diversity of the Earth's surface features reflects the perpetual spatial and temporal variations in energy exchange and transfer of matter, in their acceleration or delay, in continuous destruction and creation.

There exist landscapes with high and low intensity of various processes, with different rates of transformation as well as composed of forms of various origin and age. The complexity of some geomorphic systems may be so high that the linear dynamic model is not acceptable and the system seems to be unstable and chaotic (PHILLIPS 1996).

Among the intensively transforming landscapes are the ecotonal zones of main morphoclimatic regions and the margins of elevated morphostructures. The lithological diversity and random spatial distribution of extreme events explains the local differences.

Analysing the nature of existence (life) we touch a fundamental question of *CREATION* and *DESTRUCTION*, in which past and future are incorporated.

The old Persian holy book tells what is the *LIFE* over the Earth:

...it sleeps in stone,
wakes in a plant,
moves in an animal
and may be fully realised in the man,
sounding the question of existence...

Where is the *RELIEF* in this context?

May be ... it is the life's playground, created during the perpetual energy exchange and circulation of matter showing its resistant but slowly changing face...

.... during repeating waking up, falling asleep and waking up ....

References

BAKER, V.R. (1988): Overview. – In: BAKER V.R., R.C. KOCHEL, & P.C. PATTON (eds.): Flood Geomorphology. – 1–12, Wiley.

BAKER, V.R. (1996): Hypotheses and geomorphological reasoning. – In: RHOADS B.L. & C.E. THORN (eds.): The scientific nature of geomorphology. – 57–58, Wiley.

BARSCH, D. (1990): Geomorphology and geoecology. – Z. Geomorph. N.F., Suppl.-Bd. **79**: 39–49.
BAULIG, H. (1952): Cycle et climat en geomorphologie. – Lab.Geogr., Univ. Rennes, vol. jubil.: 215–239.
BRUNSDEN, D. (1990): Tablets of stone: towards the ten commandments of geomorphology. – Z. Geomorph. N.F., Suppl.Bd. **79**: 1–37.
BRUNSDEN, D. (1996): Geomorphological events and landform change. – Z. Geomorph. N.F.40,3: 273–288.
BÜDEL, J. (1977): Klimageomorphologie. – Gebr. Borntraeger, Berlin-Stuttgart, 304 pp.
CHURCH, M. (1996): Space, time and the mountain – how do we order what we see? – In: RHOADS, B.L. & C.E. THORN (eds). The scientific nature of geomorphology. – 147–170, Wiley.
COSTA, J.E. (1988): Rheologic, geomorphic and sedimentologic differentiation of waterfloods, hyperconcentrated flows and debris flows. – In: BAKER, V.R., R.C.,KOCHEL & P.C., PATTON (eds.): Flood Geomorphology. – 113–122, Wiley.
DAVIS, W.M. (1899): The geographical cycle. – Geogr. Journ. **14**: 481–504.
DORN, R.J. (1996): Climatic hypotheses of alluvial fan evolution in Death Valley are not testable. – In: RHOADS, B.L. & C.E. THORN (eds): The scientific nature of geomorphology. – 220, Wiley.
DURY, G.H. (1964): Principles of underfit streams. – U.S. Geol. Surv. Prof. Paper **452A**: 1–67.
DURY, G.H. (1975): Neocatastrophism. – Ann. Acad. Brasil. Cienc. **47**: 135–151.
DYLIK, J. (1967): The main elements of Upper Pleistocene paleogeography in Central Poland – Biul. Peryglac. **16**: 85–115.
FROEHLICH, W., L. STARKEL & I. KASZA (1992): Ambootia landslide valley in the Darjeeling Hills, Sikkim Himalaya, active since 1968 – Journal of Himalayan Geology, 3,1: 79–90.
GAMPER, M. (1993): Holocene solifluction in the Swiss Alps: dating and climatic implications. – In: FRENZEL, B.: ESF Project, European Paleoclimate and Man, Spec. Issue. – 1–9, Acad.Wiss, Mainz.
GILBERT, G.K. (1877): Report on the geology of the Henry Mountains, Washington. – 160 pp.
GROVE, J. (1979): The glacial history of the Holocene. – Progr. Physic. Geogr. **3**, 1: 1–54.
JAHN, A. (1956): The action of rivers during the Glacial Epoch and the stratigraphic significance of fossil erosion surfaces in Quaternary deposits. – Przegląd Geograficzny **28**, Suppl: 101–104.
KARLEN, W. (1991): Glacier fluctuations in Scandinavia during the last 9000 years. – In: STARKEL, L., K.J. GREGORY & J.B. THORNES (eds.): Temperate Palaeohydrology: 395–412, J.Wiley.
KING, L.C. (1953): Canons of landscape evolution. – Biul. Geol. Soc. Am. **64**: 721–752.
KIRKBY, M.J. (1987): The Hurst effect and its implications for extrapolating process rates. – Earth Surf. Proc. Landforms **12**: 57–67.
KIRKBY, M. (1990): The landscape viewed through models – Z. Geomorph. N.F., Suppl.-Bd **79**: 63–81.
KLIMASZEWSKI, M. (1964): On the effect of the preglacial relief on the course and the magnitude of glacial erosion in the Tatra Mountains. – Geograph. Polon. **2**: 11–21.
KNOX, J.C. (1975): Concept of the graded stream. – In: MELHORN, W.N. & R.C. FLEMAL (eds.). Theories of Landform Development. – Binghampton Symposium **6**: 169–198.
KOTARBA, A. (1995): Rapid mass wasting over the last 500 years in the High Tatra Mountains. – Question. Geogr., Spec. Issue **4**: 177–183.
KOUTANIEMI, L. (1991): Glacio-isostatically adjusted palaeohydrology, the river Ivalojoki and Oulankajoki, Northern Finland. – In: STARKEL, L., K.J. GREGORY & J.B. THORNES (eds.): Temperate Palaeohydrology. – 65–78, Wiley.
KOUTANIEMI, L. & A. RACHOCKI (1981): Palaeohydrology and landscape development in the middle course of the Radunia basin, North Poland. – Fennia **159**, 2: 335–342.
KOWALKOWSKI, A. & L. STARKEL (1984): Altitudinal belts of geomorphic processes in the Southern Khangai Mts. (Mongolia). – Studia Geomorph. Carpatho-Balcanica **18**: 95–115.
LAWRENCE, D.S.L. (1996): Physically based modelling and the analysis of landscape development. – In: RHOADS, B.L. & C.E. THORN (eds).: The scientific nature of geomorphology. – 273–288, Wiley.
LEOPOLD, L.B., M.G. WOLMAN & J.P. MILLER (1964): Fluvial processes in Geomorphology. – Freeman and Co. San Francisco-London, 522 pp.
PENCK, W. (1924): Die morphologische Analyse; ein Kapitel der physikalischen Geologie. – Stuttgart.
PHILLIPS, J.D. (1996): Deterministic complexity, explanation and predictability in geomorphic systems. – In: RHOADS, B.L. & C.E. THORN (eds.): The scientific nature of geomorphology. – 315–335, Wiley.
RHOADS, B.L. & C.E. THORN (1996): Observation in geomorphology. – In: RHOADS, B.L. & C.E. THORN (eds). The scientific nature of geomorphology. – 21–56, Wiley.

RICHARDS, K. (1996): Samples and cases: generalisation and explanation in geomorpho-logy. – In: RHOADS, B.L. & C.E. THORN (eds.): The scientific nature of geomorphology. – 171–190, Wiley.
SCHUMM, S.A. (1977): The fluvial system. – Wiley, New York, 338 pp.
SELBY, M.J. (1974): Dominant geomorphic events in landform evolution. – Bull. Internat. Assoc. Engin. Geol. **9**: 85–89.
SOERGEL, W. (1921): Die Ursachen der diluvialen Aufschotterung und Erosion. – Berlin, 1–74.
STARKEL, L. (1969): Climatic or tectonic adaptation of the relief of young mountains in the Quaternary. – Geograph. Polon. **17**: 209–229.
STARKEL, L. (1976): The role of extreme (catastrophic) meteorological events in contemporary evolution of slopes – In: DERBYSHIRE, E. (ed.): Geomorphology and Climate. – 203–246, Wiley.
STARKEL, L. (1978): First stages of relief transformation of the young uplifted mountains, – Studia Geomorphologica Carpatho-Balcanica **12**: 45–61.
STARKEL, L. (1983): The reflection of hydrologic changes in the fluvial environment of the temperate zone during the last 15000 years. – In: GREGORY, K.J. (ed.): Background to Palaeohydrology. – 213–234, Wiley.
STARKEL, L. (1984): The reflection of abrupt climatic changes in the relief and in the sequence of continental deposits. – In: MÖRNER, N.A. & W. KARLEN (eds.): Climatic Changes on a Yearly to Millenial Basis. – 135–146, Reidel Publ.Comp., Dordrecht.
STARKEL, L. (1985): The reflection of the Holocene climatic variations in the slope and fluvial deposits and forms in the European mountains. – Ecol. Mediterran. **11**: 91–97.
STARKEL, L. (1987a): The role of the inherited forms in the present-day relief of the Polish Carpathians. – In: GARDINER, V. (ed.): International Geomorphology, Part.II. – 1033–1045, Wiley.
STARKEL, L. (1987b): Long-term and short-term rhythmicity in terrestial landforms and deposits – In: RAMPINO, M.R. et al.: Climate: History, Periodicity and Predictability. – 323–332, van Nostrand Reinhold.
STARKEL, L. (1990): Fluvial environment as an expression of geoecological changes. – Z. Geomorph. N.F., Suppl.-Bd. **79**: 133–152.
STARKEL, L. (1991): Long-distance correlation of fluvial events in the temperate zone. – In: STARKEL, L., K.J.GREGORY & J.B. THORNES (eds.): 473–493, Wiley.
STARKEL, L. (1992): Zložonošć współczesnej rzezby gór î wyžyn a rekonstrukcja paleogeómorfologiczna i prognoza zmian. – Przeglad Geograficzny **64**, 1–2: 87–94.
STARKEL, L. (1995): Evolution of the Carpathian valleys and the Forecarpathian Basins in the Vistulian and Holocene – Studia Geomorphologica Carpatho-Balcanica **29**: 5–40.
STARKEL, L. (1996): Geomorphic role of extreme rainfalls in the Polish Carpathians. – Studia Geomorphologica Carpatho-Balcanica **30**: 21–38.
STARKEL ,L., K.J. GREGORY & J.B. THORNES (eds.) (1991): Temperate Palaeohydrology. – 548 pp., Wiley.
THORNES, J.B. & D. BRUNSDEN (1997): Geomorphology & Time. – 208 pp., Methuen, London.
TWIDALE, C.R. (1996): Derivation and innovation in improper geology, aka geomorphology. – In: RHOADS, B.L. & C.E. THORN (eds.): The scientific nature of geomorphology. – 361–380, Wiley.
YATSU, E. (1993): Graffiti on the wall of a geomorphology Lab. Some comments on DENYS BRUNSDEN (1990). – Handout to delegates at the 3-rd Conference of Geomorphology, Ottawa.
YOSHIKAWA, T. (1985): Landform development by tectonics and denudation. – In: PITTY, A. (ed): Themes in Geomorphology. – 194–210, Croom Helm, London.
WOLMAN, M.G. & R. GERSON (1978): Relative scales of time and effectiveness of climate in watershed geomorphology. – Earth Surf. Proc. **3**: 189–208.

Address of the author: LESZEK STARKEL, Department of Geomorphology and Hydrology, Institute of Geography, Polish Academy of Sciences, Ul.sw.Jana, 31-018 Kraków, Poland.

Z. Geomorph. N.F.	Suppl.-Bd. 115	35–50	Berlin · Stuttgart	Juli 1999

The frequency and magnitude of geomorphic processes and landform behaviour

M.J. Crozier, Wellington

with 5 figures

Summary. The parameters of frequency and magnitude and their relationship provide only a partial explanation of episodic process behaviour. Full characterization of episodic processes from a geomorphic perspective should include, initiating and ending thresholds, duration, and areal extent of process activity. Successful application of frequency-magnitude studies to engineering, management, and geomorphic problems is strongly constrained by the equilibrium conditions of the geomorphic system. Most frequency-magnitude studies involve coupling of a forcing process with a process-response. The reliability of such relationships is dependent on the recognition of thresholds, scales of time and space, and other variable system properties.

An understanding of the nature of episodic geomorphic processes and process-response systems is thus essential to the reliable assessment of geomorphic activity and landform evolution. To this end, it is useful to view process relationships from both a geomechanical and behavioural perspective. It is proposed that the stress/resistance/strain relationship can be represented by corresponding behavioural concepts of 'potency', 'susceptibility' and 'occurrence', modified by the operation of 'ambient filters'.

The extension of frequency-magnitude concepts and other aspects of episodic behaviour to the development of geomorphic terrain requires the integration of multiple processes operating on different frequency-magnitude scales. Changes in frequency-magnitude behaviour may indicate significant environmental change in either the initiating forces and/or terrain susceptibility, including the influence of human activity.

However, the signal of different geomorphic events may be diffused in time and space at different rates, making subsequent interpretation of geomorphic behaviour a difficult task.

Zusammenfassung. Die Parameter Häufigkeit und Größe und deren Verhältnis zueinander liefern nur eine teilweise Erklärung für das Verhalten episodischer Prozesse. Eine vollständige Charakterisierung von episodischen Prozessen aus geomorphologischer Perspektive sollte folgende Parameter beinhalten: auslösende und beendende Grenzwerte, Dauer und räumliche Ausdehnung der Prozeßaktivität. Eine erfolgreiche Anwendung von Häufigkeits- und Größestudien für Ingenieurwesen, Management und geomorphologische Probleme wird durch die Gleichgewichtsbedingungen des geomorphologischen Systems stark eingeschränkt. Die meisten Studien beziehen eine Koppelung von verursachendem Prozeß und Prozeßreaktion ein. Die Verläßlichkeit solcher Beziehungen ist aber von der Wahrnehmung von Grenzwerten, Zeit- und Raumdimensionen und anderen veränderlichen Systemeigenschaften abhängig.

Für die zuverlässige Einschätzung geomorphologischer Aktivitäten und der Landschaftsentwicklung ist somit das Verstehen von episodischen geomorphologischen Prozessen und Prozeßreaktionssystemen absolut notwendig. Dazu erweist es sich als nützlich, die Prozeßbeziehungen aus geomechanischer und aus verhaltenstheoretischer Perspektive zu betrachten. Es wird vorgeschlagen, daß das Verhältnis von Spannung /Widerstand/Verformung auch durch die entsprechenden verhaltenstheoretischen Konzepte 'Stärke', 'Anfälligkeit' und 'tatsächliches Ereignis' dargestellt werden kann, wobei sie durch das Wirken von 'Umgebungsfiltern' modifiziert werden.

Die Ausdehnung von Häufigkeits-Magnituden-Konzepten und anderen Aspekten episodischen Verhaltens auf die geomorphologische Landschaftsentwicklung verlangt die Integration von zahlreichen Prozessen in verschiedenen Häufigkeits- und Magnitudenskalen. Veränderungen im Häufigkeitsverhalten

0044-2798/99/0115-0035 $ 4.00

oder in der Magnitude können auf signifikante Umweltveränderungen bezüglich der auslösenden Kräfte und/oder der Erosionsanfälligkeit des Geländes, einschließlich menschlicher Einflüsse, deuten. Die Tatsache, daß Zeichen verschiedener geomorphologischer Ereignisse zeitlich und räumlich unterschiedlich gut erhalten sein können, macht deren nachfolgende Interpretation bezüglich der geomorphologischen Auswirkungen zu einer schwierigen Aufgabe.

1 *Introduction*

There are three reasons why WOLMAN & MILLER's (1960) frequency-magnitude approach has been so influential in subsequent geomorphological research:

• It provides a rationale for extrapolating short-term measurements of episodic processes over longer periods, as a way of assessing long-term rates of geomorphic processes.

• It allows the statistical identification of the most important work force (event) operating within a system, thereby providing a key variable for predicting other system qualities.

• It enables the identification of a design or planning event, for use in engineering decisions and hazard management.

However, the success of this approach in these three areas requires the system to be in a state of equilibrium: in many situations this is only a realistic assumption for short periods and small areas. The value of the approach in the wider geomorphic context is therefore open to question.

In their seminal paper, WOLMAN & MILLER used an indirect approach for determining the frequency and magnitude of sediment movement, by linking the behaviour of the transporting agent (streamflow discharge) with the volume of sediment moved. There are obviously pragmatic reasons for depending on discharge measurements to predict sediment transport. The indirect approach, however, increases the potential for error, because the relationships involved are rarely well-defined. Nevertheless, the indirect approach, where the process of interest is linked and predicted by the behaviour of the initiating process, remains a popular way of characterising frequency and magnitude; usually because the initiating process is more easily measured or offers a better quality data base than the response variable. This is particularly the case with active, high frequency processes. For example, the frequency and magnitude of soil erosion and landslides may be determined from the behaviour of rainfall conditions, eolian transport from wind speed, coastal changes from the wave regime, and fluvial sediment movement by streamflow behaviour.

The coupling of the behaviour of an initiating agent with a geomorphic response introduces a number of complexities, which are the subject of this discussion. Besides an understanding of the explicit concepts of 'process' and 'eposidicity', the coupling of episodic processes requires an appreciation of initiating forces, resistance, and strain. These are the elements that invoke and explain the relationship between the initiating process and the geomorphic response. Inevitably, such relationships involve a threshold condition. The identification of thresholds, both initiating and terminating, is fundamental to the successful employment of the indirect approach to assessing frequency and magnitude.

For frequency-magnitude studies of contemporary processes to be of wider value in interpreting landform behaviour, results of field observations must be extrapolated to wider scales of space and time. The problems involved in doing this are also addressed in this discussion. They involve questions of dissipation of response, equilibrium of systems, and the interdependence of multiple processes operating on different frequency-magnitude scales.

2 *The geomorphic process*

While the concept of 'process' is generally implicit in the appreciation of dynamic geomorphology, it is rarely defined. For the purpose of this discussion, it is defined in the following way.

A process is a pathway (sequence) of identifiable, essential, and causally linked steps leading to a designated end point. In the specific case of a geomorphic process, these steps involve movement and/or a change in mass and energy conditions within the landform system. The focus of a process study is the chosen end step – essentially, the dependent factor. The associated process pathway (Fig. 1) is thus generally invoked to explain the behaviour, controls, and/or origin of the dependent factor. The starting step, number of component steps and the end step are arbitrarily chosen. They define the process of concern and are strongly dependent on the purpose of the study and hence the spatial and temporal scales of interest. Depending on the scale, each step may be further analysed to identify a hierarchy of more detailed, nested, process pathways (Fig. 1).

Any preceding step and any subsequent step in a process pathway may be coupled as the **forcing (initiating) process** and the **process-response** respectively. The initial response to a forcing event is termed the primary response, which in turn may initiate a secondary response and other down-pathway responses.

3 *Episodicity*

It has long been recognised that geomorphic processes are episodic. Even those that are sometimes construed as continuous will be manifest as episodic on certain time scales. For example, soil creep in a given area may be considered as episodic on a diurnal or seasonal basis, continuous over a period of decades, but episodic over a period of millions of years. Eposidicity is evident at all scales. SCHUMM (1977) emphasised this point by making reference to AGER's (1973) telling observation of the stratigraphic record: 'a lot of holes tied together with sediment'.

Episodicity is therefore a fundamental concern of geomorphology, if for no other reason that suddenness can be a trauma to other parts of the ecosystem including human livelihood. The elements that characterise episodic behaviour are the start, end, duration, areal extent of operation as well as the magnitude and frequency. It is important to note that frequency and

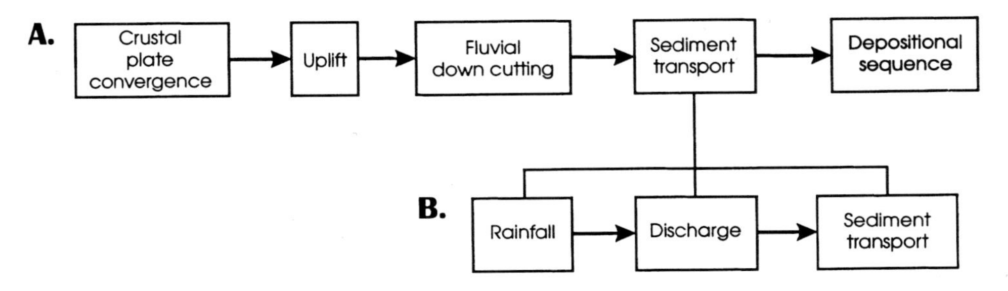

Fig. 1. Example of steps in a process pathway invoked to explain (A) a depositional sequence and (B) sediment transport. Each step can be analysed to reveal further nested process pathways.

magnitude are therefore only components of episodic behaviour and individually they provide only a partial explanation of the geomorphic process. It is not surprising, then, that in the almost 40 years since WOLMAN & MILLER (1960) established the influential tenets of frequency and magnitude, they have been revisited (THORNE & DARBY 1997), questioned and refined, if not overtly challenged (RICHARDS, this volume). The full characterisation, explanation and prediction of episodic process are a major concern for both geomorphic and social reasons.

Episodicity occurs either because eposidicity is inherent in the forcing process or because there is not a constant, proportional, relationship between the forcing process and behaviour of the factors controlling the process response. Scale is again important to this discussion because, at a given time scale, a forcing process which is viewed as continuous may produce episodic response. For example, the strain resulting from continuous crustal plate convergence may be manifest as episodic earthquake activity and coseismic uplift.

Changes in the relationship between operation of the forcing process and the consequential behaviour of the receiving system may be recognised as **thresholds** and are generally responsible for initiating and ending down-pathway steps in the process pathway. SCHUMM's (1977) characterisation of the different types of threshold still provides one of the most useful discussions on this topic.

4 *Physical and behavioural relationships between forcing events and response*

Fundamental to the relationship between a forcing process and a response are the physical concepts of applied force, resistance, and strain. For example, when force applied from streamflow overcomes resistance derived from the weight of particles resting on a stream bed, then 'strain', is manifest by sediment movement. The onset of strain indicates that a threshold has been surpassed in the relationship and another process step (sediment movement) has been initiated. However the amount and nature of that movement will relate to the properties of the system. Similarly, when shear stress on a hillslope overcomes shear strength, the strain is manifest by a process response in the form of mass movement. Again the nature and magnitude of the responding process (for example, whether a slump or earth flow) depends on the properties of the material and the hillslope system. In other words, the magnitude of the forcing process alone is not always sufficient to predict the magnitude and type of response.

As a measure of the state-of-art in frequency-magnitude studies, it is instructive to examine the findings of papers presented in 1997 to the Frequency-magnitude Symposium organised by the International Association of Geomorphologists. Many of these papers, referred to below, have identified problems in discovering a consistent relationship between the forcing process and the response. Commonly, a given magnitude of activity of the forcing event has been shown to yield different responses in space and time.

These problems do not challenge the physics of the stress/resistance/strain laws that underpin geomorphic behaviour; rather, they reflect the complexity and multivariate nature of the geomorphic system and the scale of our own interest. The greater the spatial and temporal scales involved, the more difficult it is to isolate, measure and control the important variables (BOARDMAN & FAVIS-MORTLOCK, this volume).

In moving from the confines of the laboratory and experimental plot to the more complex condition of the hillslope or catchment (BALLAIS 1997), it becomes useful to replace the considerations of **force** (stress/resistance/strain) with matching considerations of **behaviour**. In the behavioural approach, 'stress' can be represented by **potency** of the forcing process, 'resistance' as a component of **susceptibility**, and 'strain' as the **occurrence and magnitude** of geomorphic response (Fig. 2). The degree to which the implied actions and reactions are effective depends on a number of complex conditions referred to as **ambient filters**. In discussing the behavioural approach to the forcing process/response relationship, CROZIER & PRESTON (1998) define 'susceptibility' as the propensity with which the receiving system produces a response to a forcing process. 'Susceptibility' can be conveniently represented by a threshold value of the forcing process. Susceptibility is inversely proportional to the threshold value; systems with high 'susceptibility' have a low threshold.

The forcing process, however, can be treated as an independent factor. 'Potency' is defined as the ability of the forcing process to reach or exceed a given threshold. It is a measure of the energy fluctuations of the forcing system and can be characterised by the frequency with which the applied stress level exceeds a given level of susceptibility.

In geomorphic systems, however, the stress applied by the forcing process may be dissipated or concentrated by factors within the receiving system, referred to as 'ambient filters'. In other words the effectiveness of an agent may be enhanced or suppressed by ambient conditions. For example, the effectiveness of rainfall input to a hillslope may be concentrated or dissipated by topographic or stratigraphic factors; or, a chemical reaction in the weathering process may vary with ambient temperature conditions or the presence of a catalyst. Similarly, a measured stream discharge may produce different responses in the channel system, depending upon morphological characteristics of the channel, which may enhance or dissipate the

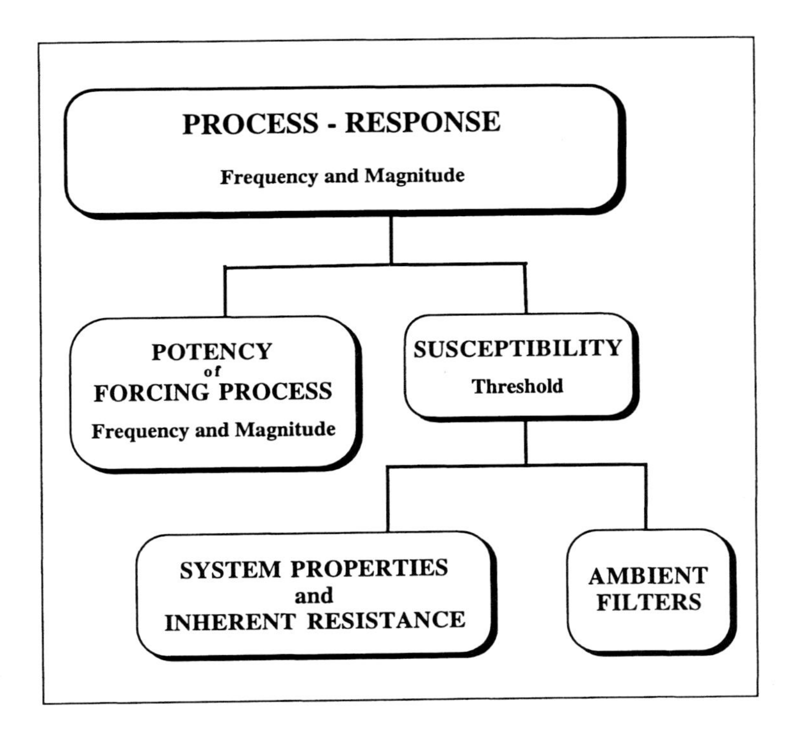

Fig. 2. Factors controlling the frequency and magnitude of process-response.

effects of discharge. These ambient filtering mechanisms in themselves may produce variance in the relationship between an externally measured input magnitude and a geomorphic process response. In this way, ambient filters may not only alter the threshold of response (susceptibility) but may also change the nature of response. Together with the inherent resistance and properties of the receiving system, ambient filters may be effective enough to dictate the type of response. For example, while a rainstorm may possess sufficient potency to initiate mass movement in a given terrain, the specific nature of the response (whether flow or slide) may vary. The nature of the response can be controlled by system properties such as clay content or the extent to which water from the triggering rainfall can be concentrated onto a specific site. GRANT et al. (1997) have shown that the down-pathway process-response from a flood and landslide event can vary dramatically based upon the degree of involvement of organic debris in the transported medium. Similarly, ALABYAN (1997) and ANDREWS (1997) conclude that the effective discharge varies in accordance with channel morphology and associated hydraulic properties of the channel. As RICHARDS (this volume) points out, the frequency and magnitude of streamflow and the degree of work accomplished is often constrained by the important system property of material availability. It is common to find that many stress-based action and reaction relationships are constrained by the supply of susceptible material.

Whereas ambient filters and system properties may obscure the relationship between the forcing process and response, other factors may also play a part. Many problems in establishing a relationship between the forcing event and a response lie simply with an inability to characterise appropriately the important stress elements of the forcing event – clearly magnitude alone is an insufficient means of characterisation in many cases.

5 *Characterising the forcing event and response*

As BOARDMAN & FAVIS-MORTLOCK (this volume) conclude from their study of soil erosion and rainfall, the characterisation of the potency of a rainstorm by its magnitude alone may only partially represent the forcing process. More direct representations of the forcing process in respect of soil erosion are attempted by examining the relationship of splash generation with magnitude and by calculating kinetic energy of rainstorms (CALLES & KULANDER 1997, BRANDÃO & FRAGOSO, this volume). Whatever the forcing process, considerable thought is required to identify and represent its effective components. For example, in the case of rainfall-triggered landslides, usually only a very broad relationship can be established between rainstorm magnitude (total rainfall) and the nature of the response – the direct effective element is actually the amount of water within the hillslope. Usually this is best approximated by comparing input rates with drainage rates and taking account also of resident storage resulting from antecedent rainfall and evapotranspiration conditions (CROZIER & GLADE, this volume) or by directly measuring soil water tensions and piezometric pressures within the slope (MACQUAIRE 1997). In the first instance, the problem of establishing consistent process-response models often amounts to identifying the effective or critical component of the forcing mechanism.

The characterisation of a process by its frequency has always taxed and fascinated geomorphologists. For a given area and process, the higher the magnitude of the event the less frequently it is manifest, certainly in time and probably also in space (COLANGELO & CRUZ 1997). The characteristics of these frequency-magnitude distributions appear to be unique to the process and to a place. HOVIUS et al. (1997) have suggested that in the case of bedrock landslides these differences may be attributed to differences between the triggering mechanism – whether seismic or climatic. COLANGELO & CRUZ (1997), on the other hand, attribute differences in the spatial frequency-magnitude distribution partly to rock type.

One of the fascinations of frequency investigations is that they provide the meeting ground between what BRUNSDEN (1993) recognises as functional process studies and historical, evolutionary research. Beyond the instrumental record the assessment of frequency and magnitude requires a myriad of dating and interpretative skills (IBSEN & BROMHEAD, this volume, HOUSE & BAKER 1997, PANIN 1997).

For a given process, frequencies are normally determined from the number of times the threshold is exceeded (occurrences) or the number of times the threshold is exceeded by a specified magnitude, in a unit of time. They are determined either stochastically, taking into account the nature of the frequency-magnitude distribution or simply given as average frequencies derived by dividing the number of events by the length of record (historical frequency). In this way frequency is expressed as a recurrence interval or return period which can also be expressed as a probability of occurrence for any specified time period.

An important property of recurrence intervals (often overlooked when comparisons are made) is that they are a function of the size of the area from which they are derived. In other words, the larger the area, all other things being equal, the more likely it will be to experience a specified event, and therefore the shorter the derived recurrence interval. Another point of caution is that the statistical derivation of frequency may obscure the clustering of events, which, in itself, may be geomorphically significant in terms of response (STARKEL this volume). Because variation of frequency through time may signal a significant environmental change (HOUSE & BAKER 1997), the identification of clustering is an important aspect of frequency-magnitude analysis. The effect on frequency-magnitude distributions of 'system memory' and 'self-organised criticality' has been addressed by BOARDMAN & FAVIS-MORTLOCK (this volume).

Another element of the forcing process besides magnitude and frequency which influences geomorphic response is duration (RICHARDS, this volume, CHIZYKOV, this volume). Clearly, in power-constrained systems, high effective streamflow, maintained for long periods, will accomplish more work than short duration events. Similarly, studies of the relationship between the magnitude and intensity of earthquakes and land deformation have shown that duration of shaking is also important in evoking a geomorphic response. Unlike frequency-magnitude distributions, little work has been published of frequency/duration distributions. Similarly, although the areal extent of process influence is also of geomorphic significance, it has received little attention.

Characterising the nature of response is also critical to the establishment of the relationship between the forcing event and the response. Geomorphic experience shows that the greater the magnitude of the forcing process above the initiating threshold, the greater the geomorphic response (NIKINOV & SERGEEV 1997, KEEFER 1984, OMURA & HICKS 1991, EYLES & EYLES 1982). That response has to be measured in terms equally as precise as those used for the forcing mechanism. Some geomorphic studies are simply confined to the recognition of occurrence; i.e., they are concerned solely with whether the forcing mechanism produces a response or not. Clearly, there needs to be a meaningful system for characterising response, because, if for no other reason, the response in one step of the process-response system may represent the forcing process for further down-pathway response.

6 *Complex response*

Adequate representation of inherent resistance, system properties and ambient filters, as well as the recognition of feedback mechanisms, is required to anticipate the nature of response throughout the process pathway. However, peculiar and often unanticipated conditions of the geomorphic system can also influence the nature of response. For example, PANIN (1997) shows that during the process of downcutting a river may encounter material of variable resistance, thereby changing the sediment yield in the course of a given process-response. Similarly, SHIMAZU (1997) describes a landslide response that became complicated as a result of topographical conditions – a landslide dam was formed which subsequently burst, producing a high magnitude debris torrent. SCHUMM (1973) in putting forward the concept of complex response, points out that different parts of a catchment system may undergo different responses to a forcing process as the responses work their way throughout the system. The appreciation of complex response is essential for accurate interpretation and deduction of causes. The interrelationships between geomorphic and ecological processes and the importance of feedback mechanisms to the nature of complex response have been addressed by CAMMERAAT (1997).

7 *What is happening to the terrain: what is normal?*

One of the significant demands placed on geomorphologists by engineers and land managers is the question of what is happening within the geomorphic system. What process behaviour can be considered normal and what changes, and rates of change need to be realised, coped with, and planned for? As previously argued, the question of not only magnitude and frequency but also duration and areal extent of geomorphic processes is fundamental to this problem. In high energy geomorphic systems there are many examples where engineers have tried to combat partially understood processes without success and at great cost. MOSLEY (1978) has clearly demonstrated that remedial efforts can be productive if the geomorphic system is appreciated or wasteful if the trends are not realised. He recognised that mountainlands in New Zealand are undergoing irreversible erosional changes, and that many headwater erosion-control measures have to work against expanding erosional activity. He concluded that mitigation measures could only be effective when applied to components of the geomorphic system where process trends were understood and amenable to control.

Clearly, it is incumbent upon geomorphologists from a social perspective to be able to answer these questions (BORA 1997); that is, to be able to represent the condition and trends within the geomorphic system in a reliable and useful way. Many purely geomorphic questions, however, encompass time scales many orders of magnitude greater than the periods of contemporary process measurement (CHEVAL 1997). At such scales, behaviour can be represented by the sorts of concepts put forward by WOLMAN & GERSON (1978) and BRUNSDEN & THORNES (1979). The fundamental parameters required to address these questions are encapsulated in Fig. 3. The difficult exercise is the translation of these concepts from an individual site to an integrated terrain at different scales. Pivotal to the question of what is happening to the geomorphic system is the 'Transient Form Ratio'. This compares the relaxation time of a geomorphic change to the frequency of those changes within the system. As SELBY (1974) and BRUNSDEN & THORNES (1979) have pointed out, some systems may have the resilience and the ability to continually resume a constant condition (various forms of equilibrium – SCHUMM

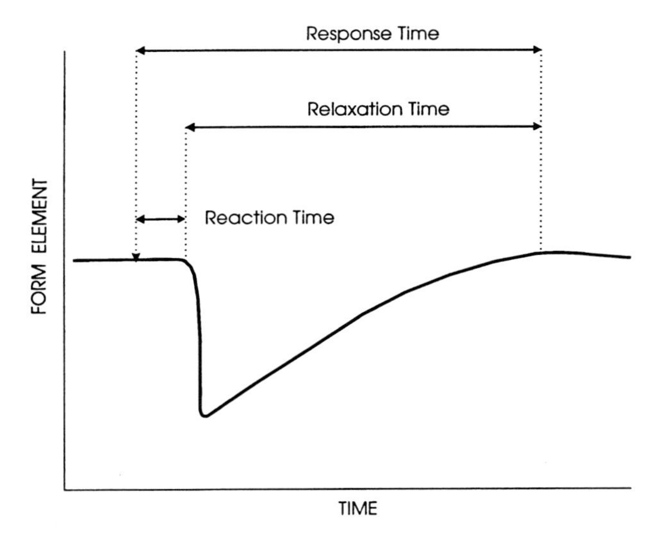

Fig. 3. Form response and recovery with time, from an initiating event.

1977). Other systems are subject to a condition of constant change and adjustment, as represented by the disequilibrium and nonequilibrium conditions described by RENWICK (1992). CROZIER & PILLANS (1991) represent this question by assessing the residence time of the effects of particular events compared with their return period or return period of other processes that would obscure the primary response. They have shown, for a large area of New Zealand, that the normal condition is a state of constant adjustment. Therefore the concept of an equilibrium condition or a characteristic geometry has little meaning to certain terrain within dynamic landscapes. Such distressed systems have been addressed in terms of geomorphic theory and have variously been classified as nonequilibrium or meta-stable conditions.

FORT'S (1997) study of the Himalayas attempts to gauge the effect of magnitude of events on the rate of landform recovery. Clearly the magnitude affects the recovery time but the question is open as to how magnitude affects the actual rate of recovery.

Thus the emphasis today in frequency-magnitude studies has moved from the original WOLMAN & MILLER (1960) focus that was directed towards assessing the effectiveness of a single process, within a limited component of the terrain, to a wider endeavour – unravelling the complexity of the whole landscape system. The challenge now is to isolate and predict the landform product of intersecting process regimes: regimes that are interrelated but working to their own frequency and magnitude agendas.

8 *Integrating the effects of multiple processes*

A geomorphic terrain is the product of the operation of numerous process-response systems, their eposidicity and their relative effectiveness. Its configuration is the net effect of the hierarchy of processes and competition amongst the formative events of multiple processes (Fig. 4).

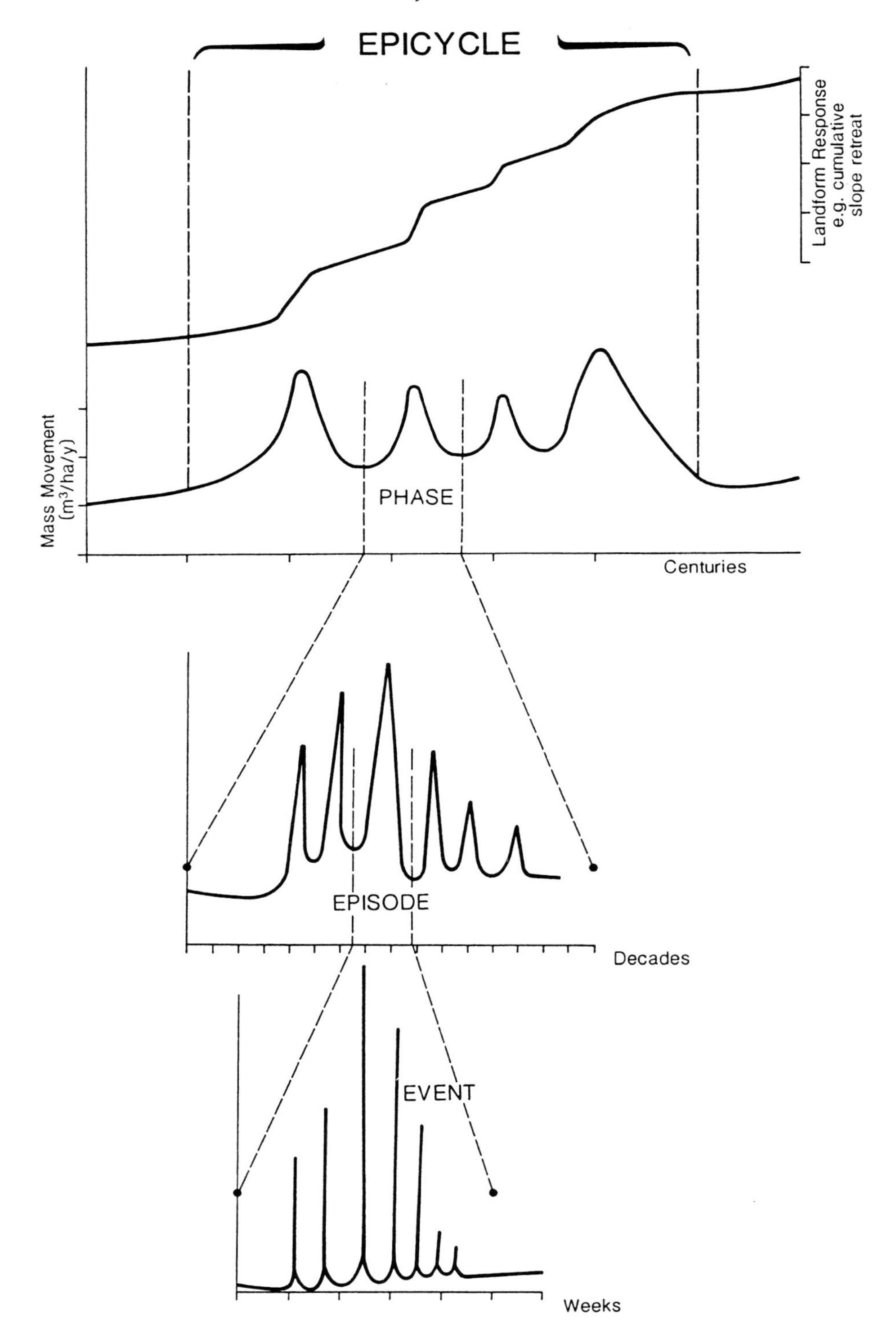
EPICYCLE
Landform Response
e.g. cumulative
slope retreat
Mass Movement
(m³/ha/y)
PHASE
Centuries
EPISODE
Decades
EVENT
Weeks

Because of the paucity of information on eposidicity through the evolutionary time of terrain, BRUNSDEN (1993) has suggested that the key to unravelling this 'palimpsest' is the identification of the persistence and rate of destruction of form elements. Interestingly, W. M. DAVIS (1899) used the same attributes to diagnose the stages of his erosion cycle, and a number of their conclusions are identical, for example; 'the exponential decrease in the rate of landform change with time'. However, BRUNSDEN argues for the importance of other factors, which do not necessarily produce a unidirectional sequential development of form through time.

Under certain geomorphic conditions, landscape processes and form can be attributed to a few dominant processes that prevent other process domains from developing. For example, HOVIUS et al. (1997) consider that uplift rates and relief development in active alpine regions mean that the occurrence of bedrock controlled landslides outpaces the development of conditions which would permit weathering constrained regolith landslides to occur. In order for this process to persist, fluvial processes need to be equally effective in order to remove the products of mass movement. Over long periods of time the effectiveness and the dominance of processes may be judged by their net effect. That we have mountains at all means, in net terms, that uplift has been greater than denudation.

On the human time scale, the interaction of process systems and their eposidicity becomes a much more practical concern. For example, predicting the frequency and magnitude of a given geomorphic process often requires considering the behaviour of processes that supply the material involved as well as the frequency-magnitude behaviour (potency) of the agent supplying power to initiate the process of concern (ANDREWS 1997). THOMPSON & ADAMS (1979), for example, can account for the offshore accumulation of sediment in New Zealand only by resorting to the episodic supply events of earthquakes and rainstorms. Similarly, the occurrence of debris flows may be viewed as a function of the frequency and magnitude of upslope events which supply a critical accumulation of debris as well as the frequency and magnitude of rainstorm triggering events (VAN STEIJN 1996). In the case of rainfall-triggered regolith slides, the occurrence is not just a function of the potency of the rainfall regime but of the frequency and magnitude of the coincidence of the dual conditions of high antecedent soil water and a given magnitude of rainfall.

The sediment delivery ratio of a catchment subject to an extreme storm event is a function of the frequency-magnitude behaviour of the external triggering regime as well as the variation in frequency-magnitude response from subcatchment to subcatchment. The effect of different intersecting processes, each with their own frequency-magnitude regime, is addressed for small catchments by BOARDMAN & FAVIS-MORTLOCK (this volume) and over large catchments by TRUSTRUM et al. (this volume). The focus of the large catchment study by TRUSTRUM et al. is the Waipaoa Catchment, North Island, New Zealand, one of the best studied and most actively eroding hill country catchments in the world. Using a multiple process approach, they contrast frequency-magnitude behaviour of this large catchment with a similar but much smaller catchment. Despite the unstable terrain, within the main drainage system of the large catchment, the magnitude and frequency of sediment movement is much as predicted by WOLMAN & MILLER (1960). By contrast, in the smaller catchment, sediment movement is dominated by episodes of slope mass movement.

Fig. 4. Hierarchy of process episodicity and landform response. Examples of the forcing processes responsible for each type of episodic behaviour are: 'epicycle', eustatic change during an orogeny; 'phase', major deforestation; 'episode', change in climatic or seismic pattern; 'event' persistent weather patterns.

9 *Back analysis*

Much geomorphology lies beyond the realms of the instrumental record. Consequently interpretation and deduction is required to establish the process pathways responsible for many geomorphic features. Critical to this endeavour is the ability to date geomorphic features (House & Baker 1997) and relate them to the forcing process-response system. This area of research has highlighted a number of questions.

- How does the longevity of evidence of a given frequency and magnitude of event compare between different processes. For example, do the results of a 100 year return period landslide event last longer in the landscape than a 100 year return period fluvial event?
- What can be said about the magnitude of the forcing event from the longevity of evidence?
- How does longevity of evidence affect the length of period over which reliable statements of frequency can be made? In other words do periods in the record lacking evidence mean no occurrence or obliteration of evidence? How quickly does the evidence of episodic behaviour diminish?
- What can be said about the nature and magnitude of the forcing event from the nature and magnitude of the response?
- What is the time lag between a given forcing event and the initiation of its response (reaction time)? Which factors control the reaction time?
- What is the nature of linkage and connectivity between systems that are involved in the process pathway? Prost (1997), for example, demonstrates a long distance linkage between coastal progradation in French Guyana and flood events from the Amazon River.
- To what extent does the response to a forcing event become diminished or enhanced as it is propagated throughout the process pathway (positive and negative feedbacks)? Is the evidence of an event lost by the 'concertina effect'? For example, a series of nickpoints caused by base level changes may merge as their headward retreat is slowed when they encounter a particularly resistant material in the channel.

Given a comprehensive understanding of process pathway behaviour, changes of process frequency and magnitude, evident in the record, can be an important key to unravelling palaeoenvironmental change. However, the role of the forcing process-response system (Fig. 2) must first be fully appreciated. Over time changes in the frequency of response may be brought about by either a change in the potency of the forcing regime or a change in susceptibility (threshold shift – Fig. 5). Temporal shifts in thresholds are not well understood and some appear to be an intrinsic response to the operation of the process itself (Anderson & Brooks 1997, Preston, this volume).

10 *The human factor*

In many studies of the forcing process/response relationship the human factor is identified as lowering the threshold of response. Prosser (1997) suggests that gully formation under conditions free from human interference may be initiated by a 100 year return period event compared with a 20 year return period event under degraded vegetation conditions. Situations must also exist where human activity increases the threshold of response but these seem to have received less attention in the geomorphic literature.

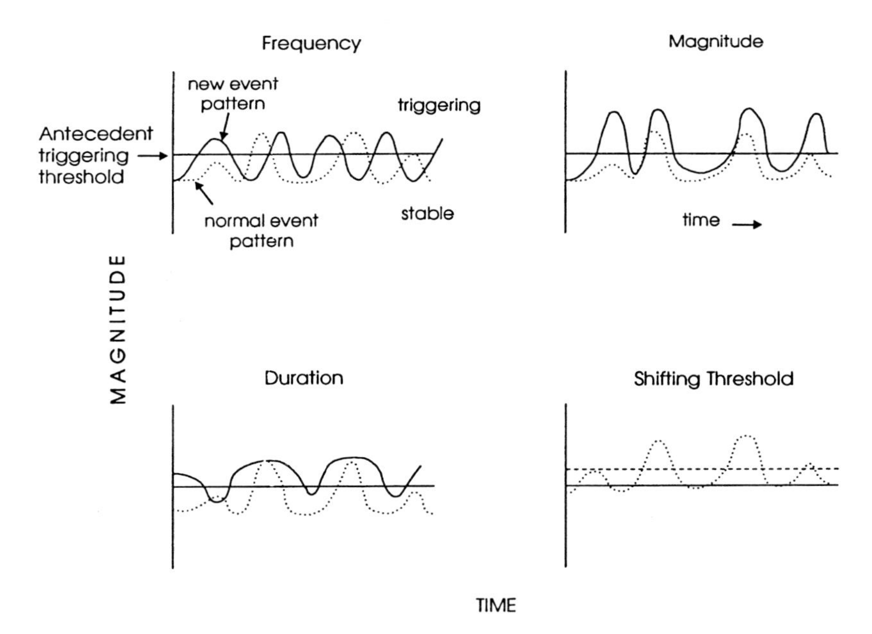

Fig. 5. Changes in episodic behaviour of the forcing (triggering) process and threshold changes, which lead to changes of frequency and magnitude of response. The diagrams illustrate how the triggering threshold can be exceeded more frequently or for a greater period of time than normal by increases in frequency, magnitude, and duration of the forcing process or by lowering of the inherent threshold.

Human activity more readily influences the susceptibility of the receiving system than the behaviour of the forcing agent. Because frequency of occurrence is a function of both the potency of the forcing mechanism as well as the susceptibility of the receiving system, it is important to be able to identify which of these factors is responsible for changes in frequency behaviour. Isolating the role of the human factor is essential before attempting reconstruction of palaeoenvironmental conditions relating to climate or tectonic activity (KASHIWAYA & OKIMURA 1997). CHIZYKOV (this volume) notes the effect of drainage of swampy soil in Belarus on increasing the frequency and magnitude of dust storms. Similarly JANICKI & ZGLOBICKI (1997) record the importance of the human effect on the vegetation cover in encouraging the appearance of ephemeral channels and their obliteration by ploughing. DANTAS et al. (1997) have identified distinct erosion phases in the hanging valleys of Southeast Brazil. They show that the phase that commenced 200 years BP, as the forest was cleared for coffee production, produced sedimentation rates almost double those of the erosion phase that occurred between 10,000 and 8,000 years BP.

11 *Conclusions*

Frequency and magnitude are key variables for characterising the behaviour of episodic geomorphic processes. The relationship between these variables provides a rationale for making engineering design decisions, establishing management strategies, and determining the important influences on landform development. However, the stability of this relationship and hence its ultimate reliability for such applications depends on the existence of equilibrium conditions within the system: in many situations, this is only a realistic assumption for short periods and small areas.

Many frequency-magnitude distributions are determined indirectly by establishing a relationship between a forcing process and a process-response. The indirect approach introduces further uncertainty into the application of frequency-magnitude relationships. This is because the relationship between a forcing process and process-response can be influenced by a number of other system properties, including ambient filters and the availability of susceptible material.

A useful way of identifying the controls on frequency and magnitude is to characterise processes in geomechanical and behavioural terms. For many geomorphic situations, properties of 'applied force', 'resistance', and 'strain' can be replaced by corresponding concepts of 'potency', 'susceptibility', and 'occurrence' (response).

Identifying episodic behaviour from the record of events remaining in the landscape presents a number of problems. These include the dissipation of event signals in space and time. Many challenges remain in explaining landform evolution resulting from multiple processes operating on different frequency-magnitude time scales.

References

AGER, D.V. (1973): The nature of the stratigraphic record. – Macmillan, London, 114pp.

BRUNSDEN, D. (1993): The persistence of landforms. – Z. Geomorph. N.F., Suppl.-Bd. **93**: 13–28.

BRUNSDEN, D. & J.B. THORNES (1979): Landscape sensitivity and change. – Trans. Inst. Brit. Geogr. **4**: 463–484.

CROZIER, M.J. & B.J. PILLANS (1991): Geomorphic events and landform response in south eastern Taranaki, New Zealand. – Catena **18** (5): 471–488.

CROZIER, M.J. & N.J. PRESTON (in press): Modelling changes in terrain resistance as a component of landscape evolution in unstable hill country. – In: HERGARTEN, S. & H. J. NEUGEBAUER (eds.): Process modelling and landform evolution. – Lecture Notes in Earth Sciences **78**: 267–284, Springer, Heidelberg.

DAVIS, W.M. (1899): The geographical cycle. – Geogr. Journ. **14**: 481–504.

EYLES, R.J. & G.O. EYLES (1982): Recognition of storm damage events. – Proceedings of the Eleventh New Zealand Geography Conference, Wellington, 1981: 118–123.

KEEFER, D.K. (1984): Landslides caused by earthquakes. – Geol. Soc. Amer., Bull. **95** (4): 406–421.

MOSLEY, M.P. (1978): Erosion in the southeastern Ruahine Range – its implication for downstream river control. – New Zealand Journ.Forestry **23**: 21–48.

OMURA, H. & D. HICKS (1991): Probability of landslides in hill country. – In: BELL, D. (ed.): Landslides. – Proceedings of the Sixth International Symposium on Landslides, Christchurch. Feb. 1992, **2**: 1045–1049, Balkema.

RENWICK, R.H. (1992): Equilibrium, disequilibrium, and nonequilibrium landforms in the landscape. – Geomorphology **5**: 265–276.

SCHUMM, S.A. (1973): Geomorphic thresholds and complex response of drainage systems. – In: MORISAWA, M. (ed.): Fluvial geomorphology – Publ. in Geomorphology **3**: 299–310.

SCHUMM, S.A. (1977): The fluvial system. – Wiley, New York, 338 pp.

SELBY, M.J. (1974): Dominant geomorphic events in landform evolution. – Bull. Int. Assn. Eng. Geologists **9**: 85–89.

THOMPSON, S.M. & J. ADAMS (1979): Suspended sediment load in some major rivers of New Zealand. – In: MURRAY, D.L. & P. ACKROYD (eds.): Physical hydrology – New Zealand experience. New Zealand Hydrol. Soc.: 213–219.

VAN STEIJN, H. (1996): Debris flow magnitude-frequency relationships for mountainous regions of Central and Northwest Europe. – Geomorphology **15**: 259–273.

WOLMAN, M.G. & R. GERSON (1978): Relative scales of time and effectiveness of climate in watershed geomorphology. – Earth Surf. Proc. Landforms **3**: 189–208.
WOLMAN, M.G. & J.P. MILLER (1960): Magnitude and frequency of forces in geomorphic processes. – Journ. Geol. **68**: 54–74.

The following references are all abstracts of papers presented to the International Association of Geomorphologists, Fourth International Symposium on Geomorphology, Bologna, 28 August – 3 September 1997: **SIMPOSIO S5 Magnitude and Frequency in Geomorphology.**

ALABYAN, A. (1997): Magnitude-frequency concept and river channel patterns.
ALLEN, P. & N. HOVIUS (1997): Sediment delivery to basins from landslide-driven fluxes.
ANDERSON, M. & S. BROOKS (1997): Identification of climatic thresholds for slope failure: application of physically-based models.
ANDREWS, E. (1997): Magnitude and frequency of bed-material transport in the Virgin River, Colorado Plateau, USA.
BALLAIS, J. (1997): Evolution of a gully in Lower Provence (France): preliminary results (October 1991 – February 1996).
BOARDMAN, J. & D. FAVIS-MORTLOCK (1997): Frequency and magnitude in soil erosion – some effects of spatial scale.
BORA, A. (1997): Floods of Assam (India): A geomorphological analysis.
BRANDÃO, C. & M. FRAGOSO (1997): Rainfall controls of soil erosion in Alentejo (Portugal).
CALLES, B. & L. KULANDER (1997): Erosive rains in Lesotho.
CAMMERAAT, E. (1997): A hierarchical approach to the geomorphological development of hillslope and catchment geomorphology: two contrastingexamples from a semi-arid and humid temperate region.
CHEVAL, S. (1997): Relation between the climate and the morphology of the south Dubroudja Plateau. Special view on the impact of the climatic hazards.
CHIZYKOV, Y. (1997): Dusty storms as a factor of modern relief-formation (by example of Belarus).
COLANGELO, A. & O. CRUZ (1997): Spatial magnitude-frequency index of mass movement event deposits in a humid tropical Precambrian plateau, and its connection with MFI of daily rainfall.
CROZIER, M. & T. GLADE (1997): Magnitude and frequency of landslide events in New Zealand.
DANTAS, M., L. DO EIRADO SILVA & A. COELHO NETTO (1997): Sediment storage on hanging valleys of SE Brazil: sedimentation and lowering rates.
FAVIS-MORTLOCK, D. & J. BOARDMAN (1997): The importance of large and infrequent rainfall events for long-term rates of soil erosion: a simulation study from the UK South Downs.
FORT, M. (1997): Large scale geomorphic events in the Nepal Himalaya and their role in the evolution of the landscape.
GRANT, G., S.I. JOHNSON, F.J. SWANSON & B. WEMPLE (1997): Anatomy of a flood: geomorphic and hydrologic controls on channel response to a major mountain flood.
HOUSE, P. & V. BAKER (1997): Unconventional methods for evaluating the magnitude and frequency of Flash Floods in ungauged desert Watersheds: an example from Arizona.
HOVIUS, N., C. STARK & P. ALLEN (1997): Landslide magnitude-frequency distributions in New Zealand, Papua New Guinea and Taiwan.
IBSEN, M. & E. BROMHEAD (1997): Head scarps and toe heaves.
JANICKI, G. & W. ZGLOBICKI (1997): The condition of development of episodic channels in loess areas of the Lublin Upland (SE Poland).
KASHIWAYA, K. & T. OKIMURA (1997): Erosional environmental change due to the Kobe earthquake inferred from pond sediments.
MAQUAIRE, O. (1997): Les glissements de Villerville-Cricqueboeuf (Calvados, France). Douze années de surveillance: frequency et magnitude.
NIKONOV, A. & A. SERGEEV (1997): Identification and quantification of seismogravitational relief disturbances: The Caucasian Mountain area as an example.
PANIN, A. (1997): Catastrophic rates of river incision: case study of Alabuga River, central Tien-Shan.
PRESTON, N.J. (1997): Event-induced changes in landsurface condition – implications for subsequent slope stability.

PROSSER, I.P. (1997): catastrophic expansion of channel networks in Holocene and Historical times.
PROST, M. (1997): Modifications de courte durée et forte intensité de la ligue du rivage de la Guyane Française analisées par télédétection.
RICHARDS, K. (1997): Events in fluvial geomorphology; auxiliary hypotheses and the normal science of magnitude and frequency.
SCHNABEL, S. & A. CEBALLOS (1997): Gully erosion and temporal variability of discharge in a small watershed in semi-arid Spain.
SHIMAZU, H. (1997): Catastrophic debris transport along the Japanese mountain river.
STARKEL, L. (1997): Role of events, phases and main climatic stages in Quaternary evolution of landscape.
THORNE, C. & S.E. DARBY (1997): Magnitude and frequency analysis of large alluvial rivers.
TRUSTRUM, N., L. REID, M. PAGE, B. GOMEZ (1997): From hillslope to floodplain: linking magnitude-frequency relations across steepland catchments.

Address of the author: M. J. CROZIER, School of Earth Sciences, Victoria University, P.O. Box 600, Wellington, New Zealand.

Z. Geomorph. N.F.	Suppl.-Bd. 115	51–70	Berlin · Stuttgart	Juli 1999

Frequency-magnitude distributions for soil erosion, runoff and rainfall – a comparative analysis

John Boardman and David Favis-Mortlock, Oxford

with 9 figures and 2 tables

Summary. The interplay between rainfall, runoff and soil loss is not simple. Responses of hydrological and erosional systems to a given runoff event are always modified by the land's surface. However, knowledge of the frequency-magnitude distributions of erosion events is essential both from the practical perspective of land management, and for an understanding of the role of small and large erosion events in landscape evolution.

This study analyses frequency-magnitude distributions of rainfall, runoff and erosion from sites in the UK, Belgium, Canada and the USA. Frequency-magnitude distributions of rainfall appear to be a poor predictor of distributions of runoff or soil loss. Site-specific influences appear to affect distributions of erosion to a greater extent than distributions of runoff: there is a greater change in the spread of absolute values between rainfall and erosion distributions than there is between rainfall and runoff distributions. For the small- to medium-sized events, soil loss frequency-magnitude distributions show a tendency toward linearity when plotted on a log-log scale. The distribution of the small- to medium-sized erosion events can be approximated by a cumulative power law function. The under-representation of larger events relative to such a distribution may be due to shifts in the balance between erosive processes, and to 'finite-size effects' which constrain the maximum size of erosion event which can occur on a given area. If the frequency-magnitude distributions of erosion events can be approximated by a power-law, then the contribution of small events to total erosion is likely to be underestimated in measured datasets.

These results suggest links with recent, but controversial, theoretical work on 'self-organised criticality'. It is important for geomorphologists to look beyond disciplinary confines and be aware of such links.

Zusammenfassung. Die Wechselwirkung zwischen Niederschlag, Abfluß und Bodenabtrag ist nicht von einfacher Natur. Die Reaktionen von hydrologischen und Erosionssystemen auf ein gegebenes (?) Abflußereignis werden immer durch das Relief modifiziert. Darum kann erwartet werden, daß die Verteilung von Abfluß und Erosion nicht auf einfache Art und Weise mit Häufigkeits- und Größenverteilungen von Niederschlag korreliert sind. Diese Studie analysiert die Verteilung von Häufigkeit und Magnitude der Niederschlags, des Abflusses und der Erosion in Gebieten Großbritanniens, Belgiens, Kanadas und den USA.

Häufigkeits- und Größenverteilungen des Niederschlags scheinen schlechte Vorhersagen für die Verteilung von Abfluß und Bodenabtrag zu liefern. Gebietsspezifische Einflüsse scheinen die Verteilung der Erosionsereignisse stärker zu beeinflussen als die Verteilung des Abflusses: bei Niederschlags- und Erosionsverteilungen tritt eine größere Veränderung in der Spannweite der absoluten Werte auf als bei Niederschlags- und Abflußverteilungen. Für kleine bis mittlere Ereignisse zeigen die Verbreitungen von Bodenabtragshäufigkeit und -größe eine lineare Tendenz, wenn sie im log-log Maßstab dargestellt werden. Die Verteilung der kleinen bis mittleren Ereignisse kann mit einer kumulierenden Potenzfunktion geschätzt werden. In diesem Fall kann durch die Kombination von kurzen Beobachtungsphasen eine Unterrepresentation großer Ereignisse im Verhältnis zu solch einer Verteilung auftreten. Dies ist begründet in einer Verschiebung im Gleichgewicht zwischen Erosionsprozessen und in ‚finite-size effects', welche die maximale Größe eines Erosionsereignisses einschränken, die auf einem Gebiet mit vorgegebener Größe auftreten können. Diese Ergebnisse legen eine Verbindung mit jüngsten, jedoch umstrittenen, theoretischen

0044-2798/99/0115-0051 $ 5.00

Arbeiten zur Selbstorganisierung von Systemen nahe. Für Geomorphologen ist es wichtig, über Disziplingrenzen hinauszuschauen und sich solcher Verbindungen bewußt zu sein.

Introduction

Soil erosion by water results from both rainfall and runoff, but their interaction is far from straightforward. Because of the way in which the land's surface modifies the responses of both hydrological and erosional systems to a given runoff event, distributions of runoff and erosion are not linked in any simple way to distributions of rainfall (cf. FAVIS-MORTLOCK & BOARDMAN 1995). These responses are due in part to temporal changes in vegetation cover and soil properties. They are also related to variation in the relative importance of erosional processes with rainfall amount and intensity, such as the increase in the contribution of splash with increased rainfall intensity, and the shift in the balance between rill and gully erosion with changed rainfall intensity (POESEN et al. 1996).

Yet knowledge of the frequency-magnitude relationships of erosion is important from both practical and theoretical perspectives. On agricultural land, infrequent large erosion events tend to attract the attention of both farmers and researchers; however frequent small events can be of more significance in moving agricultural chemicals off fields and into watercourses (HARROD 1994). An ability to quantify the probability of occurrence of both large and small erosion events is thus of considerable practical importance for the formulation of agricultural land-use policy.

Considering erosion as an agent of landscape evolution, a second issue concerns the relative geomorphic importance of large and small erosion events. In terms of geomorphic work done, does the rarity of the large erosion event disadvantage it relative to the frequent but small event? Earlier work (e.g. WOLMAN & MILLER 1960, WOLMAN & GERSON 1978, DE PLOEY et al. 1991) has provided some answers; however, considerable uncertainties remain.

The availability of data is central to both issues. Just how many small soil loss events actually occur in a landscape context? While experimental plots can be closely monitored for a wide range of event sizes, this is not possible for landscape-sized areas. Unless their numbers can be known, the relative contribution of such small – but presumably very numerous – erosion events cannot be evaluated. In this study, a comparative analysis of frequency-magnitude relationships for rainfall, runoff and soil loss at a number of sites hints at a way forward.

Analysis

Datasets from the UK, USA, Canada and Belgium (Table 1) were selected: these span a range of spatial scales, from small plots to multiple fields within a monitored area of 36 km^2. Land usage is predominantly agricultural, with one US field under natural vegetation (rangeland). Slopes range from virtually flat to 45%. The period of record varies from four years to twenty years (for the UK and Canadian datasets, the rainfall data used in this analysis covers a longer period than the soil loss and/or runoff data). Measured rainfall is available for all datasets apart from the Belgian; runoff measurements are available only for the smaller areas i.e. the US and Canadian plots/fields. Snowmelt is known to be responsible for a significant proportion of the erosion on the Canadian plots (RAMESH RUDRA, pers. commun. 1995).

Table 1. Datasets used in the analysis.

Dataset code	Description	Area (ha)	Period monitored[a]	Land use	Slope angle (%)	Soil type(s)	Reference
SD(all)	Multiple fields in monitored area of South Downs, UK	3600	1982–91	mostly winter cereals	up to 45	silt loam	BOARDMAN 1993, BOARDMAN & FAVIS-MORTLOCK 1993
B(all)	Multiple fields in central Belgium	86 fields	1981–85	winter cereals, bare	0–27	sandy loam and loam	GOVERS 1991
C4	Field at Chickasha, Oklahoma, USA	11.7	1965–76	irrigated cotton	0–1	silty clay loam and silt loam	ARLIN NICKS, pers. communi. 1995
C5	Field at Chickasha, Oklahoma, USA	5.2	1966–76	wheat	0–1	silty clay loam and silt loam	FAVIS-MORTLOCK 1998
R5	Field at Chickasha, Oklahoma, USA	9.6	1967–78	rangeland	0–8	silt loam	FAVIS-MORTLOCK 1998
CAN1	Plot at Guelph, Ontario, Canada	0.027	1962–82	minimum tillage corn	8	loam	TREVOR DICKINSON/RAMESH RUDRA, pers. communi. 1995
CAN2	Plot at Guelph, Ontario, Canada	0.027	1962–82	minimum tillage corn	8	loam	TREVOR DICKINSON/RAMESH RUDRA, pers. communi. 1995
CAN3	Plot at Guelph, Ontario, Canada	0.027	1962–82	minimum tillage corn	8	loam	FAVIS-MORTLOCK 1998
CAN4	Plot at Guelph, Ontario, Canada	0.027	1962–82	conventional tillage corn	8	loam	FAVIS-MORTLOCK 1998
CAN5	Plot at Guelph, Ontario, Canada	0.027	1962–82	conventional tillage corn	8	loam	TREVOR DICKINSON/RAMESH RUDRA, pers. communi. 1995

[a] in some cases, includes part years

Measurements of soil loss are available for all datasets. These were converted to volumetric measures where necessary by assuming a density of 1.3 t m^{-3}. Different methods were used to obtain these measurements. Soil loss data for the Canadian plots and individual US fields were obtained by sampling at an outlet; that for the multiple fields in Belgium and on the UK South Downs were obtained by measurement of rill volumes. As a result, the soil loss data should not be thought of as measuring exactly the same thing in all cases. That for the plots and individual fields represents sediment delivery, while the monitored data gives values for soil loss by rilling. Note that only some fields in a monitored area will erode in any year; and these will not necessarily be the same fields in different years.

The datasets are thus distinctly heterogenous with regard to spatial scale, length of record, site details, data definitions, and dominant erosional processes.

Some statistics for this data are given in Table 2. No assumption was made at this stage regarding distributional form, hence no measures of central tendency or spread are calculated. Since dates for erosion events obtained by monitoring can only be approximate, the decision was made to work with all datasets on an event basis rather than a daily basis. For all datasets, it was assumed that consecutive days (e.g. of rainfall) constitute a single event (cf. DE PLOEY et al. 1991). Table 2 shows the effect of aggregating daily data in this way: while the number of non-zero data items decreases, the size of the maximum data item in the dataset increases for several of the datasets.

The shapes of the distributions of rainfall, runoff and soil loss events from one dataset are shown in Fig. 1. All have a high positive skewness. The majority of data items have relatively low values, with only a few high-valued data items in the tail of the distribution. The skewness of the runoff is higher than that of the rainfall; that of the soil loss appears higher still. Such distributions would be poorly described by a Gaussian (normal) distribution; the Poisson[1] or a variant would appear more appropriate (cf. WOOLHISER & ROLDAN 1982, FAVIS-MORTLOCK 1995)

In the next stage of the analysis, the procedure adopted is based upon that of TURCOTTE (1989). Cumulative distributions for each dataset's values of rainfall, runoff and soil loss were constructed, with the number of events with a size greater than or equal to a given value plotted against that value. The distributions are then plotted on log axes (Figs. 2 to 4). This cumulative approach is in contrast to more conventional geomorphological or hydrological approaches to frequency-magnitude analysis, which consider event magnitude and recurrence interval (e.g. DE PLOEY et al. 1991) or the log of event magnitude and recurrence interval (e.g. BUFFAUT et al. 1998 in press).

One problem with this methodology concerns the smallest values. For most distributions, the smallest value plotted is that of the non-zero minimum to the dataset (Table 2). However, some of the smallest values given for runoff and soil loss were omitted on the grounds of possible unreliability: thus in all cases the smallest values plotted are 0.01 mm for runoff and 0.001 m^3ha^{-1} for soil loss. This of course decreases the range of the data somewhat.

---

[1] Which is of course discrete, in contrast to the continuous distributions produced by rainfall, runoff and soil loss.

Table 2. Some statistics for rainfall, runoff and erosion in the datasets.

Dataset code	Daily data									Event data[a]								
	rain (mm)			runoff (mm)			erosion (m^3ha^{-1})			rain (mm)			runoff (mm)			erosion (m^3ha^{-1})		
	N[b]	min.[c]	max.	N[b]	min.[c]	max.	N[b]	min.[c]	max.	N[b]	min.[c]	max.	N[b]	min.[c]	max.	N[b]	min.[c]	max.
SD(all)	6893[d]	0.1	53.8	-	-	-	-	-	-	2301[d]	0.1	223.4	-	-	-	416	0.01	234.0
B(all)	-	-	-	-	-	-	-	-	-	-	-	-	-	-	-	64	0.6	28.8
C4	760	0.3	98.8	396	<0.01	77.6	361	<0.001	2.98	482	0.3	143.8	354	<0.01	77.6	321	<0.001	2.98
C5	760	0.3	98.8	205	0.01	35.4	169	<0.001	1.38	482	0.3	143.8	106	0.01	35.4	98	<0.001	2.01
R5	821	0.3	124.0	166	0.01	45.7	125	<0.001	0.08	515	0.3	125.3	92	0.01	50.3	91	<0.001	0.08
CAN1	108	0.3	90.9	101	0.05	35.9	101	0.001	56.05	108	0.3	90.9	101	0.05	35.9	101	0.001	56.05
CAN2	108	0.3	90.9	87	<0.01	36.6	86	<0.001	60.87	108	0.3	90.9	87	<0.01	36.6	86	<0.001	60.87
CAN3	108	0.3	90.9	49	0.04	19.0	46	<0.001	5.11	108	0.3	90.9	49	0.04	19.0	46	<0.001	5.11
CAN4	108	0.3	90.9	75	<0.01	29.4	72	<0.001	43.03	108	0.3	90.9	75	<0.01	29.4	72	<0.001	43.03
CAN5	108	0.3	90.9	75	0.01	50.7	74	<0.001	46.27	108	0.3	90.9	75	0.01	50.7	74	<0.001	46.27

[a] assuming consecutive daily data to be part of same event
[b] of non-zero data items
[c] minimum non-zero
[d] rainfall data from the Southover weather station for the period 1950–90 was used

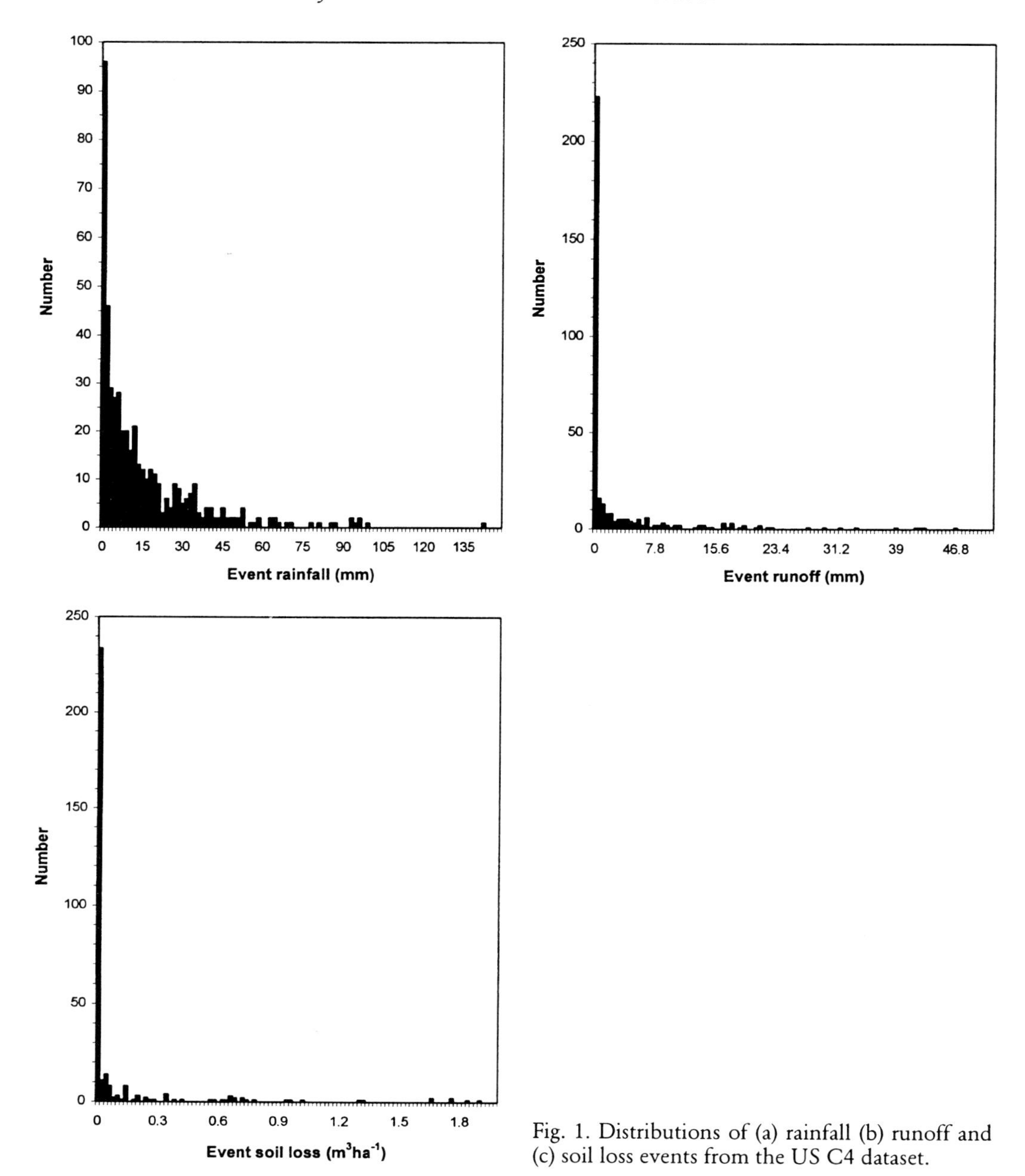

Fig. 1. Distributions of (a) rainfall (b) runoff and (c) soil loss events from the US C4 dataset.

Results

For rainfall (Fig. 2), the resulting distributions are consistently smoothly curvilinear. They differ little in shape. Any differences in distributional form are most marked for the larger rainfall events: these are likely to result (at least in part) from the small numbers of events of

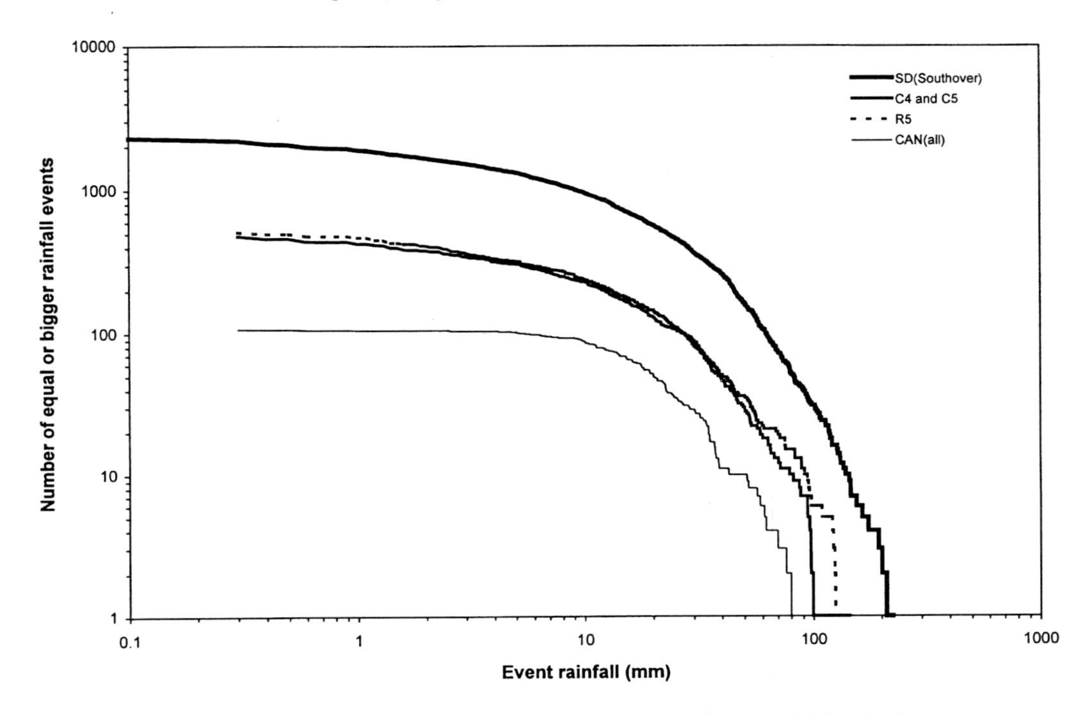

Fig. 2. Number of rainfall events greater than or equal to a given value. See Table 1 for key to datasets.

this size. The range of rainfall values for each dataset is three orders of magnitude or more. Rather less than one order of magnitude separates the maxima for each dataset. The range of numbers of rainfall events of a given size is over one order of magnitude across all datasets. The numbers of events of a given size appear to have some relationship with the number of observations in the whole dataset (Table 1), but not obviously with the period of record.

Graphs for runoff are shown in Fig. 3. These are also curvilinear, although rather less smooth than those for rainfall, presumably due to the smaller number of observations. Distributional shapes for each dataset differ more for runoff than they do for rainfall. Distributions for areas receiving the same rainfall (e.g. the Canadian plots) can be noticeably different, particularly for the larger events. The range of runoff amounts in each dataset is a little greater than the equivalent for rainfall, at over three orders of magnitude. The numbers of events of a given size for each dataset however extend across a rather smaller range than the equivalent for rainfall. Note though that runoff data for the UK South Downs and Belgium are not available; thus the range separating the highest and lowest numbers of events of a given size is broadly similar for both rainfall and runoff for the US and Canadian data. Numbers of runoff events of any size are again related to the number of observations in the total dataset (Table 1). This is however only broadly true: compare the distributions of the US C5 and Canadian CAN1 sites, each of which have almost the same total number of runoff events. There is no clear relationship between the number of events of a given size and the length of record.

Fig. 4 shows the soil loss frequency-magnitude distributions. These are noticeably different in shape from those for rainfall or runoff. There is a clear tendency to straightness for the

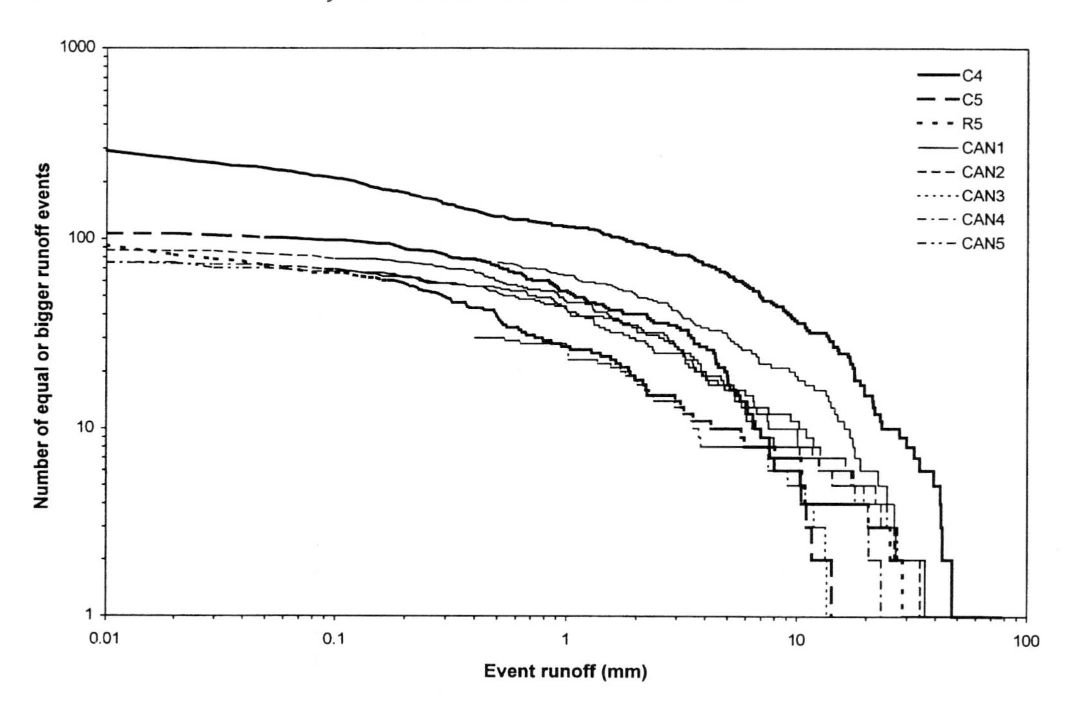

Fig. 3. Number of runoff events greater than or equal to a given value.

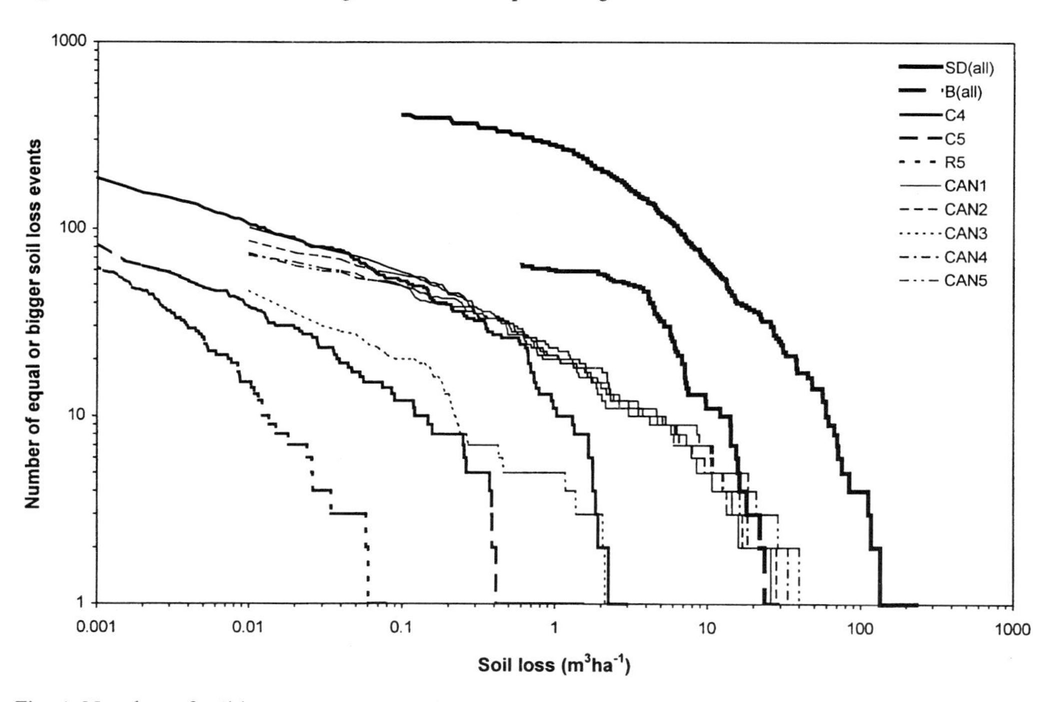

Fig. 4. Number of soil loss events greater than or equal to a given value.

small to medium soil loss events for the US and Canadian datasets (i.e. those from plots and individual fields). For these datasets, the distributions for higher values of soil loss have a 'downturn' which preserves the curvilinearity of the rainfall and runoff distributions. There is some tendency to straightness for medium-sized events on the South Downs data, although the general form here is curvilinear. The Belgian data has an irregular curvilinear form. As with runoff, all soil loss distributions are less smooth than those for rainfall, reflecting the smaller number of events.

A notable difference between the soil loss distributions and those for runoff and rainfall concerns the increased range of maximum values for soil loss across all datasets: over three orders of magnitude. The range of values for any individual dataset varies from about two to over four orders of magnitude; the total range is over five orders of magnitude. The range of numbers for a given size of soil loss event has also generally increased relative to rainfall or runoff. The number of erosion events of a given size is less obviously related to the total number of events (e.g. the US C4 dataset has more events than the Belgian dataset, but these are smaller in magnitude). As with runoff, the number of events of any size is not clearly related to the length of record.

A similar analysis was also carried out assuming that each day represents a separate event (cf. De Ploey et al. 1991): results are not shown here, but are broadly similar.

Discussion

Rainfall, runoff and erosion

Given the heterogeneity of the datasets, the similarity of their cumulative frequency-magnitude distributions for runoff and erosion is surprising. All runoff distributions have a very similar shape; soil loss distributions differ more from each other, but again have a broadly similar shape.

On the evidence of these datasets, it appears that knowledge of the frequency-magnitude distribution of rainfall at any locality can supply little information on the frequency-magnitude distributions of the runoff and erosion which result from it. The Canadian plots all receive the same rainfall (Fig. 2); but site-specific influences act to produce noticeably different distributions for runoff (Fig. 3) and for soil loss (Fig. 4: note the widely divergent response of plot 3). The same applies to the US C4 and C5 fields. The rainfall for US R5 differs little from that for C4 and C5, but its runoff and soil loss distributions are different again. This conclusion appears to be true both for the plot-sized Canadian areas and for the field-sized US areas, with their differing covers and slope angles. Frequency-magnitude analysis of rainfall or runoff is therefore a poor predictor of erosion.

In terms of absolute values of order of magnitude, the separation between the datasets is much wider for soil loss than for runoff or rainfall. This implies that the link between rainfall distributions and soil loss distributions is 'weaker' than is that between rainfall and runoff; thus the influence of site-specific factors would appear to be greater for soil loss than for runoff.

A power-law relationship

The tendency to linearity of the lower-magnitude portions of the soil loss distributions for plots and fields is suggestive: can these be fitted by a straight line? Linear log-log frequency-magnitude relationships are known to underlie distributions of a variety of other natural phenomena as diverse as the scaling relationships of large-scale channel networks (CRAVE & DAVY 1997) and soil aggregates (CRAWFORD et al. 1997), the magnitudes of earthquakes (e.g. TURCOTTE 1989) and the sizes of species extinctions during geological time (RAUP 1991). Relationships of this kind may be described by a power law function of the form $F = c\,M^{-D}$ where F is frequency, M is magnitude, c is a constant and D is the fractal dimension (e.g. GOODCHILD & MARK 1987, EVANS & MCCLEAN 1995).

The difficulty with this assumption is that the 'straight-line' portions of these soil loss distributions are relatively short: those for the Canadian plots are the longest, at about three orders of magnitude. The tendency of datasets of limited length to approximate a straight line on a log-log graph is well-known (Ian Evans, personal communication 1997). Fig. 5 shows the results of linear regressions on portions of the US C4 and UK South Downs datasets. For the C4 data, values for soil loss events between 0.001 and 0.650 m^3ha^{-1} were used in the regression (i.e. the whole of the lower- and middle-valued portion, up to the 'downturn'). For the South Downs data, only the portion for soil loss values from 0.1 to 100.0 $m^3ha^{-1}$was used in the regression (i.e. an arbitrarily-chosen central portion only).

For the straight-line portions of the erosion distributions in Fig. 4, the cumulative frequency-magnitude distributions are well fitted by a power-law relationship. Such relationships are

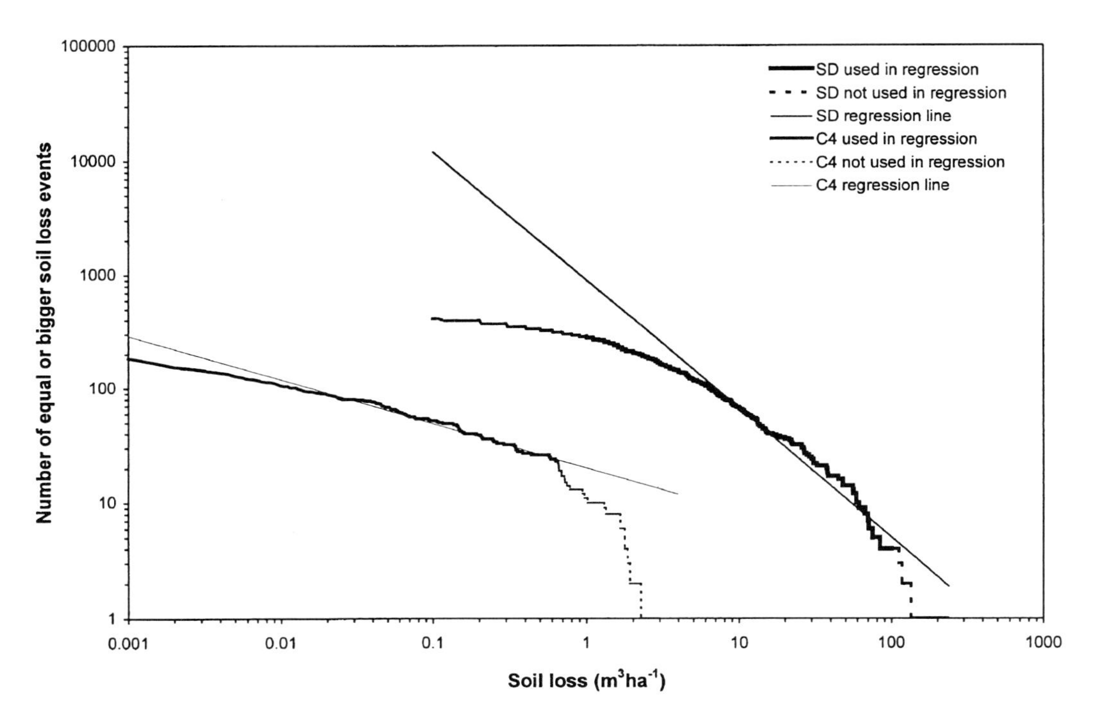

Fig. 5. Linear regression on portions of the US C4 and UK South Downs datasets. See text for explanation.

scale-invariant, so that in this case the logarithms of the numbers of events of any two magnitudes and above form a constant ratio to the logarithms of these magnitudes, irrespective of the values of the two magnitudes.
Thus:

$$\log A = c \cdot \log B \quad (1)$$

where:

$$A = \sum_{m=M_1}^{m=M_{max}} F_m \quad (2)$$

$$B = \sum_{m=M_2}^{m=M_{max}} F_m \quad (3)$$

M_1 and M_2 are the magnitudes of two erosion events
M_{max} is the magnitude of the largest erosion event in the dataset
F_m is the number of erosion events of magnitude m
c is a constant.

Why cumulative distributions?

As stated earlier, all frequencies have been plotted cumulatively i.e. as the number of events with magnitude equal to or greater than some given magnitude. At first glance, it might appear that the reverse approach could have been adopted, i.e. using the number of events with magnitude equal to or less than a given value. This is, however, not possible in practice since the total numbers of events of smaller magnitudes must always be unknown. It is still possible to consider the data in this 'reverse' way, however. The results show that to get (for example) ten events of 80 m^3ha^{-1} or above, two events of 100 m^3ha^{-1} or above are 'necessary'. This may be turned around – two events of 100 m^3ha^{-1} or above 'require' ten events of 80 m^3ha^{-1}. It must be emphasised however that such a statement implies nothing regarding the temporal sequence of the small and large erosion events.

Extrapolation to smaller and larger events

If this power-law relationship is also assumed to be valid for smaller events than those plotted, the straight-line portion of each plot may be extrapolated in this direction. This would permit an estimate of the number of smaller erosion events which occurred on that site, but which are too tiny to be measured; and hence the contribution of such events to total erosion (see below).

Under-reporting of small events may be the explanation for the differences between the data for plots and single fields, and the data for multiple fields, in terms of the straightness of the distributions for the smaller events (Fig. 4). Since the datasets for soil loss on plots and individual fields were measured as sediment delivery at an outlet, they will include data for all erosion events which were of sufficient size to deliver sediment to that outlet. However, measurement of rill volumes will necessarily omit the smallest events (cf. EVANS & BOARDMAN 1994, EVANS 1995), which will therefore be under-represented in the Belgian and UK South Downs datasets.

However, extrapolation beyond the smallest events measured is likely to be indefensible. The relative importance of erosional processes varies over a range of rainfall amounts and intensities, e.g. the contribution of splash increases with increased rainfall intensity. Therefore the balance between erosional processes may well shift for events which are much smaller than those graphed.

What about the larger events, which clearly cannot be fitted to an extrapolation of the straight-line portion of the distribution? In every case (Fig. 4), there is a 'downturn' of the distribution for larger events: in other words, there is a greater under-representation of large events as the size of the event increases, compared with the linear distribution assumed for the smaller events. Other non-geomorphological work on such 'downturned' power-law relationships (e.g. RAUP 1991, KAUFFMANN 1995) has suggested that this kind of under-representation of larger events is the result of some constraint(s) on the occurrence of larger events. A number of such constraints may be operating on the soil loss data considered here (see next section).

Nonetheless, upward extrapolation may have some value in defining an upper limit for the frequency-magnitude relationships of the largest erosion events which are to be expected in a given area. The difficulty here is that this upper limit will – as common sense suggests – itself increase as the period of observation lengthens.

Constraints upon the occurrence of larger events

Length of dataset. One such constraint is the period of observation. Clearly, the longer the record, the greater the probability that it will include low-frequency, high-magnitude events. Fig. 6 shows the UK South Downs data disaggregated temporally, so that the first line represents events which occurred in the first year of monitoring, the second line events during the first two years, and so on up to the full ten years. As expected, the distributions are smoother for the longer-period distributions: increased jaggedness is most in evidence for the larger events in the short-period datasets. However, in this analysis the number of events plotted for any size class depends both upon the number of events in that size class and upon the number of events in all greater size classes. As a result, the addition of new larger events will affect the numbers plotted for all smaller magnitudes. Thus additional large events tend to shift the whole line upwards and to the right. This can be seen in Fig. 6.

However, there is no evidence of any change in the shape of the 'downturn' between long-period and short-period distributions.

Changed balance of erosional processes. As with the smallest events, there may be a shift in the relative importance of erosional processes for the largest events. For example, a shift in the balance between rill and gully erosion with changed rainfall intensity was noted by POESEN et al. (1996). At both ends of the event-size spectrum, the importance of these processes does not necessarily vary smoothly across the whole range of rainfall magnitudes: there appear to be broad thresholds separating the domains in which each dominates. Thus the 'downturn' in the plots of Figs 4 and 5 may also in part be due to a shift to a new domain, with a differing balance between erosional processes.

Effects of spatial scale. The size of the area on which the erosion events occur is another constraint upon the numbers of large events. Fig. 7 plots the full ten years of the South Downs data for a range of field size categories. As in Fig. 6, the subsets with the smallest numbers of events show the greatest jaggedness. However, there is a suggestion here that the most marked

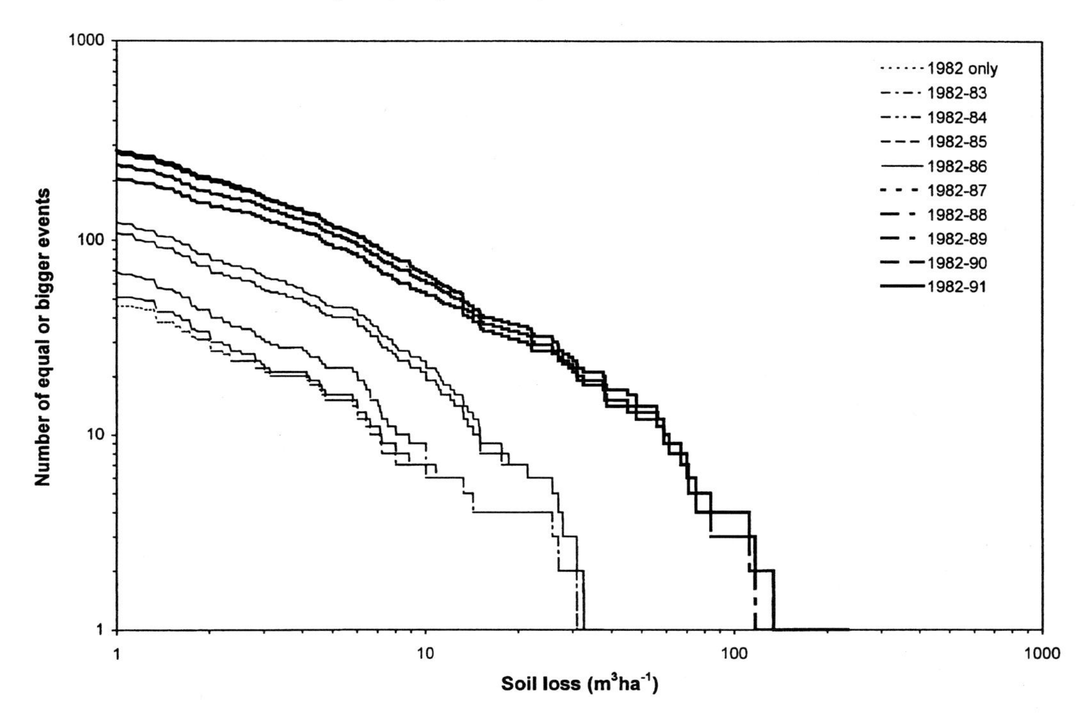

Fig. 6. Number of soil loss events on the South Downs greater than or equal to a given value for a single year, two years etc.

'downturns' occur on the distributions for the smallest areas: i.e. that the smallest areas are most deficient in large events.

There are a number of reasons why such a negative-feedback effect might be expected. The first is that common sense suggests that there is a top limit to the maximum size of erosion event which can ever occur on a given area. For example, if a uniform soil depth of 20 cm is assumed for a 1 ha South Downs field, then an erosion event of 2000 m^3ha^{-1} will remove all soil on the field. While it is difficult to envisage such an event actually occurring on a 1 ha field – the highest recorded erosion rate observed on the South Downs for a whole field is 234 m^3ha^{-1} over 11 hectares in 1987–88 (BOARDMAN 1988) – it is possible that rates approaching these might, during the largest events and given ideal conditions for erosion, be approached on small areas within a field (cf. EVANS & BOARDMAN 1994). (Note that very high rates of past erosion may have occurred even on the small fields formerly in use on the South Downs. However, this is likely to be the result of higher contemporary soil erodibilities compared with those of the present-day stony soils: see FAVIS-MORTLOCK et al. 1997). An analogous 'finite size effect' is invoked by KAUFMANN (1995) to explain the 'downturn' in RAUP's (1991) data for the frequency-magnitude distribution of species extinctions.

A second negative-feedback effect of increased spatial scale is increased opportunity for long-term sediment storage, which manifests itself as a decrease in sediment delivery ratio with increase in area. However, this may be subsequently released during large events (TRIMBLE 1983, SLATTERY & BURT 1997). A central issue here is the degree of connectivity between land-

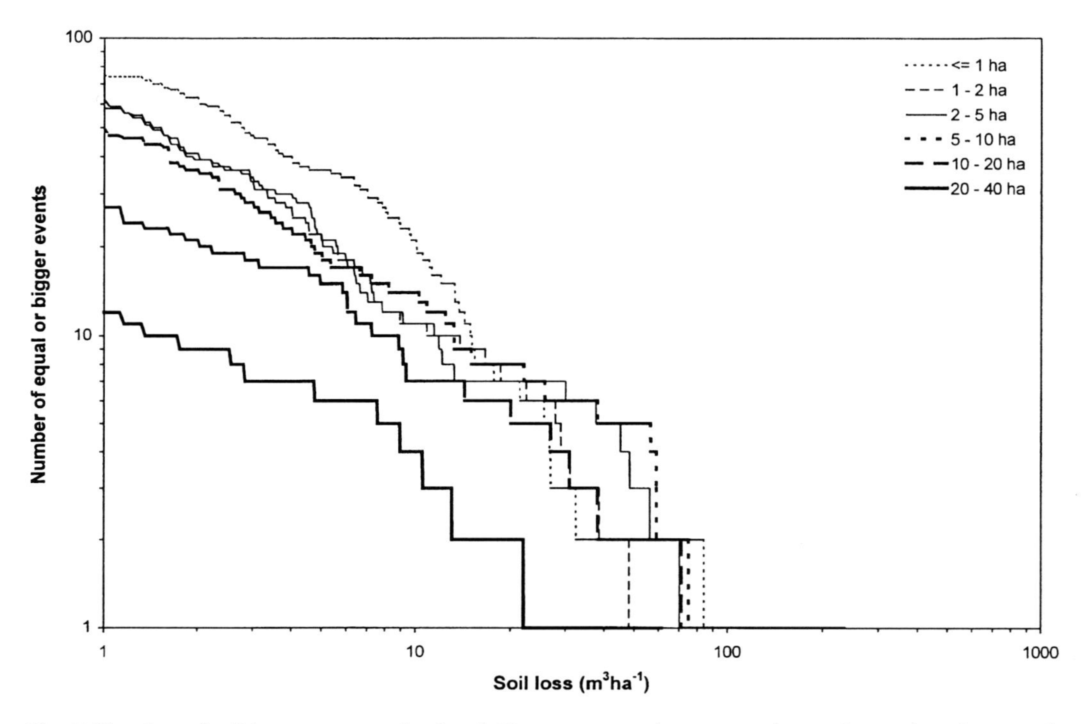

Fig. 7. Number of soil loss events on the South Downs greater than or equal to a given value, for several categories of field size.

scape elements. In the largest erosion events, normally separate erosional subsystems may begin to link up, so increasing the erosional efficiency of the whole system. This varies through time and in relation to the magnitude of runoff events. It is also affected by topography, sometimes quite subtly: for example, it appears that valley bottoms in the South Downs with a gradient of less than 0.04 m m^{-1} are not at risk of gullying under present conditions (BOARDMAN 1992), so that hillslope rills in such valleys would not be expected to become part of a larger system even during a large event.

The issue is complicated still further by increased heterogeneity of vegetation cover as one moves from hillslope plots to small catchments. For example, while it is clear that on the agricultural South Downs most erosion occurs as the result of a few storms, these are only effective when and where there is low crop cover. While storms of >1000 year return period have little erosive effect in summer, storms with a 25 year return period may give rise to erosion rates >200 m^3ha^{-1} on individual fields with little crop cover in the autumn period. However, on the South Downs at least, the year-to-year vagaries of land use appears to have little influence on the frequency-magnitude distributions of soil loss events: Fig. 8 shows individual distributions calculated for each year of the period of monitoring. While there is great variability in amounts eroded from year to year (cf. BOARDMAN & FAVIS-MORTLOCK 1993), the slopes of the distributions vary little. It may be that there is some cancelling of opposed random influences on erosion rate on individual fields here, which – given a sufficient number of fields – from year to year tends to preserve a similar ratio between the numbers of events of different

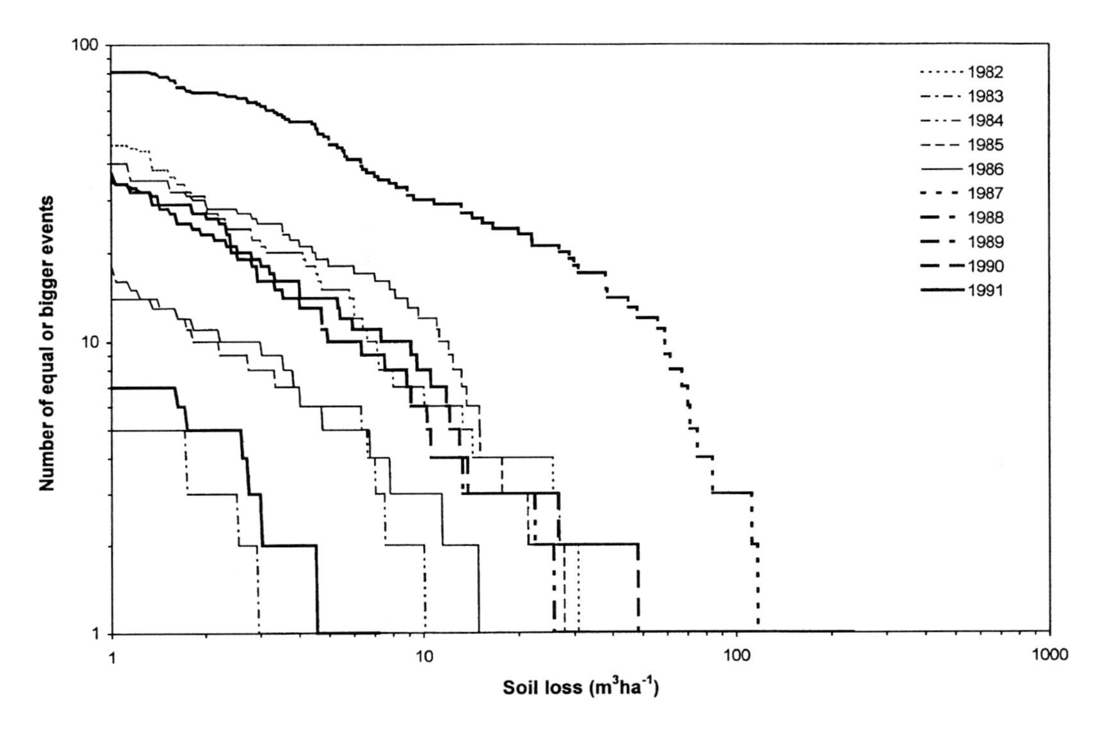

Fig. 8. Number of soil loss events on the South Downs greater than or equal to a given value for each year of the period of monitoring.

size classes. Such broad-scale order arising from finer-scale randomness is discussed by PHILLIPS (1997) in the context of equifinality and 'simplexity'.

Work which uses experimental plots fails to address these scale issues (EVANS & BOARDMAN 1994, EVANS 1995).

The contribution of large and small erosion events to total soil loss

Fig. 9 shows the contribution of soil loss events of different sizes to total soil loss on the US C4 field and the monitored area of the South Downs. Note that this total is expressed on a notional per-hectare basis, i.e. the spatial extent of erosion for each event is not considered. This means that, for the South Downs, there is no simple relationship between this total and field measurements of total soil loss during the period of monitoring (cf. for example BOARDMAN 1993).

For the South Downs, the smallest category of erosion events (< 5 m^3ha^{-1}) makes the largest contribution (about 15%); this is due to the large numbers of events in this category. However several other sizes of event also contribute notably: the single largest event (234 m^3ha^{-1}) contributes about 7%. Erosion rates are in all cases much lower on the C4 field, but here the smallest category of events does not make the greatest contribution: this comes from relatively moderate-sized events.

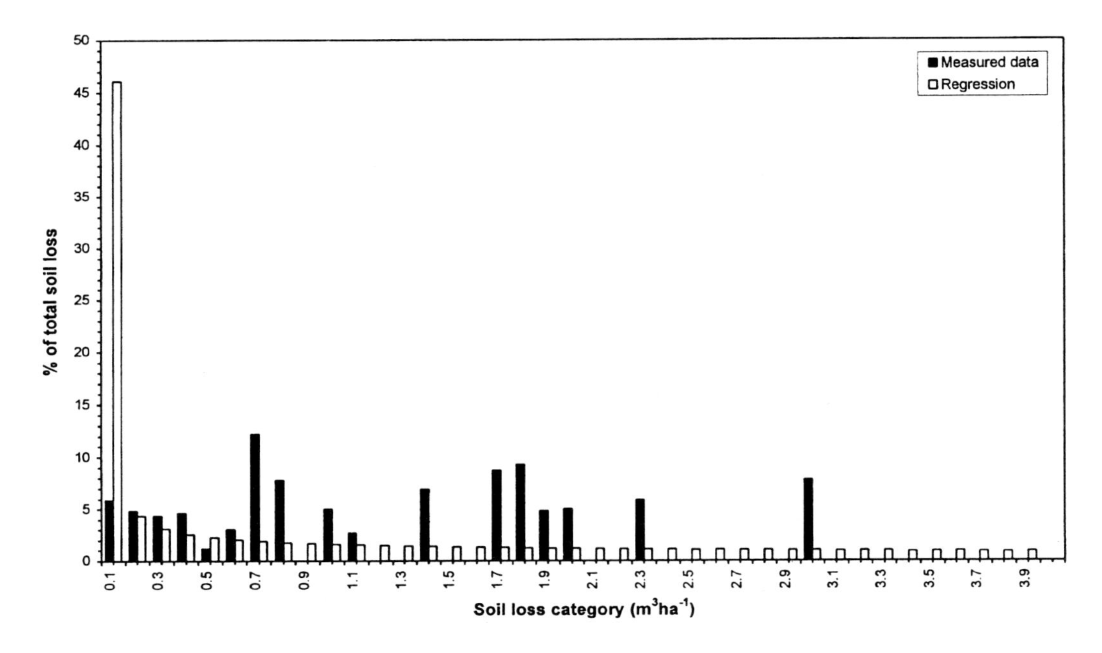

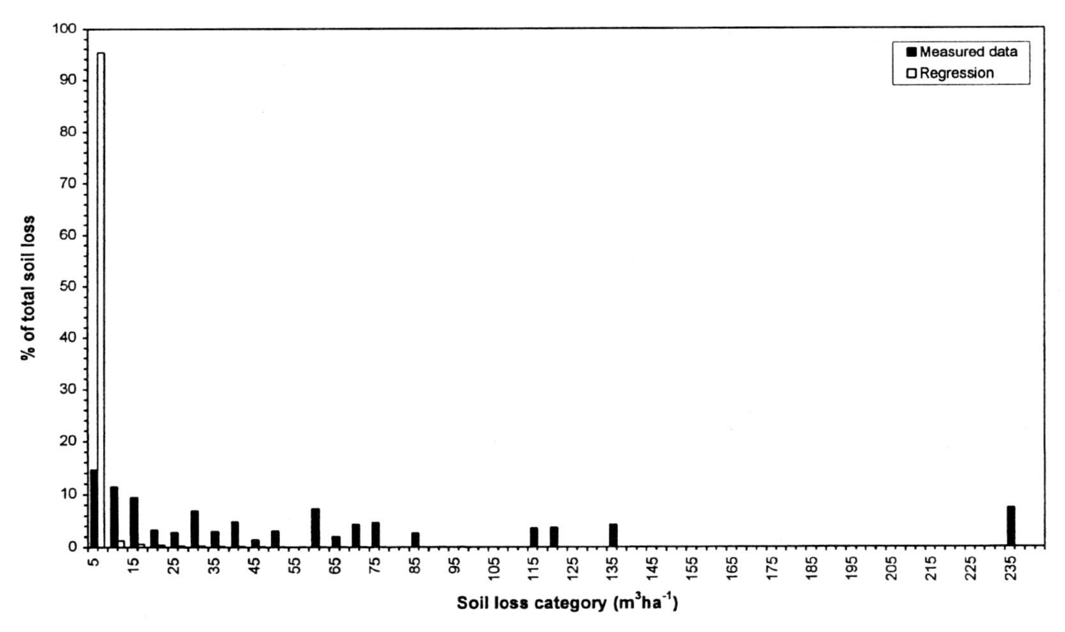

Fig. 9. The contribution of erosion events of different sizes to total soil loss for US C4 (upper) and the South Downs (lower), for both measured data and the regressions of Fig. 5. Note that both x- and y-axis scales differ.

Fig. 9 also shows the results of a similar analysis for the power-law regression lines shown in Fig 5. For both sites, the relative contribution of the smallest events increases. For the South Downs, this increase is very marked: events under 5 m^3ha^{-1} now contribute over 95% of per-hectare erosion. For the C4 site, the smallest events now contribute about 46%.

If it is true that the frequency-magnitude distributions of these erosion events do approximate to a power-law, then these results suggest that the contribution of the smallest erosion events to total erosion is underestimated in both measured datasets .

Erosion and self-organised criticality

Is there any theoretical reason why a power law relationship might underlie frequency-magnitude distributions of erosion? BAK & CHEN (1988) and BAK (1997) suggest that a power-law frequency-magnitude distribution is one of the hallmarks of complex systems which exhibit 'self-organised criticality'. Such systems possess internal dynamics which result in their having a 'memory' – they will react in different ways at different times to the same external stimulus depending on the system's current state, which in turn will depend on its past history. The canonical example is a developing sandpile: dropping a single grain onto it can result in the grain finding a secure resting place, triggering a major avalanche, or anything in between. According to BAK & CHEN (1988), the sizes and frequencies of avalanches from such a sandpile will follow a power-law distribution (see also HOVIUS et al. 1997, HERGARTEN & NEUGEBAUER 1998).

There is still debate regarding the definition and limits of applicability of self-organised criticality (e.g. CAPRA 1996, SAPOZHIKOV & FOUFOULA-GEORGIU 1996, LEWIN 1997, MEHTA 1997). However, it seems possible that, from the results of this analysis, erosional systems can (within limits) exhibit power-law frequency-magnitude distributions – one of the traits of self-organised criticality (cf. FAVIS-MORTLOCK 1996, 1998 in press, HERGARTEN & NEUGEBAUER 1996). Geomorphology may have a role to play in the further development – and particularly the validation – of concepts such as self-organised criticality.

Conclusions

From an analysis of the frequency-magnitude distributions of rainfall, runoff and erosion from sites in the UK, Belgium, Canada and the USA, a number of conclusions may be drawn.

- Frequency-magnitude distributions of rainfall are a poor predictor of distributions of runoff or soil loss. Site-specific influences appear to operate more strongly on distributions of erosion than runoff since rainfall and erosion distributions differ more in absolute terms than do rainfall and runoff distributions.
- On a log-log graph, soil loss frequency-magnitude distributions show a tendency toward linearity for the small- to medium-sized events, in comparison with distributions of rainfall and runoff. If it is assumed that the frequency-magnitude distribution of the small- to medium-sized erosion events can be validly described by a power law (i.e. by a straight line on a log-log graph), then this may offer a way of quantifying the numbers of events which are too small to be measured, and hence of determining their contribution to total erosion.

- A number of possible reasons may be advanced for the 'downturn' (i.e. under-representation) of the distribution for the larger events relative to such a power-law. One of these is that the periods of observation of the datasets used for the analysis are too short to include an adequate sample of the low-frequency high-magnitude events. Analysis of erosion data for the South Downs for varying sample lengths suggests that this is a possibility. Other possibilities are a shift in the balance of erosive processes with event size, and the operation of a 'finite-size' effect which constrains the maximum size of erosion event which can occur on an area of given size. Experimental plots fail to address these scale issues.
- If the 'downturned power-law' distributional form is found to applicable to other erosion datasets – in particular, datasets for landscape-sized areas – then it could offer a way of estimating the numbers of erosion events which are too small to be adequately recorded. However, it promises to be less useful for estimating the numbers of large erosion events, since the presence of the 'downturn' complicates extrapolation of the distribution in the direction of bigger events.
- The results of this study suggest links with recent theoretical work on self-organised criticality. It is important for geomorphologists to be aware of such links.

Acknowledgements

We are grateful to Trevor Dickinson and Ramesh Rudra (University of Guelph) for supplying the Canadian data, and to the late Arlin Nicks (USDA-ARS) for supplying the US data; both were originally used in the GCTE validation of field-scale erosion models. Thanks also to Nel Caine for discussion, and to Peter Hutchinson, Richard Tol and Bob Evans for their comments upon an earlier draft. This paper is a contribution to the Soil Erosion Network of GCTE, a Core Research Project of the International Geosphere-Biosphere Programme.

References

BAK, P. (1997): How Nature Works: the Science of Self-organised Criticality. – Oxford University Press, Oxford, UK. 212 pp.

BAK, P. & K. CHEN (1988): Self-Organised Criticality. – Phys. Rev. Letters A **38**: 364–374.

BOARDMAN, J. (1988): Severe Erosion on Agricultural Land in East Sussex, UK, October 1987. – Soil Technology **1**: 333–348.

BOARDMAN, J. (1992): Current Erosion on the South Downs: Implications for the Past. In: BELL, M. & J. BOARDMAN (eds), Past and Present Soil Erosion, Oxbow Monograph 22, Oxbow Books, Oxford, UK. pp. 9–19.

BOARDMAN, J. (1993): The Sensitivity of Downland Arable Land to Erosion by Water. – In: THOMAS, D.S.G. & R.J. ALLISON (eds.): Landscape Sensitivity. – Wiley, Chichester, UK., pp. 211–228.

BOARDMAN, J. & D.T. FAVIS-MORTLOCK (1993): Simple Methods of Characterizing Erosive Rainfall with Reference to the South Downs, Southern England. – In: WICHEREK , S. (ed.): Farmland Erosion in Temperate Plains, Environments and Hills. – Elsevier, Amsterdam, The Netherlands. pp. 17–29.

BUFFAUT, C., M.A. NEARING & G. GOVERS (1998): Statistical distributions of daily soil loss from natural runoff plots and the WEPP model. – Soil Sci. Soc. Amer. Journ. **62** (3): 756–763.

CAPRA, F. (1996): The Web of Life. – HarperCollins, London, UK, 320 pp.

CRAVE, A. & P. DAVY (1997): Scaling Relationships of Channel Networks at Large Scales: Examples from Two Large-Magnitude Watersheds in Brittany, France. – Tectonophysics **269**: 91–111.

CRAWFORD, J.W., S. VERRALL & I.M. YOUNG (1997): The Origin and Loss of Fractal Scaling in Simulated Soil Aggregates. – Europ. Journ. Soil Sci. **48**(4): 643–650.

De Ploey, J., M.J. Kirkby & F. Ahnert (1991): Hillslope Erosion by Rainstorms – a Magnitude-Frequency Analysis. – Earth Surf. Proc. Landforms **16**: 399–409.
Evans, I.S. & C.J. McClean (1995): The Land Surface is not Unifractal: Variograms, Cirque Scale and Allometry. – Z. Geomorph. N.F. Suppl.-Bd. **101**: 127–147.
Evans, R. (1995): Some Methods of Directly Assessing Water Erosion of Cultivated Land – a Comparison of Measurements made on Plots and in Fields. – Prog. Phys. Geogr. **19**(1): 115–129.
Evans, R. & J. Boardman (1994): Assessment of Water Erosion in Farmers' Fields in the UK. – In: Rickson, R.J. (ed.): Conserving Soil Resources: European Perspectives. – CAB International, Wallingford, UK. pp. 13–24.
Favis-Mortlock, D.T. (1995): The Use of Synthetic Weather for Soil Erosion Modelling. – In: McGregor, D.F.M. & D.A. Thompson (eds.): Geomorphology and Land Management in a Changing Environment. – Wiley, Chichester, UK. pp. 265–282.
Favis-Mortlock, D.T. (1996): An Evolutionary Approach to the Simulation of Rill Initiation and Development. – In: Abrahart, R.J. (eds.): Proceedings of the First International Conference on GeoComputation (Volume 1). – School of Geography, University of Leeds. pp. 248–281.
Favis-Mortlock, D.T. (1998): Validation of Field-Scale Soil Erosion Models using Common Datasets. In: Boardman, J. & Favis-Mortlock, D.T. (eds.): Modelling Soil Erosion by Water. – Springer-Verlag NATO-ASI Global Change Series, Heidelberg, Germany, pp. 89–127.
Favis-Mortlock, D.T. (1998): A self-organising dynamic systems approach to the simulation of rill initiation and development on hillslopes. – Computers and Geosciences **24**(4): 353–372.
Favis-Mortlock, D.T. & J. Boardman (1995): Nonlinear Responses of Soil Erosion to Climate Change: a Modelling Study on the UK South Downs. – Catena **25**(1–4): 365–387.
Favis-Mortlock, D.T., J. Boardman & M. Bell (1997): Modelling Long-Term Anthropogenic Erosion of a Loess Cover: South Downs, UK. – The Holocene **7**(1): 79–89.
Goodchild, M.F. & D.M. Mark (1987): The Fractal Nature of Geographic Phenomena. – Ann. Associ. Amer. Geogr. **77**(2): 265–278.
Govers, G. (1991): Rill Erosion on Arable Land in Central Belgium: Rates, Controls and Predictability. – Catena **18**: 133–155.
Harrod, T.R. (1994): Runoff, Soil Erosion and Pesticide Pollution in Cornwall. – In: Rickson, R.J. (ed.): Conserving Soil Resources: European Perspectives. – CAB International, Wallingford, UK, pp. 105–115.
Hergarten, S. & H.J. Neugebauer (1996): A Physical Statistical Approach to Erosion. – Geol. Rdsch. **85**: 65–70.
Hergarten, S. & H.J. Neugebauer (1998): Self-Organized Criticality in a Landslide Model. – Geophys. Res. Lett. **25**(6): 801–804.
Hovius, C., C.P. Stark, & P.A. Allen (1997): Sediment Flux from a Mountain Belt derived by Landslide Mapping. – Geology **25**(3): 231–234.
Kauffman, S. (1995): At Home in the Universe. – Penguin, London, UK, 321 pp.
Lewin, R. (1997): Critical mass. – New Scientist, 23rd August 1997.
Mehta, A. (1997): The World in a Grain of Sand? – Review in The Times Higher Educational Supplement, 12th December 1997, p. 21.
Phillips, J.D. (1997): Simplexity and the Reinvention of Equifinality. – Geogr. Anal. **29**(1): 1–15.
Poesen, J., K. Vandaele & B. Van Wesemael (1996): Contribution of Gully Erosion to Sediment Production on Cultivated Lands and Rangelands. – In: Walling, D.E. & B.W. Webb (eds.): Erosion and Sediment Yield: Global and Regional Perspectives. – IAHS Publication no. 236, Wallingford, UK, pp. 251–266.
Raup, D.M. (1991): Extinction: Bad Genes or Bad Luck? – Norton, New York, USA.
Sapozhikov, V.B. & E. Oufoula-Georgiu (1996): Do the Current Landscape Evolution Models show Self-Organised Criticality? – Water Resources Res. **32**(4): 1109–112.
Slattery, M.C. & T.P. Burt (1997): Particle Size Characteristics of Suspended Sediment in Hillslope Runoff and Stream Flow. – Earth Surf. Proce. Landforms **22**(8): 705–719.
Trimble, S.W. (1983): A Sediment Budget for Coon Creek Basin in the Driftless Area, Wisconsin, 1853–1977. – Amer. Journ. Sci. **283**: 454–474.
Turcotte, D.L. (1989): Fractals in Geology and Geophysics. – Pure Appl. Geophys. **131**(1/2): 171–196.
Wolman, M.G. & J.P. Miller (1960): Magnitude and Frequency of Forces in Geomorphic Processes.– Journ. Geol. **68**: 54–74.

Wolman, M.G. & R. Gerson (1978): Relative Scales of Time and Effectiveness of Climate in Watershed Geomorphology. – Earth Surf. Proce. **3**: 189–208.
Woolhiser, D.A. & J. Roldan (1982): Stochastic Daily Precipitation Models 2. A Comparison of Distributions of Amounts. – Water Resources Res. **18**: 1461–1468.

Addresses of the authors: Dr John Boardman, School of Geography and Environmental Change Unit, University of Oxford, 5 South Parks Road, Oxford OX1 3TB, U.K., Dr. David Favis-Mortlock, Environmental Change Unit, University of Oxford, 5 South Parks Road, Oxford OX1 3TB, U.K.

Z. Geomorph. N.F.	Suppl.-Bd. 115	71–86	Berlin · Stuttgart	Juli 1999

Sediment production, storage and output: The relative role of large magnitude events in steepland catchments

N.A. TRUSTRUM, Palmerston North, B. GOMEZ, Terre Haute, M.J. PAGE, Palmerston North, L.M. REID, Arcata and D.M. HICKS, Christchurch

with 4 figures

Summary. Magnitude-frequency analysis provides an approach to assess the effectiveness of rare events for generating and transporting sediment in the 32 km² Tutira catchment and the 2205 km² Waipaoa River basin, located in the erodible softrock hill country of the eastern portion of New Zealand's North Island. Shallow landsliding, triggered by high-intensity storms, is an important erosion process throughout the region. The depositional record of Lake Tutira reveals the relationship between event magnitude and sediment yield of the Tutira catchment during the last 100 years, and illustrates the extent to which climate and land use change have affected the amount of sediment generated by landsliding over the past 2250 years. Magnitude-frequency relationships for hillslope erosion and for sediment deposition are closely related in the Tutira catchment, and large-magnitude, low-frequency landsliding events have been responsible for much of the deposition in Lake Tutira since European settlement. In the Waipaoa River basin, processes that affect sediment output are examined using a sediment budget approach. There, large storms appear to play a lesser role compared with the cumulative influence of more frequent, lower magnitude events. The difference in the effectiveness of large magnitude storms is probably attributable to the lower relative contribution of landsliding to catchment sediment yields, compared with other erosion processes such as gully and stream bank erosion where sediment can be generated by more frequent, lower-magnitude storms. Furthermore, the depositional response of large-magnitude storms may be buffered by transport lags, temporary storage, and a general diffusion of the magnitudes and frequencies of sediment contributions from the spatially variable distribution of erosion processes within the various tributaries. Certainly, the long-term record of sedimentation in Lake Tutira suggests that the magnitude and frequency of the erosional response varies with land use, vegetation type and climatic regime.

Zusammenfassung. *Sedimentproduktion, -speicherung und -austrag: Die anteilige Rolle von Ereignissen mit hoher Magnitude in steilen Einzugsgebieten.* – Die Magnitude-Frequenz Analyse stellt eine Methode dar, im 32 km² großen Tutira-Einzugsgebiet und dem 2205 km² großen Flußeinzugsgebiet des Waipaoa die Effektivität seltener Ereignisse für die Erzeugung und den Transport von Sediment abzuschätzen. Beide Einzugsgebiete liegen im mit leicht erodierbaren Gesteinen ausgestatteten Hügelland des östlichen Teils der Nordinsel von Neuseeland. Flachgründige Hangrutschungen, ausgelöst durch Niederschlagsereignisse mit hoher Intensität, sind ein wichtiger Erosionsprozeß in der gesamten Region. Die Sedimentfolge des Tutira-Sees offenbart die Beziehung zwischen der Magnitude eines Ereignisses und der Sedimentproduktion des Tutira-Einzugsgebietes während der letzten 100 Jahre und veranschaulicht das Ausmaß, zu dem Veränderungen des Klimas und der Landnutzung die Menge des durch Hangrutschungen produzierten Sedimentes während der letzten 2250 Jahre beeinflußt haben. Prozesse, die den Sedimentaustrag des Waipaoa beeinflussen, werden mit Hilfe eines Sedimentbudget-Ansatzes untersucht. Es existiert eine starke Beziehung zwischen Magnitude und Frequenz von Hangerosion und Sedimentablagerung im Tutira-Einzugsgebiet, und Hangrutschungsereignisse mit hoher Magnitude und niedriger Frequenz waren verantwortlich für einen Großteil der Ablagerungen im Tutira-See seit Beginn der Europäischen Besiedlung. Im Einzugsgebiet des Waipaoa scheinen starke Niederschlagsereignisse im Vergleich mit den kumulativen

0044–2798/99/0115–0071 $ 4.00

Effekten von Ereignissen mit höherer Frequenz und niedrigerer Magnitude eine geringere Rolle zu spielen. Der Unterschied in der Effektivität von Niederschlagsereignissen hoher Magnitude ist wahrscheinlich dem niedrigeren relativen Beitrag zum Sedimentaustrag von Hangrutschungen im Vergleich mit anderen Erosionsprozessen wie Gully- und Ufererosion zuzuschreiben, wo Sediment durch höherfrequente Niederschläge mit niedrigerer Magnitude produziert werden kann. Außerdem kann die Ablagerungsreaktion auf Stürme mit hoger Magnitude durch Transportverzögerungen, vorübergehende Speicherung und ein allgemeines Verschwimmen der Magnituden und Frequenzen der Sedimentbeiträge der räumlich variabel verteilten Erosionsprozesse in den verschiedenen Subeinzugsgebieten gepuffert werden. Die langzeitliche Sedimentationsfolge im Tutira-See legt außderdem nahe, daß die Magnitude und Frequenz der Erosionsreaktion mit Landnutzung, Vegetationstyp und Klimabedingungen variiert.

Introduction

Increased public awareness of the consequence of erosion and the promulgation of legislation designed to mitigate environmental degradation and facilitate resource management (e.g., New Zealand's Resource Management Act, 1991), have rekindled interest in the routing of sediment through catchments (REID & TRUSTRUM, in press). A major research question facing geomorphologists and resource managers is how to evaluate the downstream impact of land use changes that have occurred, or are occurring, on upland hillslopes. Although the relation between sediment production in upland areas and the sediment yield at a basin outlet has been the subject of investigation for over half a century (GLYMPH 1954, WALLING & WEBB 1996), an understanding of the connections between cause and response remains far from complete. In the short term, a considerable proportion of the sediment generated by erosion does not exit the basin (ROEHL 1962, DEDKOV & MOZZHERIN 1996). This undelivered sediment is of indeterminate origin, since neither the path by which it makes its way to the basin outlet nor the length of time it remains in transport are well known.

Magnitude-Frequency Concepts

In the long term, the amount of work accomplished by a series of events of given magnitude is the product of the effect of a discrete event of that magnitude and the frequency with which it recurs. WOLMAN & MILLER (1960) used magnitude-frequency analysis to assess the effectiveness of rare events for transporting sediment. The discharge frequency was assumed to be log-normally distributed and the sediment transport rate to be a power of discharge; though NASH (1994) found that the power function of discharge was not always a good predictor of sediment transport rate at high discharges.

Defining work as the amount of suspended sediment transported by a river, WOLMAN & MILLER (1960) concluded that large floods transported only a minor proportion of the annual suspended sediment load and that most work was accomplished by frequent events of moderate intensity. However, NOLAN et al. (1987) observed that, in five northern California catchments, most suspended sediment transport was accomplished by high magnitude, low frequency discharges. HICKS (1994) found that this was the case in an urbanizing basin in New Zealand's North Island, but that in basins with more stable land use patterns, most suspended sediment transport was accomplished by frequent events of moderate intensity. Wolman and Miller also observed that the relative amount of work done during different events was not necessarily synonymous with the relative importance of these events in forming a landscape or a particular feature of the landscape. The latter issue was addressed by WOLMAN & GERSON (1978), who noted that effectiveness cannot be related simply to frequency and magnitude of force. Both the timing and location of erosion differ from place to place within a river basin and the products of erosion are not uniformly distributed across the landscape. Material moves sequentially from place to place (WOLMAN & GERSON 1978), and the effectiveness of a flood event of given magnitude depends on the effects of previous floods (NEWSON 1980, BEVIN 1981, GOMEZ et al. 1995).

Attempts to assess the relative effectiveness of events of similar magnitude may additionally be complicated by spatial and temporal variations in sediment storage and mobilization within a catchment (PEARCE 1986). The sensitivity of the landscape, as defined by the threshold value that must be reached or surpassed before motion is initiated, which in itself depends on the type of erosion process involved, may also have a bearing on the effectiveness of an event. For example, PAIN & HOSKING (1970) argued that although events of moderate magnitude and frequency may be more important than large-magnitude, low-frequency flood events in environments where sheetwash erosion contributes most of the sediment to the stream channel, the reverse may be true in areas where mass movements, such as landslides, provide the dominant erosion control (cf. HACK & GOODLETT 1960, O'LOUGHLIN et al. 1978, OKUNISHI & IIDA 1983, PAGE et al. 1994a, BURBANK et al. 1996, HOVIUS et al. 1997). Thresholds of landscape sensitivity can shift as successive rainfall events strip regolith from a hillslope. Without sufficient recovery time for soils to reform the rainfall threshold becomes higher with each successive event. This process is referred to as event or terrain resistance (CROZIER 1996). Furthermore, as PROSSER & SLADE (1994), FROEHLICH & STARKEL (1995), and BROOKS & BRIERLY (1997) emphasize, it is the physical characteristics and state of landscape evolution that set the catchment scale boundary conditions and are the primary determinant of landscape sensitivity to change and the effectiveness of floods. However, PHILLIPS (1997) cautions that while one may be able to assess the sensitivity of landforms to causal events, one cannot necessarily make inferences about recurrence and relaxation times based on the stability of landforms and infer the nature of the forcing factors (tectonic, climatic and anthropogenic) from the geomorphic response (cf. BRUNSDEN & THORNES 1979).

Spatial and temporal variations in erosion thresholds, process dominance, and effects due to prior events make it particularly difficult to determine the role that events of a given magnitude play in the production and dispersal of sediment in large drainage basins. Indeed, assessments of an event's effectiveness typically are made on the basis of at-a-station sediment transport data, or by assessing the scale of a particular erosional or depositional response at a restricted number of locations in the landscape. However, the prediction of future conditions and the reconstruction of past circumstances both require information about the relation between hillslope changes and downstream responses. In the former case, the problem often is one of tracing the response to changes in hillslope conditions downstream through an increasingly complex system of tributaries, sediment sources and sediment sinks (cf. GILBERT 1917). In the latter case, the inverse is true, since typically the responses are known but the multitude of factors influencing those responses must be disentangled before the upbasin conditions can be inferred from the downstream effects (cf. PHILLIPS 1997). The tractability of either problem depends on the extent to which the linkages between hillslope and channel processes, and the changes in the magnitude-frequency response of the sequence of processes are understood.

A sediment budget can be used to connect the spatial and temporal distribution of erosion processes which generate sediment with the processes that transport sediment and govern the sediment yield at the drainage basin outlet (REID & DUNNE 1996). Most sediment budgets describe the processes and rates of sediment production in a catchment, often with the objective of predicting sediment yields. The accuracy with which prediction of sediment yield can be achieved depends also on an understanding of processes of sediment loss and remobilisation that regulate sediment delivery. Information is required about erosion and sediment transport rates and fluxes of sediment, and the acquisition of such information requires an understanding of the relative importance of events of different magnitude and frequencies in the production and transport of sediment.

In this paper we address the role magnitude-frequency considerations play in the elucidation of links between hillslopes and channels, and describe some of the problems inherent in such evaluations. Our discussion employs examples from the results of ongoing work on catchment processes in two river basins of different size in the erodible softrock hill country of New Zealand. Shallow landsliding is an important erosion process in this area, and we focus on the contribution that this process makes to the sediment yield of these watersheds. In particular we examine the role that storms of different magnitude play in the production, storage and output of sediment in these two catchments.

Study catchments and research approach

The two catchments discussed in this paper are the Tutira catchment and Waipaoa River basin, located in the eastern portion of New Zealand's North Island (Fig. 1). High-intensity storms that induce severe erosion are an important component of the rainfall regime of New Zealand's east coast. Both the Tutira catchment and Waipaoa River basin are representative of the erodible hill country in which these events characteristically trigger shallow landslides (PAGE et al. 1994b). In the historic period the largest storm event was Cyclone Bola, which occurred in March, 1988 and, depending on the locality, generated between 300 and 900 mm of rain over a 3-day period. Rainfall records have been kept at Tutira since 1894, while in the Waipaoa River basin records began in 1890 at the coast and in 1913 inland. In the Tutira catchment the mean annual precipitation is 1438 mm and, in the Waipaoa River basin, depending on altitude, it varies from 1000 to 3000 mm.

In the Tutira catchment the sediment sources and the depocentre (Lake Tutira) are closely coupled (Fig. 2), but the links in the much larger Waipaoa River basin are more complex. First, using the depositional record of Lake Tutira, we examine the importance of event magnitude as evidenced by the contribution that sediment generated by landslides makes to the sediment yield of the Tutira catchment in the historic period, and the extent to which land use and climate change over a 2250-year period have affected the amount of sediment generated by landsliding. Using a sediment budget approach, we then examine the processes that regulate sediment output from the Waipaoa River basin in contemporary times.

Tutira catchment study area and methods

The 32 km^2 Tutira catchment is underlain by late Pliocene and early Pleistocene sandstones and siltstones interbedded with conglomerates (LOWE 1987). The local rate of uplift was between 0–2 mm yr^1 during the last 10 000 years (WELLMAN 1967). Lake Tutira, which is currently

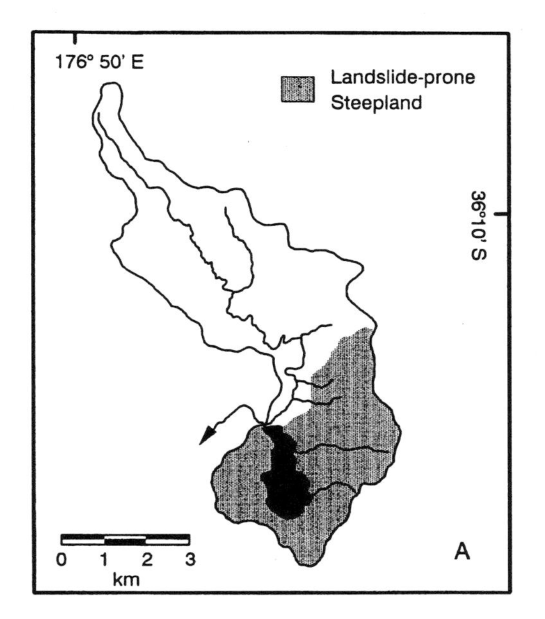

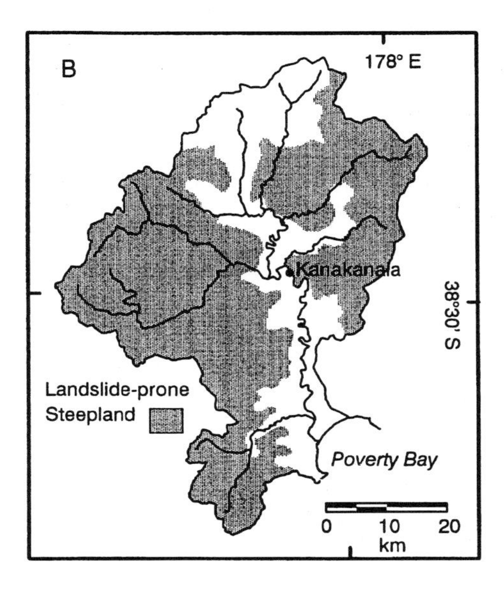

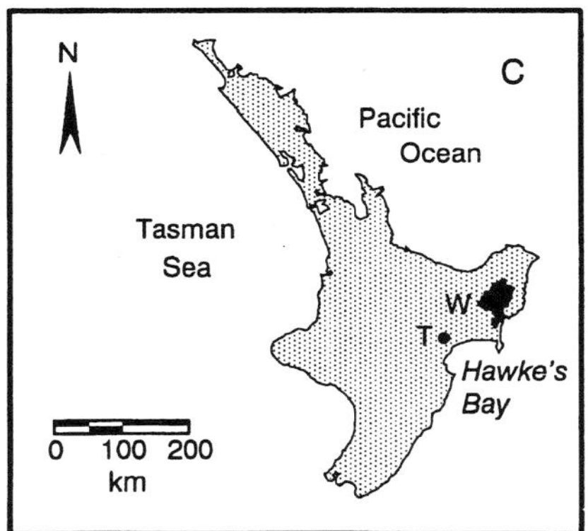

Fig. 1. A) Tutira catchment; B) Waipaoa River basin; and C) North Island, New Zealand, showing location of Tutira catchment (T) and Waipaoa River basin (W).

the sink for sediment generated within the catchment, has an area of 1.79 km^2. The lake formed as a result of a landslide which occurred about 6500 years ago. (Trustrum & Page 1992, Page et al. 1994a). North of the lake, where the late Pliocene sediments crop out, the gently undulating terrain consists of a series of NE-SW trending asymmetrical ridges with long inclined dip slopes and shorter, steeper scarps. Ridges are separated by vales, and the stream system is entrenched for most of its length. To the south of the lake where the terrain is steeper, a series of narrow, infilled alluvial valleys drain to the lake, and the steep hillslopes formed of early Pleistocene sediments are landslide-prone (Fig. 2). Prior to Polynesian (Maori) settlement about 490 cal years B.P. the catchment was covered with indigenous forest (Wilmshurst 1997, Page & Trustrum 1997). By 1873, when the first Europeans arrived, most of the forest had been destroyed by fire and replaced by bracken and scrub (GUTHRIE-SMITH 1953). European settlers converted the entire catchment to pasture.

A series of 53 cores was obtained in and around Lake Tutira (PAGE et al. 1994a and b). Two modified Mackereth piston corers were used to obtain 25 of the cores, 20 cores were obtained using a freezebox corer, and 8 cores were taken from valley floors. One 16.6 m valley floor core was obtained from the swamp at the northern end of the lake. Nine radiocarbon ages were obtained from the latter core, which also contained 14 tephra layers, and cores obtained from the deeper part of the lake typically exhibited well-defined laminations that delineate individual storm sediment pulses. A composite stratigraphy for the period of European settlement was constructed using a combination of Mackereth cores, which extended into the period since European arrival but contained disturbed or incomplete upper layers, together with the freezebox cores, which contained an undisturbed record of the uppermost sediments. A chronology for the period of European occupation was established using the local rainfall records, ^{137}Cs, pollen and diatom analysis, evidence of eutrophication and the presence of tephra. The laminated sediments also provide a high resolution record of storm events over the past 2250 years (EDEN & PAGE 1998). The longer term depositional chronology was constrained using tephrochronology in conjunction with pollen and diatom analysis. A total of 340 rainfall events were identified in the pre-European record, and, in the period following European settlement, storm sediment pulses were correlated with the local rainfall record.

Fig. 2. Lake Tutira and surrounding landslide-prone steepland terrain, showing the 1988 (Cyclone Bola) storm-induced landslides and floodplain sedimentation. (Photo: N. A. TRUSTRUM, taken in September 1998, six months after Cyclone Bola).

Erosional response of the Tutira catchment

The relation between sediment thickness derived from the lake cores and the historical storm rainfall record suggests that high-magnitude events produce disproportionally large amounts of sediment in comparison with low-magnitude events (Fig. 3A). However, the variance associated with the relationship suggests that this may not be a simple relationship. For example, a variation in the threshold of terrain resistance (which influences sediment availability) can produce different landslide frequencies for storms with similar antecedent precipitation and seasonal occurrence. The susceptibility of the landscape to landsliding increased following removal of

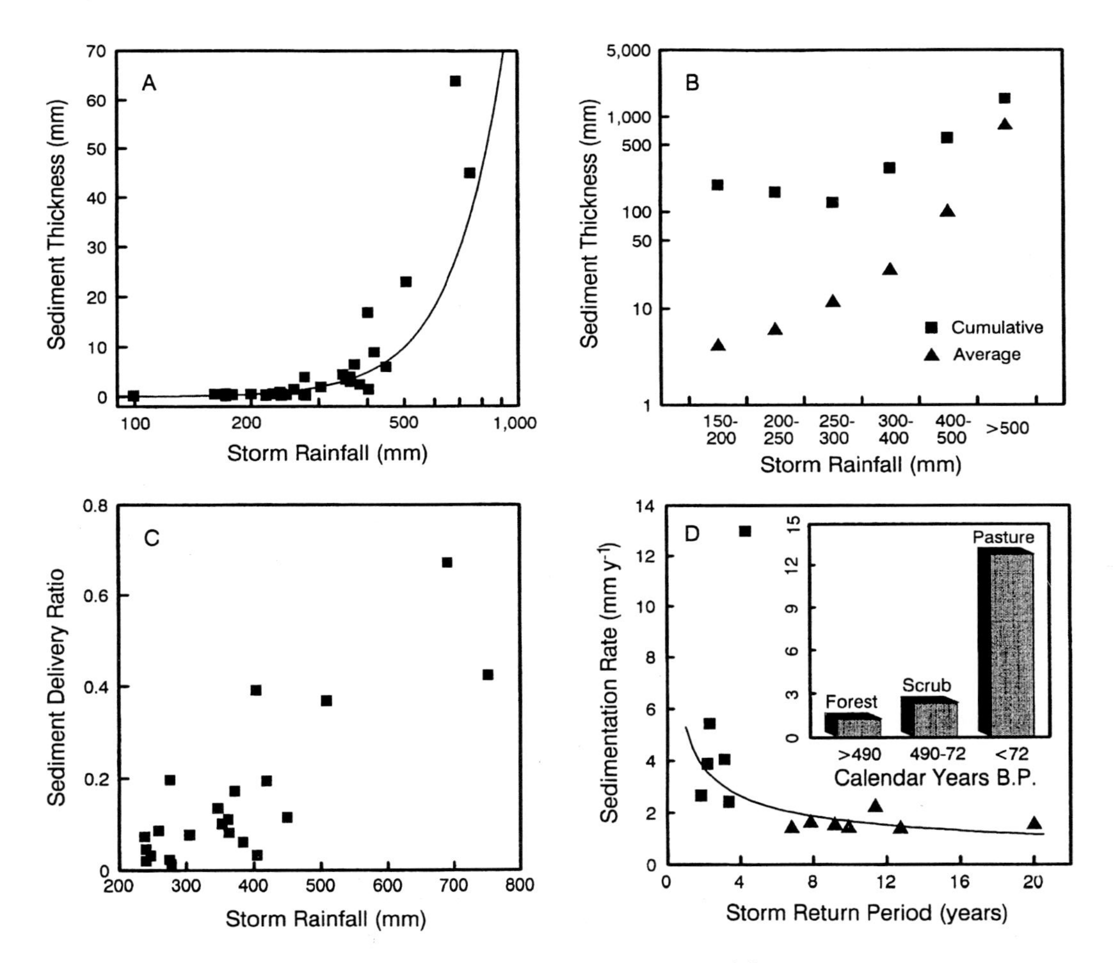

Fig. 3. A) Relation of sediment thickness in Lake Tutira to storm rainfall.
B) Cumulative and average sediment thickness in Lake Tutira for storms occurring within the specified rainfall class (after PAGE et al. 1994a).
C) Relation of sediment delivery ratio to storm rainfall (after DYMOND, pers. comm.).
D) Variation of sedimentation rate in Lake Tutira with paleoclimatic regime (solid squares relate to periods of increased storminess and solid triangles represent intervals between periods of storminess) and (inset) land use/vegetation type (after PAGE & TRUSTRUM 1997, EDEN & PAGE 1998).

the indigenous vegetation, and it is likely that earlier storms, such as that of 1938, removed a large proportion of the sediment that had accumulated under the indigenous forest cover. Once these storage sites were evacuated, sediment had to be derived from other locales. In effect, later storms would be expected to yield less sediment because terrain resistance had likely increased (cf. CROZIER & PRESTON 1998). This trend is also supported by FRANSEN & BROWNLIE (1995) who showed that for the Tamingimingi catchment, 30 km south of Lake Tutira, the area of landsliding in the larger-magnitude 1988 rainfall event was 77% of the area of landsliding in the 1938 event. Sediment availability is also constrained by the rates of regolith stripping and soil recovery between storms (cf. SHIMOKAWA 1984, TRUSTRUM & DEROSE 1988). Differences in individual storm sediment delivery ratios, antecedent conditions, within-storm intensities and land management induced changes in storage capacity may also increase the variance about the rainfall:sediment-yield relationship. During small storms that are below the threshold of landsliding, terrain resistance has a diminished influence. Sediment generated during these storms is flushed from storage sites which are replenished during higher magnitude events.

The thickness of sediment yielded by storms of given magnitude during the historic period is a measure of geomorphic effectiveness. Although the average thickness of sediment yielded by storms increases with increasing rainfall (Fig. 3B), the high frequency of low magnitude events (47 storms with rainfalls between 150 and 250 mm) produces a greater cumulative sediment thickness than moderate-frequency, moderate-magnitude rainfall events (11 events with rainfalls between 250–300 mm). While landsliding accounts for the majority (89%) of the sediment mobilised during large-magnitude events in the Tutira catchment (PAGE et al. 1994b), for smaller storms a greater percentage of sediment is derived by other erosion processes, including fluvial erosion from low-order channels and valley floor storage sites. Evidence for this change in process is provided by the presence of storm sediment pulses correlated to the 150 to 250 mm rainfalls. These storms are below the threshold for significant landsliding and it is likely that the majority of this sediment is generated by sheetwash erosion and channel erosion processes. Because much of this sediment may be excavated from redeposited sediments, the thickness of sediment produced by small storms is less affected by timing of previous storms. Above the threshold for landsliding the effectiveness of individual storms becomes increasingly more pronounced, and the two largest (1938 and 1988) storms generated about half the storm-related sediment in the 93 year period (1895–1988) for which rainfall records are available. The disproportionate influence of large storms results because the spatial frequency of landsliding (numbers of slides/unit area) and the sediment delivery ratio both increase with increasing storm magnitude (Fig. 3C). The former effect is well documented and reflects the influence that elevated soil moisture levels have on soil strength (cf. MONTGOMERY & DIETRICH 1994). The latter likely relates to the connectivity that an increasing number of landslides scars have to an extended stream channel network during rare, high-magnitude events, and by increased transmission of debris-tail sediment by overland flow. In order to construct Fig. 3C, a surrogate measure of the extent to which increasing rainfall influences sediment delivery was obtained by combining magnitude-frequency curves for landsliding derived for an analogous landtype in the Waipaoa catchment (REID & PAGE, in press) with the lake deposition record (Fig. 3A). Thus, during events which generate between 200 and 400 mm of rainfall, sediment delivery ratios are ~0.1, whereas during large-magnitude events that generate >600 mm of rainfall, sediment delivery ratios are between 0.4 and 0.7 (JOHN DYMOND, pers. comm.). Site-specific differences in depositional rates on hillslope and valley-floor storage sites may also decrease during high magnitude events (as discussed below).

The incidence of landsliding, and hence the magnitude of the erosional response, also vary with land use and vegetation type, and climatic regime (PAGE & TRUSTRUM 1997, EDEN & PAGE 1998). Clusters of sediment pulses in the depositional record suggest that, under the near-complete cover of indigenous forest prior to Polynesian settlement, and under the fern/scrub cover before European settlement, periodic increases in high-rainfall storms (and presumably also in the incidence of landsliding) were associated with La Niña phases of the El Niño-Southern Oscillation (ENSO) that have been identified from New Zealand and Southern hemisphere palaeoclimatic evidence (EDEN & PAGE 1998). Five of the six clusters of sediment pulses were related to increased storm frequency, the longest period lasting more than 200 years, while the other periods generally were of much shorter duration. On average, each such period is associated with a ~400% increase in storm frequency and a ~100% increase in the sedimentation rate, though the sedimentation rate varies greatly from one period to another (Fig. 3D). Compared with the sedimentation rate under the indigenous forest cover, the sedimentation rate increased by ~60% in response to the fire-induced vegetation changes that occurred following Polynesian settlement. There was a massive increase in landsliding following the catchment's conversion to pasture during the period of European settlement (Fig. 3D), when the sedimentation rate increased by an order of magnitude (PAGE & TRUSTRUM 1997). These examples demonstrate the utility of a continuous climate-landslide record to quantify the extent to which vegetation and climatic changes cause shifts in landscape sensitivity and erosion triggering thresholds (CROZIER 1996).

Waipaoa study area and methods

The 2205 km^2 Waipaoa River basin drains the eastern flanks of the Raukumara Ranges. In the headwaters the rate of uplift is 5–10 mm yr^{-1}, but the coastal plain is subsiding in the southwest (MAZENGARB et al. 1991). Different combinations of lithology, structure and topography in the basin give rise to distinct landform combinations that exhibit varying degrees of stability (GAGE & BLACK 1979, BLACK 1980). For example, conspicuous (up to 0.2 km^2) amphitheatre-like gully complexes develop where crushed and weathered argillite rocks crop out in the headwaters (PEARCE et al. 1981., DEROSE et al. 1998), while the hills surrounding the Poverty Bay Flats that are underlain by Miocene Pliocene sandstone and mudstone are prone to landsliding. Maori settlements on the lower floodplain date from about 700 years ago (JONES 1988), but wholesale clearance of the indigenous forest did not begin until the early 1830s, following the arrival of European settlers (PULLAR 1962). By the 1880s much of the land in the basin's lower reaches had been cleared and conversion to pasture in the headwaters was underway. The latter phase of deforestation was completed by 1920, and only 3% of the basin now remains under primary forest. Reforestation of portions of the headwaters began in 1960, and commercial timber harvesting began in 1990. Deforestation during the late 19th and early 20th centuries caused particularly intense erosion of the incoherent rocks in the Waipaoa River basin headwaters (O'BYRNE 1967, ALLSOP 1973). Based on measurements of water and suspended sediment discharge made since 1960, the mean annual suspended sediment load of the Waipaoa River at Kanakanaia is 10.7 $x10^6$ tons yr^1, or 6750 tons km^{-2}-yr^{-1}, which is high by global standards (HICKS et al., in press).

Sediment budgets are being constructed for each of the 16 land types represented in the Waipaoa River basin. The contribution that the principal sediment-generating mechanisms (i.e., shallow landslides, gully erosion, sheetwash and earthflows) make to sediment production has been locally determined by a variety of methods, including area and elevation differences derived from high-resolution digital elevation models constructed from aerial photographs (e.g., DEROSE et al. 1998). In the case of shallow landslides, information derived from conventional photogrammetric interpretation and field measurements has been used to obtain erosion rates for twenty 1–15 km^2 subcatchments which are representative of the principal landslide-prone land types present in the basin (REID & PAGE, in press). Aerial photographs were used to determine the number of landslides generated by storms of known magnitude. The resulting correlation with storm magnitude permitted landslide frequency to be estimated from storm rainfall. This relation was then combined with storm magnitude-frequency relations derived from rainfall records to yield a long-term magnitude-frequency relation for landsliding. Average scar volumes were determined from field measurements. The average rate of sediment production by landsliding for a given land type was determined by multiplying the mean landslide volume by the average landslide frequency. A delivery ratio was assessed on the basis of the proportion of landslides that conveyed sediment directly to channels, as determined from aerial photographs.

Upstream and downstream links in the Waipaoa River basin

Connections between erosion in the headwaters and the downbasin response are often obscured in large river basins. Nevertheless, the magnitude-frequency characteristics of suspended sediment transport in the Waipaoa River basin were examined for evidence of sediment sources. Estimates of event sediment loads at the Kanakanaia gauging station, derived by integrating the suspended sediment rating curve with water discharge over discrete quickflow events, were used to compile an event-specific magnitude-frequency distribution of sediment loads (Fig. 4A). For the 23 year period of record, ~50% of the suspended sediment load of the Waipaoa River was transported during events that re-occur at least once per year, while 86% of the load was transported by events with return periods less than 10 years. The sediment load during the 1988 storm was ~26 x 10^6 t, which is nearly 3 times the mean annual sediment load and was the largest event load in the record (HICKS et al., in press). The 1988 storm plots as an apparent outlier above the Extreme Value type II trend (see Fig. 4A) displayed by smaller magnitude events, and it is also responsible for the discontinuity on the relation of cumulative sediment yield to event frequency (Fig. 4B). It is unclear whether this storm is a true outlier, which might reflect a change in sediment regime that could be related to a variation in the type and intensity of the important erosion processes and/or to the proportion of sediment delivered to the channel system, or whether it is simply a storm with larger return period that happened to occur during the period of record (HAAN 1977, p. 138). If it were actually a 100 year event, for example, there is a 20% chance that it would occur during the 23 year record period. The return period estimated for the peak flow of the 1988 storm, based on an extended historical record of peak flood levels, was 100 years, which suggests that the sediment yield event may have been of similar return period.

There was a 100% increase in the suspended sediment concentration for a given water discharge in the three years following the 1988 storm as compared with the three years before the storm (Fig. 4C). Elevated levels of gully, earthflow, and landslide activity combined to increase the amount of sediment supplied directly to river channels and riparian storage areas during the 1988 storm. However, the magnitude of the shift in the suspended sediment rating

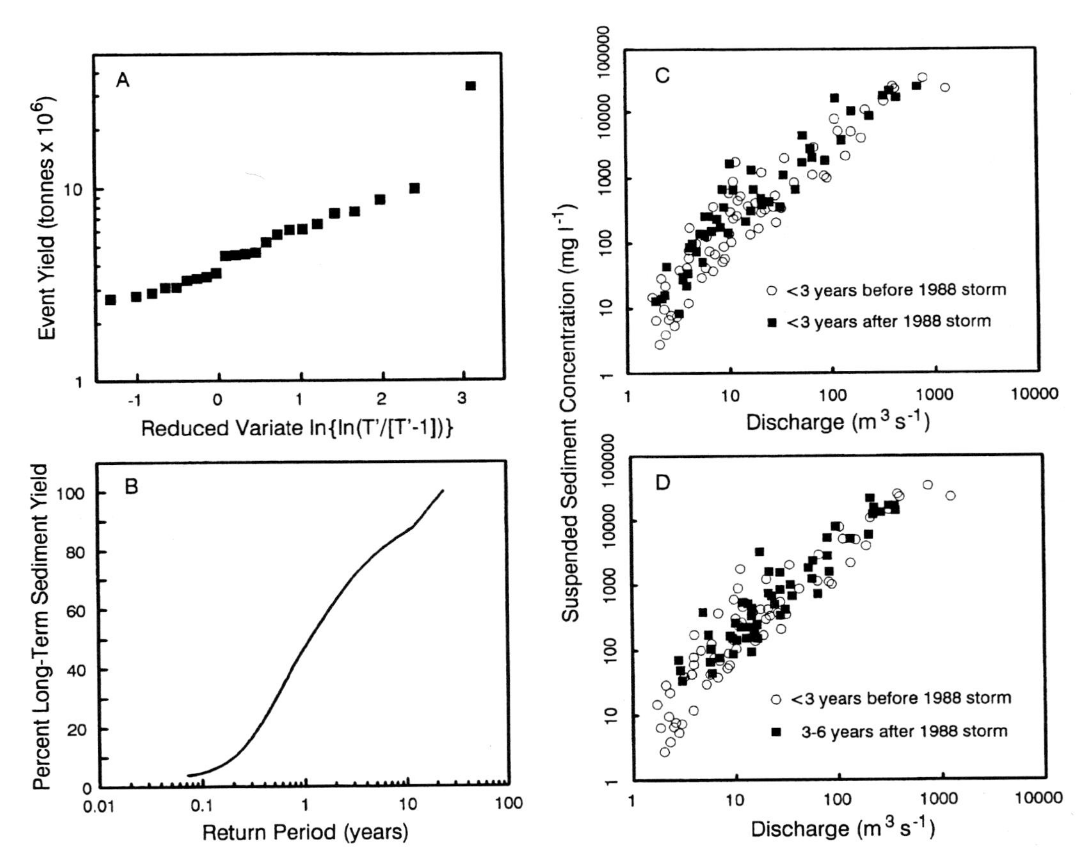

Fig. 4. A) Relation of suspended sediment yield in the Waipaoa River, at Kanakanaia, to event magnitude (reduced variate function, where: T' is the return period in months).
B) relation of cumulative sediment yield to event frequency. In Figures 4A & B the Cyclone Bola sediment yield event plots at about a return interval of 20 years, but this is probably an artifact of the 23 year period of record, and more likely has a return period of 100 yr on the basis of the historical flood record.
C and D) Relation of suspended sediment concentration in the Waipaoa River at Kanakanaia with discharge for specified periods before and after the 1988 storm.

curve after the storm suggests that either the eroded surfaces continued to erode after the storm or considerable volumes of sediment were left in temporary storage by the storm and were available for transport during ensuing smaller storms, albeit temporarily, as is indicated by the subsequent return of the rating curve to pre-storm levels 3 years after the storm (Fig. 4D). Thus the relative influence of large storms is somewhat larger than per-storm yields would suggest.

Virtually all the sediment contributed to stream channels by shallow landslides is transported in suspension; soils in the landslide-prone terrain are predominately fine-grained and landslide failure planes typically coincide with the base of the soil profile. It is estimated that shallow landslides typically contribute about 10% to 19% of the Waipaoa River's annual suspended sediment load at Kanakanaia; ~50% of all landslides occur during storms with a recurrence interval <7 years; and ~75% during storms with a recurrence interval of <25 years (Reid

& PAGE, in press). Gully erosion, earthflows, sheetwash erosion and channel erosion contribute the balance. Assuming a sediment delivery ratio of 0.5, PAGE et al. (in press) estimate that during the 1988 storm about one-half of the suspended sediment load at Kanakanaia was contributed by shallow landslides. Sediment is removed from landslide scars and debris tails by sheet and rill erosion both during the initiating event and subsequent storms. Though landslide scars were still visible 9 years after the 1988 storm, surface sealing and revegetation cause the rate of surface erosion to decrease exponentially as the bare surfaces heal (cf. TRUSTRUM & DEROSE 1988). The increased variance of the magnitude-frequency relation between storm rainfall and landsliding caused by factors such as antecedent precipitation and terrain resistance, was only considered to be minor compared to the influence of storm magnitude (REID & PAGE, in press).

In addition to suspended sediment transport, variations in the rate of sediment deposition on the floodplain show how the overbank environment, which functions as a significant sediment sink in many large river basins (cf. MARRON 1992, MERTES 1995), acts to regulate sediment output from the Waipaoa River basin during floods of differing magnitude (GOMEZ et al., in press). The rate of vertical accretion on a 0.5 km^2 area of the floodplain bordered by a 2.5 km long bend was determined by relating the floodplain stratigraphy to the 1948–1995 flood history. Overbank flow occurred across the bend at discharges >1800 $m^3 s^{-1}$. During events with recurrence intervals of 5 to 30 years (peak discharge between 1905 and 3965 $m^3 s^{-1}$), the rate of vertical accretion was typically 14 to 18 mm h^{-1} (GOMEZ et al. 1988). The measured rate of vertical accretion during a flood in 1996 (peak discharge 2284 $m^3 s^{-1}$) was 15 mm h^{-1}. However, because flow velocities across this section of the floodplain were too high to permit deposition from suspension, the rate of vertical accretion during the 1988 storm (Cyclone Bola: return period >100 years, peak discharge 5287 $m^3 s^{-1}$) was only 6 mm h^{-1}. Subsequent measurements of vertical accretion from two other cores and repeat surveys of transects across the floodplain, show that although rates of vertical accretion are high, the net loss to the system is comparatively small. For an eleven year period between 1979 and 1990 floodplain storage accounted for only 5% of the total suspended sediment load, and 16% of the sediment load during events that exceeded bankfall discharge (GOMEZ et al., in press).

Discussion and conclusions

WOLMAN & MILLER'S (1960) pioneering concept of geomorphic work encompassed both event frequency and magnitude. They and many other researchers have developed these concepts further, even to the point where some now question whether the original concept has outlived its usefulness as a research tool (RICHARDS, this issue). Results from the Tutira and Waipaoa studies further illustrate these concepts, demonstrating that the interpretation of depositional records and reconstruction of the magnitude-frequency relationships is not straightforward. In the 32 km^2 Tutira catchment, magnitude-frequency relationships between hillslope erosion and sediment deposition are closely related and, as in other small, highly disturbed, steepland catchments (cf. NOLAN et al. 1987), large-magnitude, low-frequency landsliding events were responsible for much of the deposition in Lake Tutira during the historic period. By contrast, in the 2205 km^2 Waipaoa River basin, landsliding makes a smaller relative contribution to catchment suspended sediment yield than does the combination of other erosion processes (e.g., gully, earthflow, and streambank erosion). Furthermore, suspended sediment data suggest that 86% of the sediment load is transported by events with return periods less than 10 years. Hence, large storms have a diminished role compared with the cumulative influence of more frequent, lower magnitude events. This conclusion matches that of Wolman and Miller's magnitude and frequency analysis of the effectiveness of rare events for transporting sediment. Large-magnitude, low-frequency events generate and mobilize large amounts of sediment from primary hillslope source areas, but much of this material goes directly into storage (for comparison, ~43% in the Tutira catchment during the 1988 storm). However, small events, which derive sediment predominately from secondary storage areas and sources with high sediment delivery ratios (MARUTANI et al., in press, – e.g., gullies and streambanks), have a high enough frequency of recurrence to cumulatively outweigh the large, rare events as transporters of sediment.

In large river basins where diverse terrain types are present, the effectiveness of large magnitude storms also varies with the spatial distribution of catchment characteristics (JACOBSON et al. 1989). The change in effectiveness is probably attributable to spatial variations in the distribution of erosion processes and erosion thresholds. In the Waipaoa River basin, the temporarily increased suspended sediment load at given water discharges after major storms indicates a temporary up-shift in the magnitude-frequency distribution of event sediment loads. This shift may occur because of the role that event sequencing plays in determining the magnitude of suspended sediment transport in large river basins, and because the suspended sediment yield is regulated by the sediment sinks that are interposed between the hillslopes and the basin outlet. Because these sinks do not behave in a homogeneous manner, the influence they have on suspended sediment yield varies in both space and time. Similar discontinuities between the magnitude and frequency of sediment mobilised on hillslopes, and channel response, were noted by FROEHLICH & STARKEL (1995) for temperate areas subject to extreme rainfalls.

Many magnitude-frequency analyses are divorced from the historical context of landscape evolution, but the long term (c. 2250 year) record of sedimentation in Lake Tutira suggests that the incidence of landsliding, and hence the magnitude of the erosional response, also varies with land use, vegetation type, and palaeoclimatic regime. The recognition of 5 periods of increased sedimentation, related to increased storm magnitude and frequency, provides evidence for late Holocene changes in climate regime. However, vegetation change appears to have been an even more important forcing factor. Sustained increases in landsliding, by up to an order of magnitude, occurred following the catchment's conversion to pasture during the period of European settlement. Tectonism is another important factor influencing sediment delivery. Compared with Lake Tutira, rates of tectonic uplift in the Waipaoa catchment are between 5 to 10 times higher. Several flights of remanent terraces dated at about 11500 years B.P. (BLACK 1980), above the Waipaoa floodplain indicate that this catchment has undergone significant landscape change during several major erosion periods in the Holocene and late Pleistocene. But it is unclear how tectonism and climate have interacted to shape this landscape or whether one factor has predominated. Insufficient time has elapsed to evaluate the effects that anthropogenically-induced increases in erosion and sedimentation will have on landscape morphology, and how these effects might be modified by tectonism and potential climate shifts.

It is postulated that future advances in addressing the role that magnitude-frequency considerations play in the elucidation of links between hillslopes and channels will require multidisciplinary studies of the relative roles of climatic, tectonic, and anthropogenic forcing factors. Furthermore, these studies should be undertaken in sedimentary systems were there are sufficient depositional records to unequivocally link upbasin cause with downbasin response.

Acknowledgements

This research was funded by the NZ Foundation for Research, Science and Technology under Contract C09612. Gisborne District Council provided logistical support and other resources. Numerous colleagues have also assisted our research efforts, but we especially thank Ron DeRose, Dennis Eden, Mike Marden, Dave Peacock, Ted Pinkney, Donna Rowan, Janet Wilmshurst, John Dymond, Ian Lynn and the grape growers of the Hawke's Bay and Gisborne regions for enhancing the quality of the field experience.

References

ALLSOP, F. (1973): The Story of Mangatu. – Government Printer, Wellington, New Zealand, 100p.

BEVIN, K. (1981): The effect of ordering on the geomorphic effectiveness of hydrologic events. – Internat. Assoc. Hydrolog. Sci. Publ. **132**: 510–526.

BLACK, R.D. (1980): Upper Cretaceous and Tertiary geology of Mangatu State Forest, Raukumara Peninsula, New Zealand. – N.Z. Jl. Geol. Geophys. **23**: 293–312.

BROOKS, A. & G.J. BRIERLY (1997): Geomorphic responses of lower Bega River to catchment disturbance, 1851–1926. – Geomorphology **18**: 291–304.

BRUNSDEN, D. & J.E. THORNES (1979): Landscape sensitivity and change. – Transactions of the Institute of British Geographers **4**: 463–484.

BURBANK, D.W., J. LELAND, E. FIELDING, R.S. ANDERSON, N.BROZOVIC, M.R. REID & C. DUNCAN (1996): Bedrock incision, rock uplift, and threshold hillslopes in the northwestern Himalayas. – Nature **379**: 505–510.

CROZIER, M.J. (1996): The climate-landslide couple: a Southern Hemisphere perspective. – Paleoclimate Res. 19 ESF Spec. Issue **12**: 329–350.

CROZIER, M. J. & N.J. PRESTON (1998): Modelling changes in terrain resistance as a component of landform evolution in unstable hill country. – In: HERGARTEN, S. & H. J. NEUGEBAUER (Eds.): Process Modelling and Landform Evolution. – Lecture Notes in Earth Sciences, Vol. 78, Springer-Verlag, Heidelberg.

DEDKOV, A.P. & V.I. MOZZHERIN (1996): Erosion and sediment yield on the Earth. – Internat. Assoc. Hydrol. Scis. Publ. **236**: 29–33.

DEROSE, R.C., B. GOMEZ, M. MARDEN & N.A. TRUSTRUM (1998): Gully Erosion in Mangatu Forest, New Zealand, Estimated From Digital Elevation Models. – Earth Surf. Proc. Landforms **23**: 1045–1053.

EDEN, D.N. & M.J. PAGE (1998): Palaeoclimatic implications of a storm erosion record from late Holocene lake sediments, North Island, New Zealand. – Palaeogeogr., Palaeoclimat. Palaeoecol. **139**: 37–58.

EDEN, D.N., P.C. FROGGATT, N.A. TRUSTRUM & M.J. PAGE (1993): A multi-source Holocene tephra sequence from Lake Tutira, Hawke's Bay, New Zealand. – N.Z. Journ. Geol. Geophys. **36**: 233–242.

FRANSEN, P. & R. BROWNLIE (1995): Historical slip erosion in catchments under pasture and radiata pine forest, Hawke's Bay hill country. – New Zealand Forestry, November 1995, 29–33.

FROEHLICH, W. & L. STARKEL (1995): The response of slope and channel systems to various types of extreme rainfall: A comparison between the temperate zone and humid tropics. Geomorphology **11**,4: 337–345.

GAGE, M. & R.D. BLACK (1979): Slope-stability and geological investigations at Mangatu State Forest. – N.Z. Forest Service Techn. Paper No. 66, New Zealand Forest Service, Wellington, 37 p.

GILBERT, G.K. (1917): Hydraulic mining debris in the Sierra Nevada. – U.S. Geol. Surv. Prof. Paper **105**, 154 p.

GLYMPH, L.M. (1954): Studies of sediment yields from watersheds. – Internat. Assoc. Hydrol. Scis. Publ. 36: 173–191.

GOMEZ, B., L.A.K. MERTES, J.D. PHILLIPS, F.J. MAGILLIGAN & L.A. JAMES (1995): Sediment characteristics of an extreme flood: 1993 upper Mississippi River valley. – Geology **23**: 963–966.

GOMEZ, B., D.N. EDEN, D.H. PEACOCK & E.J. PINKNEY (1998): Flood Plain Construction by Rapid Vertical Accretion and Channel Change: Waipaoa River, New Zealand. – Earth Surf. Proc. Landforms **23:** 405–418.

GOMEZ, B., J.D. PHILLIPS, F.J. MAGILLIGAN & L.A. JAMES (1997): Floodplain sedimentation and sensitivity: summer 1993 flood, upper Mississippi River valley. – Earth Surf. Proc. Landforms **22**: 923–936.

GOMEZ, B., D.N. EDEN, D.M. HICKS, N.A. TRUSTRUM, D.H. PEACOCK & J. WILMSHARST (in press): Contribution of floodplain sequestration to the sediment budget of the Waipaoa River, New Zealand. – In: ALEXANDER, T. & S.B. MARRIOT (eds.): Floodplains: Interdisciplinary Approaches. – Geological Society of London, Special Publication.

GUTHRIE-SMITH, W.H. (1953): Tutira: The Story of a New Zealand Sheep Station. – 3rd Edition, William Blackwood and Sons Ltd., London. 464 p.

HANN, C.T. (1977): Statistical methods in hydrology. The Iowa State University Press, Ames, 378pp.

HACK, J.T. & J.C. GOODLETT (1960): Geomorphology and forest ecology of a mountain region in the central Appalachians. – U.S. Geol. Surv. Prof. Paper **347**, 64p.

HICKS, D. M. (1994): Landuse effects on magnitudefrequency characteristics of storm sediment yields: some New Zealand examples. – Internat. Assoc. Hydrol. Scis. Publ.n **224**: 395–402.

HICKS, D.M., B. GOMEZ & N.A. TRUSTRUM (in press): Erosion thresholds and suspended sediment yields: Waipaoa River basin, New Zealand. – Water Resources Res.

HOVIUS, N., C.P. STARK & P.A. ALLEN (1997): Sediment flux from a mountain belt derived by landslide mapping. – Geology **25**: 231–234.

JACOBSON, R.B.I., A.J. MILLER & J.A. SMITH (1989): The role of catastrophic geomorphic events in central Apalachian landscape evolution. – Geomorphology **2**: 257–284.

JONES, K.L. (1988): Horticulture and settlement chronology of the Waipaoa River catchment, East Coast, North Island, New Zealand. – N. Z. Journ. Archaeol. **10**: 19–51.

LOWE, D.A. (1987): The geology and landslides of the Lake Tutira-Waikoau area, northern Central Hawke's Bay. – M.Sc. Thesis, Victoria University of Wellington, New Zealand, unpubl.

MARRON, D.C. (1992): Floodplain storage of mine tailings in the Belle Fourche River system: a sediment budget approach. – Earth Surf. Proc. Landforms **17**: 675–686.

MARUTANI, T., M. KASAI, L.M. REID & N.A. TRUSTRUM (in press): Influence of storm-related sediment storage on the sediment delivery ratios of tributary catchments in the upper Waipaoa River, New Zealand. – Earth Surface Processes and Landforms.

MAZENGARB, C., D.A. FRANCIS & P.R. MOORE (1991): Sheet Y1 Tauwhareparae 1:50,000 scale. – Dept. Scient. Industr. Res., Wellington, New Zealand, 1 sheet and notes, 52 p.

MERTES, L.A.K. (1995): Rates of flood-plain sedimentation on the central Amazon River. – Geology **22**: 171–174.

MONTGOMERY, D.R. & W.E. DIETRICH (1994): A physically based model for the topographic control on shallow landsliding. – Water Resources Res. **30**: 1153–1171.

NASH, D.B. (1994): Effective sedimenttransporting discharge from magnitudefrequency analysis. – Journ. Geol. **102**: 79–95.

NEWSON, M. (1980): The geomorphological effectiveness of floods a contribution stimulated by two recent events in midWales. – Earth Surf. Proc. **5**: 116.

NOLAN, K.M., T.E. LISLE & H.M. KELSEY (1987): Bankfull discharge and sediment transport in northwestern California. – Internat. Assoc. Hydrol. Scis. Publ. **165**: 439–449.

O'BYRNE, T.N. (1967): A correlation of rock types with soils, topography and erosion in the Gisborne – East Cape region. – N.Z. Journ. Geol. Geophys. **10**: 217–231.

O'LOUGHLIN, C.L., L.K. ROWE & A.J. PEARCE (1978): Sediment yields from small forested catchments, North Westland-Nelson, New Zealand. – Journ. Hydrol., New Zealand, **17**: 1–15.

OKUNISHI, K. & T. IIDA (1983): Evolution of hillslopes including landslides. – Transact. Japan. Geomorph. Union **2**: 191–200.

PAGE, M.J., L.M. REID & I.H. LYNN (in press): Sediment production from Cyclone Bola landslides, Waipaoa Catchment. – Journ. Hydrol., New Zealand.

PAGE, M.J. & N.A. TRUSTRUM (1997): A late Holocene lake sediment record of the erosional response to land use change in a steepland catchment, New Zealand. – Z. Geomorph., N.F. **41**, 3: 369–392.

PAGE, M.J., N.A. TRUSTRUM & R.C. DEROSE (1994a): A high resolution record of storm induced erosion from lake sediments, New Zealand. – Journ. Paleolimnol. **11**: 333–348.

PAGE, M.J., N.A. TRUSTRUM & J.R. DYMOND (1994b): Sediment budget to assess the geomorphic effect of a cyclonic storm, New Zealand. – Geomorphology **9**: 169–188.

PAIN, C.F. & P.L. HOSKING (1970): The movement of sediment in a channel in relation to magnitude and frequency concepts a New Zealand example. – Earth Sci. Journ. **4**: 1723.

PEARCE, A.J. (1986): Geomorphic effectiveness of erosion and sedimentation events. – Journ. Water Resources **5**: 551–569.
PEARCE, A.J., R.D. BLACK & C.S. NELSON (1981): Lithologic and weathering influences on slope form and process, eastern Raukumara Range, New Zealand. – Internat. Assoc. Hydrol. Scis. Publ. **132** (Proceedings of Christchurch Symposium): 95–122.
PHILLIPS, J.D. (1997): Humans as geologic agents and the question of scale. – Amer. Journ. Scis. **297**: 98–115.
PROSSER, I.P. & C.J. SLADE (1994): Gully formation and the role of valley-floor vegetation, southeastern Australia. – Geology **22**: 1127–1130.
PULLAR, W.A. (1962): Soils and agriculture of Gisborne Plains. – N.Z. Dept. Scient. and Industr. Res. Soil Bur. Bull. **20**, 92 p.
REID, L.M. & T. DUNNE (1996): Rapid evaluation of sediment budgets. – Catena Verlag, Reischierken, Germany, 164p.
REID, L.M. & M.J. PAGE (in press): Magnitude and frequency of landsliding in a large New Zealand catchment. – Geomorphology.
REID, L.M. & N.A. TRUSTRUM (in press): Sediment budgets and land management planning: Examples from New Zealand. – Journ. Environmental Planning and Management.
RICHARDS, K.S. (1999): The magnitude-frequency concept in fluvial geomorphology: a component of a degenerating research programme? – Z. Geomorph.N.F., Suppl.-Bd. **115**: 1–18.
ROEHL, J.E. (1962): Sediment sources, delivery ratios and influencing morphological factors. – Internat. Assoc. Hydrol. Scis. Publ. **59**: 202–213.
SHIMOKAWA, E. (1984): A natural recovery process of vegetation on landslide scars and landslide periodicity in forested drainage basins. – In: O'LOUGHLIN, C.L. & A.J. PEARCE (eds.): Symposium on effects of forest land use on erosion and slope stability. – Honolulu, Hawaii, May 1984, 99–108.
TRUSTRUM, N.A. & M.J. PAGE (1992): The longterm history of Lake Tutira watershed:implications for sustainable land use management. – In: HENRIQUIS, P.R. (ed.): The Proceedings of the International Conferece on Sustainable Land Management. – Napier, Hawke's Bay, New Zealand, 17–23 November 1991: 212–215.
TRUSTRUM, N.A. & R.C. DEROSE (1988): Soil depth – age relationship of landslides on deforested hillslopes, Taranaki, New Zealand. – Geomorphology **1**: 143–160.
WALLING, D.E. & B.E. WEBB (1996): Erosion and sediment yield: a global overview. – Internat. Assoc. Hydrol. Scies. Publ. **236**: 3–19.
WELLMAN, H.W. (1967): Report on studies relating to Quaternary diastrophism in New Zealand. – Minutes of the working group meeting for the Neotectonic study of the Pacific Regions, Appendix 5. Quat. Res. (Japan) **6**: 34–36.
WILMSHURST, J.M. (1997): The impact of human settlement on vegetation and soil stability in Hawke's Bay, New Zealand. – N.Z. Journ. Botany **35**: 97–111.
WOLMAN, M.G. & R. GERSON (1978): Relative scales of time and effectiveness of climate in watershed geomorphology. – Earth Surf. Proc. **3**: 189–208.
WOLMAN, M.G. & J.P. MILLER (1960): Magnitude and frequency of forces in geomorphic processes. – Journ. Geol. **68**: 54–74.

Address of the authors: N.A. TRUSTRUM and M.J. PAGE, Landcare Research, Private Bag 11052, Palmerston North, New Zealand (trustrumn@landcare.cri.nz & pagem@landcare.cri.nz), B. GOMEZ, Department of Geography and Geology, Indiana State University, Terre Haute, IN 47809 USA (bgomez@indstate. edu), L.M. REID, USDA Forest Service, Pacific Southwest Research Laboratory, 1700 Bayview Drive, Arcata, CA 95521, USA (lreid/psw_rsl@fs.fed.us) and D.M. HICKS, National Institute of Water & Atmospheric Research, P.O. Box 8602, Christchurch, New Zealand (m.hicks@niwa.cri.nz).

Z. Geomorph. N.F.	Suppl.-Bd. 115	87–111	Berlin · Stuttgart	Juli 1999

Atmospheric triggering and geomorphic significance of fluvial events in high-latitude regions

Martin Gude, Jena, and Dieter Scherer, Basel

with 7 figures and 4 tables

Summary. Strong annual variability and seasonal snow covers determine hydrological processes and related sediment displacements in high latitude regions. The most significant period for runoff and fluvial sediment transport is the snowmelt period in early summer. Atmospheric events during this time initiate a broad range of fluvial events, comprising runoff events with daily or seasonal frequency but also minor sluhflows with annual frequency, and high-magnitude slushflows (slush torrents) which are characterized by frequencies of several years to decades. To assess magnitude and frequency of these atmospheric, hydrologic, and geomorphic processes, a discussion on general aspects of magnitude-frequency relations is preceded. Based on this discussion, atmospheric triggering and resulting hydrologic and geomorphic events in periglacial environments are analyzed with special focus on drainage basins in NW-Spitsbergen and in N-Sweden. The examples display the moderate variability in magnitude of hydrologic events, but a high variability in the geomorphic consequences. Generally, geomorphic impact is characteristic for certain fluvial event types, but the connection to atmospheric triggering is more complex, and our today's knowledge of these interrelations is still limited.

Zusammenfassung. Hydrologische Prozesse und die von ihnen verursachten Sedimenttransporte zeichnen sich in den Höheren Breiten durch eine hohe jährliche Variabilität aus und sind in starkem Maße durch saisonale Schneedecken beeinflußt. Die wichtigste Phase des Abflusses und der Sedimenttransporte ist die Schneeschmelzperiode im Frühsommer. In dieser Zeit auftretende atmosphärische Prozesse initiieren eine Vielzahl fluvialer Ereignisse, die von täglichen und saisonalen Abflussereignissen, bis hin zu Sulzströmen (slushflows) reichen. Deren Frequenzen reichen von jährlichen Ereignissen bei kleineren Sulzströmen bis zu Sulzstromereignissen mit grossen Magnituden (Sulzmuren), die nur mit niedrigen Freqenzen auftreten und zu Jährlichkeiten von mehreren Jahren bis Jahrzehnten führen. Eine einleitende Diskussion zu Magnitude-Frequenz-Relationen atmosphärischer, fluvialer und geomorphologischer Ereignisse beleuchtet allgemeine Aspekte dieser Thematik. Basierend auf diesen Überlegungen werden dann die atmosphärischen Auslöseprozesse sowie die daraus resultierenden fluvialen und geomorphologischen Ereignisse in periglazialen Räumen anhand von Beispielen aus NW-Spitzbergen und N-Schweden analysiert. Diese zeigen, daß die Magnituden fluvialer Ereignisse nur eine beschränkte Variabilität aufweisen, während die daraus resultierenden geomorphologischen Konsequenzen eine hohe Variabilität besitzen. Letztere sind im wesentlichen vom jeweiligen fluvialen Ereignistyp abhängig, aber die Zusammenhänge zu den atmosphärischen Auslöseprozessen ist komplex, und unser heutigen Wissen darüber ist immer noch begrenzt.

Introduction

In high-latitude regions, fluvial processes are significantly influenced by seasonal snow covers and frozen ground. Snow deposits possess a high potential for strong or even extreme events since snow acts as a water storage prone to release within short periods. This is mainly due to

0044-2798/99/0115-0087 $ 6.25

the annual cycle of atmospheric processes in high latitudes causing water storage in snowpacks throughout winter, and a short but pronounced snowmelt period in spring. Furthermore, frozen ground increases the amount of water delivered to surface flow due to the prevention of infiltration.

The occurrence of seasonal snow covers, in combination with a strong annual pattern of atmospheric processes and either seasonal ground frost or permafrost, offers an adequate method to define the spatial context in which fluvial events in high-latitude regions are discussed. The above definition coincides with that given by CLARK (1988) in his comprehensive overview on periglacial hydrology, where he refers to processes in high latitudes as well as in high altitudes. In a more restrictive manner, KANE et al. (1992) refer to the existence of permafrost in defining the spatial extent of problems of hydrological processes with respect to climate change. In order to evaluate the geomorphic effects of fluvial events, which are also in the scope of this paper, periglacial areas should be defined as regions, where frost-dynamic processes are prevailing and generating detectable results (WASHBURN 1980, THORN 1991). By including the conditions of the strong annual pattern of climate, the term 'periglacial' is spatially defined for both geomorphology and hydrology.

In this context, a restriction to small and mid-size basins is appropriate, since large basins often are not entirely within the periglacial domain and therefore reflect a mixed flow pattern in the river. In general, it can be stated that the intensity of a signal of a hydrological response in a river increases with decreasing size of its basin. Smaller basins reflect more purely the control by a single process-response complex like net radiation causing snowmelt-runoff. In larger basins, an overlap of different discharge regimes is prominent and obscures evidence for flood generation analysis, since snowmelt may occur in one part of the basin and rainfall in other parts.

In analyzing magnitude and frequency of fluvial events, the triggering atmospheric conditions are of major importance, together with the snow conditions and the geomorphic setting. Fluvial events are generated either by snowmelt or by rain (or both simultaneously) and by snowpack conditions, while terrain features, soils characteristics and the thermal regime of the ground modify water movements within snow covers and open channels. The interaction of these conditions controls magnitude and frequency of fluvial events. On the other hand, event magnitude and frequency effect geomorphic change, a fact first stressed in detail by WOLMAN & MILLER (1960). In recent decades, one focus of research in fluvial geomorphology was directed towards process geomorphology with special attention to the evaluation of formative events

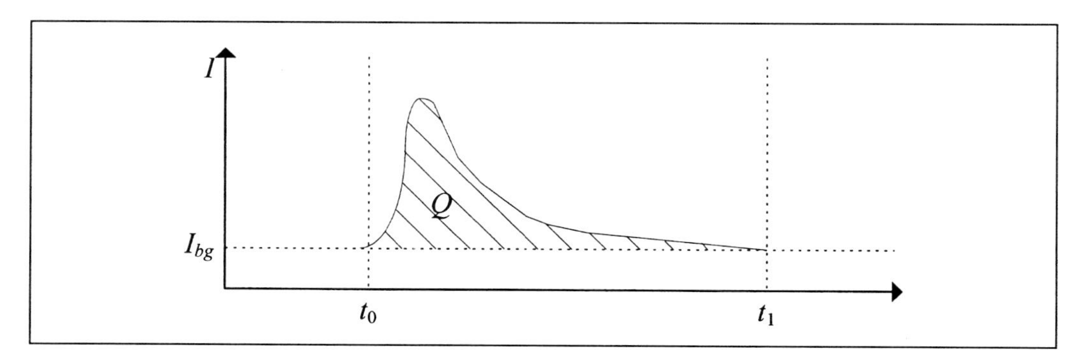

Fig. 1. Event definition by a local maximum of the process intensity (see text).

and event sequences (cf. BRUNSDEN 1996). Although the significance of paleo-data, such as sediment stratigraphies and former channels is obvious, a lack of information on processes forming these landscape features in periglacial environments was noted by CLARK (1988). Research to date offers a comprehensive overview on hydrological processes, but geomorphic evidence is sparsely investigated. This is mainly due to the fact that data on high-magnitude/low-frequency fluvial events are rare, and arctic landscapes have a wide variety of topography, former and actual glaciation and climate. Evidently, the evaluation of the geomorphic effects of events and the identification of correlating erosional and depositional features cause difficulties.

In this context, the discussion of magnitude and frequency of fluvial events includes atmospheric and snow conditions, soil characteristics, soil thermal regime and geomorphic features. After a theoretical discussion on the general magnitude-frequency concept of events, the particularities of fluvial events at high latitudes will be discussed.

1 *General considerations on magnitude-frequency*

1.1 *Concepts and problems of magnitude-frequency relations*

The concept of magnitude-frequency relations is based on a stochastic treatment of dynamic processes. The magnitude Q can be defined as the energy or mass exchange connected with a process. Q can also be expressed by the mean intensity $I \equiv \bar{\dot{Q}}$ over a time interval dt by

$$Q(t) = \bar{\dot{Q}}(t) \cdot dt \equiv I(t) \cdot dt. \quad (2.1)$$

The total effect S of a process in the time interval $[t_0, t_1]$ can be defined by

$$S = \int_{t_0}^{t_1} dt \cdot Q(t). \quad (2.2)$$

The magnitude-frequency relation $\tilde{Q}(\upsilon)$ describes the percentage of the total effect over an unlimited time period in each frequency interval between υ and $\upsilon + d\upsilon$. The frequency υ is usually given in d^{-1} for faster processes (e.g. atmospheric processes) or in a^{-1} for the slower ones (e.g. geomorphological processes).

In reality however, we are dealing with data sets over limited time periods using discrete time intervals and measurement devices of limited precision. The measured magnitudes are grouped in l classes $C \equiv \{C_0 \ldots, C_{l-1}\}$ covering the magnitude interval $[Q_{min}, Q_{max}]$ both to enable statistical analyses and to reduce the problem of data errors. The number l of classes depends both on the observed maximum Q_{max} and on the accuracy of the measurements. The classes are usually scaled logarithmically for processes spanning over many orders of magnitude.

Let N be the number of time intervals Δt, for which magnitudes Q_i or mean intensities I_i were measured and grouped in the classes C_i. The probability density function $P(C)$ is then estimated from

$$P(C_i) \cong \frac{N(C_i)}{N}, \quad (2.3)$$

where $N(C_i)$ is the number of elements in class C_i. The higher the total number of observations N, i.e. the sum of all $N(C_i)$, the better is $P(C)$ approximated by formula (2.3). The mean frequency $\bar{\upsilon}(C_i)$ of each magnitude class C_i is given by

$$\bar{\upsilon}(C_i) = \frac{P(C_i)}{\Delta t}, \tag{2.4}$$

which leads to the mean return period $\bar{T}(C_i)$ for magnitudes of non-zero probability

$$\bar{T}(C_i) = \frac{1}{\bar{\upsilon}(C_i)}. \tag{2.5}$$

With the help of the probability distribution $\Phi(C)$

$$\Phi(C_i) = \sum_{k=0}^{i} P(C_k) \tag{2.6}$$

it is possible to compute the probability of observing extreme magnitudes $Q \in C > C_{crit}$ by

$$P(C > C_{crit}) = 1 - \Phi(C_{crit}), \tag{2.7}$$

where C_{crit} is the critical class defining the magnitude threshold. This finally leads to the mean frequency $\bar{\upsilon}(C > C_{crit})$ for extreme magnitudes

$$\bar{\upsilon}(C > C_{crit}) = \frac{P(C > C_{crit})}{\Delta t} = \frac{1 - \Phi(C_{crit})}{\Delta t}. \tag{2.8}$$

An example is given to demonstrate the terms and concepts introduced above. Looking at a fluvial system, a measured runoff of 1 $m^3 \cdot s^{-1}$ is a mean intensity within a certain time interval Δt, e.g. 1 h. The total mass exchange (transported volume) of water is 3600 m^3 within this hour. If runoff is assumed to be constant over the time interval, the magnitude Q at a certain time t would be 1 $m^3 \cdot s^{-1} \cdot dt$. The total effect S in this hour is then 3600 $m^3 \cdot h$, implying a constant effect rate $\dot{S}$ of 1 m^3, which is identical to $\bar{Q}$. The introduction of an integration time Δt shows the problem of discrete time intervals, for which magnitudes can be determined individually by measurements. In field experiments, an additional time discretization error may result from the sampling interval, if no (quasi-)continuous measurement is possible. Using magnitude-class intervals {1 l, 10 l, ..., 1 m^3, 10 m^3, ...,10^6 m^3}, runoff measurements of one year would result in a total number N of 8760 observations. Assuming that 8 observations reached hourly averaged magnitudes $\bar{Q}$ between 10^4 and 10^5 m^3, and 2 measurements exceeded even 10^5 m^3, then the probability densities for those magnitudes would be 0.091% respectively 0.023%. The resulting frequency for magnitudes higher than 10^4 m^3 would then be 10 a^{-1}.

Unfortunately, there are several unavoidable drawbacks in the determination and application of magnitude-frequency relations. First, the limited observation period does not lead to the exact probability density function, since the probability density computed by formula (2.3) of magnitudes not observed so far would be zero, although such magnitudes may have occurred prior to the observations, or may occur in the future. By means of theoretical considerations there is a chance to extrapolate the probability density function for magnitudes higher than the observed maximum. The second problem arises from the statistical treatment of the underlying process. Mean values for frequencies respectively return periods and cannot be used for prognostic purposes, since the events of a certain magnitude interval are not uniformly distributed in time. This is due to the episodic nature of most geosystem processes (cf. WOLMAN & MILLER 1996, CROZIER, this volume). The third problem is strongly related to the problems already mentioned. The processes, for which the magnitude-frequency relations are determined, are connected with state transitions of dynamic systems, which not only depend on internal but also on external forces driving the system. As long as the system is in a dynamic equilibrium with its environment represented by the external forces (e.g. by steady, periodic or cyclic boundary conditions), the observed magnitude-frequency relation remains valid. However, shifts in the environmental conditions are able to cause dramatic changes in the magnitude-frequency relation (e.g. HOUSE & BAKER 1997) depending on the sensitivity of the system to the external forces.

1.2 *Magnitude-frequency relations of events*

In the context of this paper, magnitude-frequency relations of fluvial events in periglacial environments are subject of discussion. Therefore, a general definition of events is needed, which is suitable for practical research work. Events can be explicitly defined on discrete time scales as the result of singular processes that have a duration of one time interval or less. Their magnitude Q_i within the time interval t_i of length Δ is the mean energy or mass (volume) exchange rate $\bar{I}_i$ multiplied with Δt

$$Q_i \equiv Q(t_i) = \bar{I}_i \cdot \Delta t = \int_{t_i}^{t_i+\Delta t} dt \cdot I(t). \tag{2.9}$$

Using this definition, it is a philosophical question whether long-term processes should be treated as separate phenomena, or could be regarded as event sequences (cf. BRUNSDEN 1996). The second viewpoint makes it much easier to use the mathematical treatment of magnitude-frequency relations as described above. The only complication for the determination of magnitude-frequency relations is the necessity of composing them by using different time scales for different parts of the spectrum. For example, the high frequency part, where $\bar{\upsilon}$ is greater than 1 d^{-1} could be based on hourly observations, while the low frequency part with frequencies of 1 a^{-1} and less may result from yearly budgets. The lower the frequency the longer is the maximum duration of the events contributing to the magnitude-frequency relation. Therefore, this definition of an event makes it prerequisite to adopt the length of the time intervals adequately.

A different approach to events is to define them by a local maximum of the process intensity $I(t)$ (cf. also Fig. 1). The magnitude Q of an event is then given by

$$Q = \int_{t_0}^{t_1} dt \cdot I(t) - I_{bg}. \tag{2.10}$$

I_{bg} is a background intensity level (e.g. a small non-zero mass or energy exchange rate, or by the sensitivity of the measurement system); the times t_0 and t_1 are given by the fact that $I(t)$ must exceed I_{bg} during the event period. By this definition of an event, its time interval $\Delta t = t_1 - t_0$ is defined implicitly by the underlying process.

The methods for the determination of the magnitude-frequency relation as described above can also be applied to events following this definition. An example may be a runoff event that shows increased but different hourly runoff values (intensities) for a certain day. Assuming a baseflow runoff of 0.1 $m^3 \cdot s^{-1}$ and measured hourly runoff values of 0.1, 1.0, 3.0, 2.0, 1.5, 1.0, 0.5, 0.2 and 0.1 $m^3 \cdot s^{-1}$ following a convective rainstorm event, the duration of this fluvial event would be 7 h, resulting in a magnitude of $8.5 \cdot 7 \cdot 3600$ m^3, i.e. $2.142 \cdot 10^5$ m^3.

In geomorphology, the interest in fluvial events is given by their impact on landscape development, but this demands additional information on the geosystem dynamics. Assuming that $\tilde{Q}_f(\upsilon)$ is the known magnitude-frequency relation of a certain type of fluvial events, then the question arises how this relation is connected to the magnitude-frequency relation $\tilde{Q}_g(\upsilon)$ of a specific geomorphologic process, i.e. we are interested in the relation $g(\tilde{Q}_f)$

$$\tilde{Q}_g(\upsilon) = g\left(\tilde{Q}_f(\upsilon)\right). \tag{2.11}$$

g depends not only on the processes involved, but also on specific drainage basin features and on the environmental conditions. Atmospheric processes are triggering fluvial events, particularly in periglacial environments. In the case where the magnitude-frequency relation of the formative fluvial events is not known, or the sensitivity of a geomorphological process on the atmospheric boundary conditions is the aim of a study, we need to know the relation $f(\tilde{Q}_a)$

$$\tilde{Q}_f(\upsilon) = f\left(\tilde{Q}_a(\upsilon)\right), \tag{2.12}$$

where $\tilde{Q}_a(\upsilon)$ is the magnitude-frequency relation of a certain atmospheric process. The composed relation $G(\tilde{Q}_a)$

$$\tilde{Q}_g(\upsilon) = G\left(Q_a(\upsilon)\right) = g\left(f\left(\tilde{Q}_a(\upsilon)\right)\right) \tag{2.13}$$

finally gives the chance to carry out sensitivity studies, which allow an assessment of the geomorphic impacts of shifting magnitude-frequency relations of atmospheric processes caused by regional or global climate changes.

However, the determination of the relation between atmospheric processes and geomorphic impacts is seriously complicated by the fact that several distinct atmospheric processes are responsible for triggering one type of fluvial events, and also that several types of fluvial event cause comparable geomorphic impacts in periglacial environments. The total impact of different processes can not be determined by a simple superposition of the magnitude-frequency relations of each process, since the processes are partly correlated resulting in non-linear interactions of complex structures. This will be discussed in detail in the following chapters, and is also discussed by CROZIER (this volume). In addition, it will be shown that further information on the couplings of atmospheric, fluvial and geomorphic processes are necessary for the reconstruction of the relations given in equations 2.11, 2.12 and 2.13. In particular, quantitative data on intensity thresholds (e.g. for runoff values that are sufficient for initiating bed load transports) or on preconditioning processes (e.g. thawing processes that make sediments disposable for fluvial erosion) are most relevant for a full mathematical-physical treatment. Nevertheless, the conceptual framework presented in this chapter forms the background for practical field research on this topic.

Processes leading to events are not only limited in time but also in the spatial domain. Intensity, magnitude and impact of an event are therefore integral values of the affected area. Computing areal densities of these variables is a common approach to compare areas of different size. But in many cases, e.g. when measuring energy fluxes into the snow cover, the acquired data are already areal density values. Therefore, an integration over the area of interest, e.g. a certain drainage basin, is necessary. However, this integration is difficult, since point measurements may not represent the average value of the area. This problem makes spatially-distributed studies of the involved processes essential.

2 *Atmospheric triggering and fluvial event types*

Beside the influence of atmospheric conditions, snow cover, soil characteristics and frost regime within the soil control timing, duration, and magnitude of events in high-latitude fluvial environments. Additionally, water routing is significantly influenced by topography. A schematic interaction diagram of this complex system displaying the process connections between triggering event types and fluvial event types is given in Fig. 2. The relations between the factors can be defined as causing a certain type of event, but knowledge about the relations is conceptual rather than quantitative in many cases (Fig. 2).

Three general types of fluvial events at high latitudes can be distinguished: minor slushflows, slush torrents and flood waves. Although these events display nonlinearity concerning their magnitudes, they represent a consistent event sequence as defined e.g. by BRUNSDEN (1996). In addition, other event types also occur in connection with fluvial activity, such as lake outburst floods from landslide or glacier damming (e.g. GLOFs) or jökulhlaups. These events are highly irregular in frequency and, more important, their time of occurrence can merely be addressed by an analysis of the meteorological and hydrological conditions. Therefore, the focus is directed to events closely coupled with atmospheric conditions causing snowmelt and rainfall.

2.1 *Atmospheric triggering by snowmelt and rainfall*

Seasonal snow covers are characteristic elements of periglacial environments. As long as the snow cover is present and stable, fluvial events are limited to areas where snow is already removed or was never existing. But prior to the opening of channels, meltwater is prevented from being drained effectively, which frequently results in meltwater accumulation and a more or less dramatic break-up of the channel.

Due to energy influx into the snow cover caused by atmospheric processes, liquid water becomes available by snowmelt. Meltwater is able to trigger fluvial events of different types and magnitudes. The total energy exchange ΔU between the snow cover and its environment in the area ΔA is given by

$$-\frac{\Delta U}{\Delta t} = \Delta A \cdot \left(R_{top} - R_{base} + H_{top} + \lambda E_{top} - G_{base} + \Delta \dot{Q}_{top} - \Delta \dot{Q}_{base} \right), \quad (3.1)$$

where R is the net radiation, H the sensible heat flux, λ the latent energy of vaporisation ($2.5 \cdot 10^6$ J·kg^{-1}), E the evaporation, G the conductive and $\Delta \dot{Q}$ the convective heat flux. The

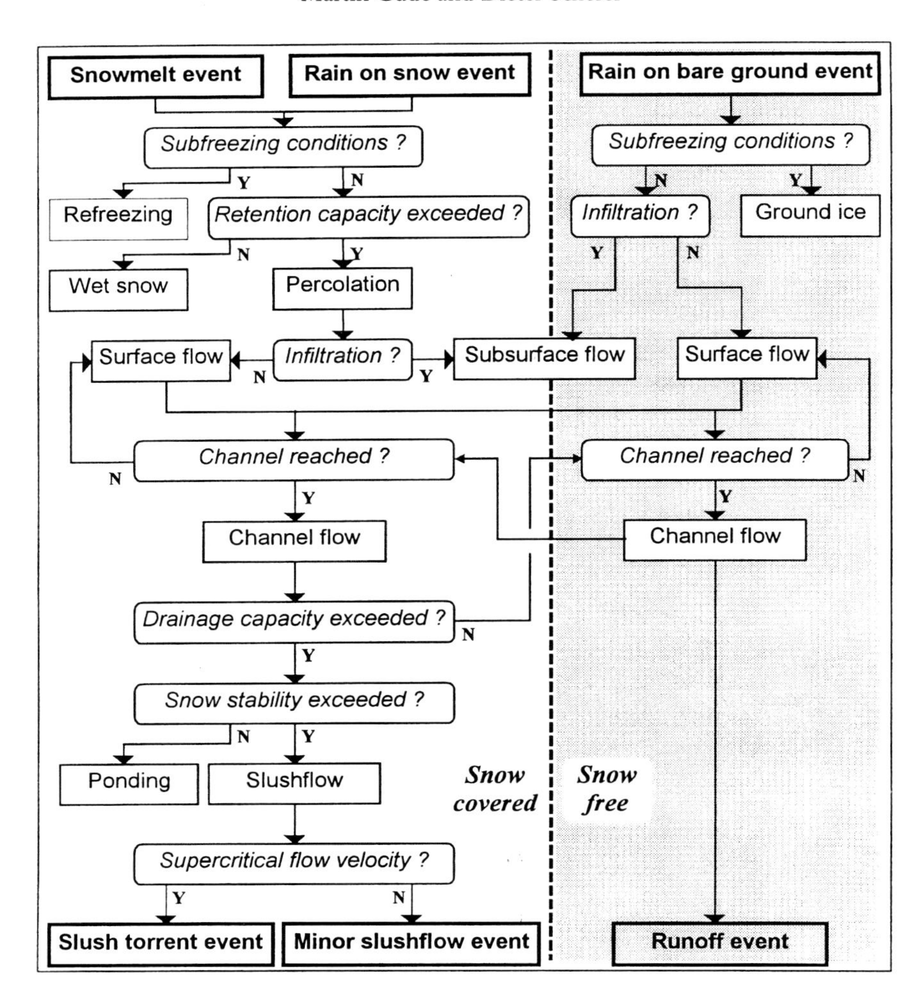

Fig. 2. Schematic interaction diagram displaying the relations between triggering atmospheric events and resulting fluvial events. Processes in snow covered and snow free areas principally are distinguished due to the retention capacity of snow. Interaction between both area types is common when meltwater from snow covered areas reaches snow free areas (e. g. channel sections) and vice versa.

subscripts *top* and *base* indicate the upper and lower boundary of the snow cover. The signs in equation (3.1) result from the mathematical definition of the vertical axis, i.e. all downward fluxes are counted negative. All energy fluxes are given as mean energy flux densities averaged over the time interval Δt and the area ΔA, while the evaporation is the mean mass flux density in Δt and ΔA. As long as the mean temperature of a snowpack is less than 0 °C, a positive ΔU will be used to heat the snow, either directly or by melt-percolation-refreeze cycles. Reaching isothermal snow conditions at 0 °C, meltwater is produced that no longer refreezes. As soon as the gravity potential exceeds the matrix potential, meltwater percolates towards the snow base. When liquid water forms a saturated layer above an impermeable layer, meltwater follows the hydraulic gradient, accumulates in areas of horizontal meltwater convergence and generates fluvial events.

There are several specific conditions in periglacial environments which can be used to assess the relative importance of the different terms in equation (3.1) with respect to the magnitude-frequency relation of meltwater production and their potential for triggering high-magnitude fluvial events (cf. also CHURCH 1988). If the snow cover has a depth of at least a few decimetres, then R_{base} and G_{base} can be disregarded. The same holds for $\Delta \dot{Q}_{base}$ due to the existence of frozen ground preventing significant infiltration rates of meltwater into the subsurface layer. The convective heat flux $\Delta \dot{Q}_{top}$ caused by 'warm' rainfall is negligible with respect to high-magnitude fluvial events. This results from the fact that the amount of meltwater generated by the convective heat flux is small compared to amount of rainwater directly added to the snow cover.

With respect to high-magnitude fluvial events caused by snowmelt, three terms of equation (3.1) remain important: net radiation, sensible, and latent heat flux at the top of the snow cover. As many energy balance studies in periglacial environments demonstrate (e.g. MALE & GRANGER 1981, OHMURA 1981, BLÖSCHL et al. 1988, MARKS & DOZIER 1992, PLÜSS & MAZZONI 1994, SCHERER 1994), the main energy supply for snowmelt is given by net radiation, followed by sensible heat flux. Latent heat flux is of minor importance, although in periods of strong warm air advection the vertical gradient of the specific humidity is frequently directed towards the atmosphere causing a significant energy flux (and a minor water flux) into the snow cover.

Net radiation is given by

$$R_{top} = (1-\alpha)\cdot {}^{\downarrow}R_{sw} + \varepsilon \cdot {}^{\downarrow}R_{lw} - \varepsilon \cdot \sigma \cdot T_{top}^{4}, \tag{3.2}$$

where α is snow albedo (typically 0.5 to 0.95), ${}^{\downarrow}R_{sw}$ is the incoming solar radiation, ε is the emissivity of snow (about 0.99), ${}^{\downarrow}R_{lw}$ is the incoming long-wave radiation from the thermal emission of the atmosphere, σ is the Stefan-Boltzmann constant ($5.67 \cdot 10^{-8}$ W·m^{-2}·K^{-4}) and T_{top} is the surface temperature of the snow cover. During periods of intensified snowmelt, thermal emission of snow, the last term in equation (3.2), is fixed to approx. 310 W·m^{-2} due to a surface temperature of 0 °C. The effect of incoming short-wave radiation strongly depends on snow albedo, which decreases during the snowmelt period causing a positive feedback on snowmelt. However, both short- and long-wave incoming radiation are limited due to astronomic and atmospheric constraints. Under extreme meteorological conditions, i.e. strong advection of warm and moist air and a cloud coverage that does not reduce incoming short-wave radiation significantly but increases the incoming long-wave radiation, energy loss by net long-wave radiation diminishes, so that net short-wave radiation directly controls net radiation.

Sensible heat flux is controlled both by air temperature T_a, which is directly related to the vertical temperature gradient at the snow surface, and by horizontal wind velocity u, which strongly influences the exchange coefficient for sensible heat k_H

$$H = \rho_a \cdot c_p \cdot k_H \cdot \frac{\partial T}{\partial z} = \rho_a \cdot c_p \cdot f(u) \cdot \frac{\partial T}{\partial z}. \tag{3.3}$$

The density of the air is given by ρ_a (1.29 kg·m^{-3} at 0 °C and 1013 hPa), while c_p denotes the specific heat capacity of air at constant pressure (1.005 J·K^{-1}·kg^{-1}). Both u and T_a can vary over a wide range causing strong temporal and spatial variations of the sensible heat flux. A superposition of high air temperatures ($T_a > 10$ °C) and high wind velocities ($u > 10$ m · s^{-1}) may result in sensible heat flux values higher than net radiation during clear sky conditions close to the summer solstice (SCHERER 1994).

Fluvial events generated by rainfall in periglacial environments can be separated into rain on bare ground and rain on snow events (cf. CHURCH 1988 for a comprehensive discussion of floods in cold climates triggered by rainfall). In the first case, the effectiveness of rainfall for the generation of fluvial events depends on thickness and water content of the active layer, since it provides an intermediate storage increasing retention of rainwater. Rain on snow events depend on thickness and mean temperature of the snow cover. If the snow cover is thick and cold, liquid water is not able to travel a long distance downslope without refreezing. However, each rainfall event that is not able to generate a fluvial event provides a significant influx of energy in the snow cover due to the released energy of $3.34 \cdot 10^5$ J·kg^{-1} when refreezing. By this process, rain on snow events cause a successive destabilization of the snow cover.

2.2 *Fluvial event types*

2.2.1 *Break-up phenomena*

Break-up of streams in high latitudes is one of the major events during the hydrological year. The event is most pronounced in high arctic rivers, where during winter period water flow normally is completely diminished in small rivers, and huge ice covers are developed in large rivers. Both snow in the channel and ice cover on the river cause a damming, but display different processes concerning the effects on event development.

In channels choked with snow, the first meltwater is forced to flow through the matrix of the snowpack, which significantly delays runoff. Inflow of meltwater to the channel from adjacent slopes frequently results in a saturation of the snowpack, at least in low gradient sections of the channel. Then, removal of the snowpack commences mainly through the following processes:

- successive erosion of snow by flowing water on the surface of the snowpack (typically caused by full saturation),
- successive erosion of snow by water flow within the snowpack, e.g. on ice layers, or at the base (frequently effects the formation of open pipes),
- abrupt mobilization of the whole snowpack moving in the channel as a slushflow (preferably preconditioned by a hydrostatic pressure gradient, cf. e. g. WOO & SAURIOL 1980, SCHERER 1994, SCHERER & PARLOW 1994, GUDE & SCHERER 1995).

Whereas the first two cases lead to a gradual opening of the channel without remarkable event dynamics the third process can produce extreme events consisting of slush waves, i.e. water-saturated snow, and typically also of sediment. These slushflows were first described as break-up phenomena by WASHBURN & GOLDSWAITH (1958). More recently, e.g. WOO & SAURIOL (1980) compiled detailed investigations on the hydrologic significance of slushflows in a catchment in Resolute (NWT, Canada). Slushflow initiation is discussed in relation to meteorological factors and the effects of slushflows on landscape in further studies (NOBLES 1966, HESTNES 1985, ONESTI 1985, CLARK & SEPPÄLÄ 1988, RAPP 1960). The most comprehensive work in this context is compiled by NYBERG (1985), who investigated recent and ancient slushflow activity in Northern Sweden. But only since 1994 the significance of slushflows for event chronologies in fluvial systems has been stressed (Barsch et al. 1994a, Gude & Scherer 1995).

The original term slushflow was supplemented by the term slush avalanche to account for high energy events that should be separated from normal break-up processes (cf. e.g. Nyberg 1985). Since the usage of the term slush avalanche may imply confusion with other avalanche types, the term slush torrent is proposed instead (Barsch et al. 1993). This also stresses the confinement of the processes to the fluvial environment.

Slushflows can reasonably be separated from frequently quoted flood waves accompanying the onset of snowmelt. Unfortunately, very little information is available on discharge rates of slushflows, whereas the first flood waves have been measured in many cases (cf. Church 1988, Clark 1988). Consequently, comparable data on the significance of slushflows for event chronologies in high-latitude regions are rare, although slushflows sometimes exceed the peak values of runoff event after opening of the channel by a factor far more than 1'000 (Table 1). This deficit in knowledge is mainly caused by the difficulties of measuring discharge volume fluxes, and by the fact that slushflows were not regarded as fluvial events, although this was already mentioned by the initial description by Washburn & Goldthwait (1958), and also stressed in the comprehensive overview published by Nyberg (1985).

The dominating reason for magnitude differences between runoff events and slushflows is the water storage capacity of a channel snowpack. Thick snowpacks are able to delay runoff effectively leading to a high potential for an abrupt release of the stored water. Once the snow is removed from the channel, i.e. the drainage network is established, retention decreases significantly. Consequently, slushflow discharge is a result of a cumulative effect of meltwater production over a certain time period, while runoff events in snow-free channels display the actual water supply to the channel.

2.2.2 *Runoff events*

Flood waves are the most prominent type of runoff events, and they are well documented in a number of high-latitude basins in North America, Scandinavia, and Siberia. The significant increase of water stage above mean discharge level for a certain period (hours to few days) in

Table 1. Magnitudes and maximum intensities of fluvial events in Liefdefjord (NW-Spitsbergen) and Kärkevagge (N-Sweden). (after: Barsch et al. 1994 and Gude & Scherer 1995).

	Runoff events		**Slushflow events**			
			Minor slushflows		Slush torrents	
	m^3	$m^3 \cdot km^{-2} \cdot s^{-1}$	m^3	$m^3 \cdot km^{-2} \cdot s^{-1}$	m^3	$m^3 \cdot km^{-2} \cdot s^{-1}$
Kvikkåa	50,000	0.5	3,000	4	20,000 –	1,200 –
(Liefdefjorden)					40,000	2,400
Kärkerieppe	20,000	1.5	no data	no data	10,000 –	75 –
(Kärkevagge)					20,000	150

cold climates is not only caused by rainfall events but also frequently by snowmelt or mixed events (e. g. rain on snow). A comprehensive overview on floods in cold climates is given by CHURCH (1988), including high altitude periglacial environments. Furthermore, he refers to the characteristics of floods in order to separate different river regimes in the arctic (cf. also CLARK 1988). Taking into account that channel water flow in many catchments originates from different source areas, a clear separation of different regimes appears obvious only in rare cases (cf. MARSH & WOO 1981, CLARK 1988). For example, the analysis of the spring flood characteristics of a wetland area detected no significant difference to basins with incised channels (GLENN & WOO 1997). Furthermore, many basins are composites of wetland areas, steep slopes with rock outcrops or talus material, firn fields, and channels with variable incision and gradient. Therefore, a general separation of rivers supplied by nival or by glacial meltwater seems more suitable particularly in respect to the problem of flood generation (MARSH & WOO 1981). A more detailed separation of basin characteristics requires an analysis that defines hydrological response units within the basins in order to assess the high variability in hydrological regime patterns (cf. CLARK 1988). WOO (1986) provides a comparative statistical analysis of discharge parameters for northern Canadian rivers. The size of their basins range between 1'500 and 68'000 km^2, and hence these basins probably represent mixed type regimes, in which event peaks are smoothed and variability is much less than in small basins.

It is most common that flood waves in the very beginning of snowmelt are cited to reach the highest runoff values during the summer season (CLARK 1988, CHURCH 1988, KANE et al. 1992). Between 30 to 90% of the annual runoff is delivered within the first 2 or 3 weeks of open water flow in the channels (CLARK 1988). Beside these major floods, the diurnal course of meteorological parameters effects a periodical variability, which is most pronounced in alpine relief. Main driving force is the subsequent insolation on different slope areas in a basin, each of which representing a separate source area for meltwater. As the angle of direct solar radiation varies only moderately in high latitudes during the main discharge period, strong diurnal fluctuations in discharge must be attributed to the melting of snow on different slopes and not only to the angle of the sun. Data displayed in Fig. 3 stress the importance of certain slopes for discharge variability. The weak, but significant peak in daily discharge occurs in response to net radiation peak on W and NW exposed slopes and not in response on net radiation peak on horizontal ground. This is calculated taking into account flow velocities of meltwater. Recent research on snowmelt flood generation especially in the context of snowmelt-runoff-models emphasizes the control by relief features (BRAUN et al. 1994, KIRNBAUER et al. 1994).

In general, flood waves represent the liquid water supply in the basin generated by intensified snowmelt, rainfall or both. Once water supply in the basin is increasing, a combination of different factors control the stage in the river (cf. Fig. 2). The snow cover itself can cause an effective retention when rain or meltwater refreezes within a snowpack, which frequently is accompanied by the formation of ice layers. In the same way, meltwater reaching the ground surface may refreeze on top or in the upper parts of the ground (WOO & HERON 1981). In inclined areas, ice layers within the snowpack and on the ground act as deflecting strata that direct water flow from vertical to downslope flow (cf. COLBECK 1975, WOO & HERON 1981). Saturation above the ice layer and subsequent snow metamorphism increases the permeability significantly. Furthermore, concentration of flow along depression lines allows progressive erosion and the development of open pipes (GUDE & SCHERER 1998). These processes can account for effective and rapid drainage of meltwater to the main channel, even in areas where snow coverage is 100%. This case is typical for the first runoff period, and is revealed by the

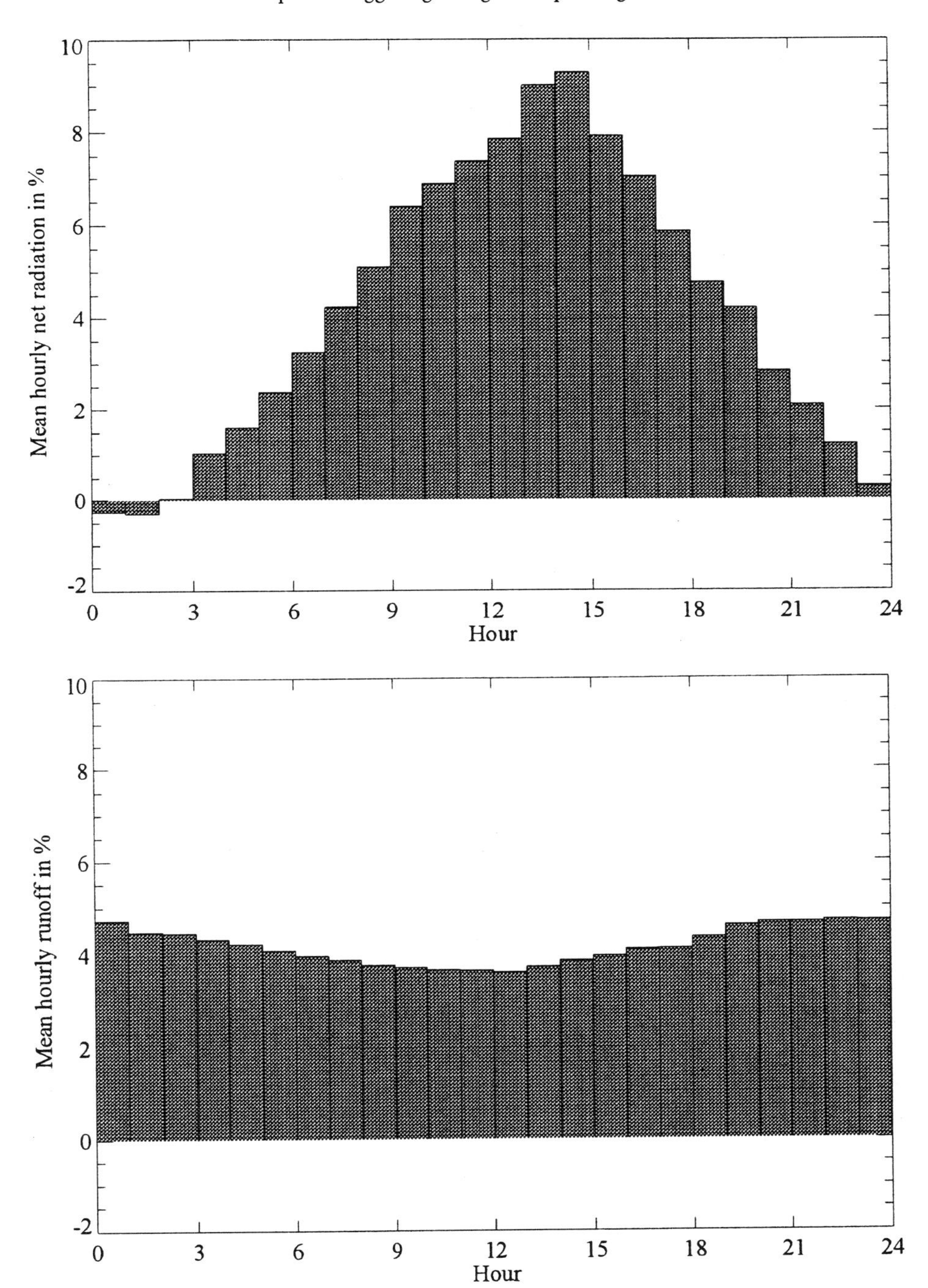

Fig. 3. Daily cycles of net radiation and runoff in the Kvikkåa drainage basin (Liefdefjorden, NW-Spitsbergen) during the runoff season 1991. The hourly means (Central European Summer Time) are given as percentages of the total values.

domination of atmospheric compared to ground originated ions in the discharged meltwater. Depending on the snow conditions in the channel, water inflow from the slopes and tributary chutes causes either an accumulation in the snow-filled channel, or flow in the open channel. In the latter case, retention depends only on channel geometry and roughness as well as on reservoir effects, e.g. on floodplains.

3 *Geomorphic impacts of fluvial events*

Prior to the recent consideration of the significance of event magnitude and frequency for geomorphic development, a more general discussion on the influence of fluvial processes in high latitudes has taken place for some decades. The overall presence of valleys in elevated periglacial environments displays the active role of linear erosion factors, i.e. valley glaciers and rivers. Despite the fact that many valleys obviously possess features deriving from fluvial formation, the variety of their shapes frequently stresses a multi-process development. This includes erosional forms of former glaciated periods as well as solifluction and other mass movements affecting the fluvial environment. Furthermore, the variety of processes within the fluvial environment result in a number of different erosional and depositional landform types ranging from incised channels to meanders and fans.

In this context, it is not surprising that geomorphologists state the overall dominance of valleys, but consensus is weak concerning the dominating valley shapes in high-latitude areas (e.g. WASHBURN 1979, CARTER et al. 1987, RUDBERG 1988, FRENCH 1996). Whereas some authors (e.g. MCCANN & COGLEY 1978, RUDBERG 1988) define the presence of alluvial fans and gravel-filled valley bottoms as typical, CARTER et al. (1987) state that deeply incised valleys with narrow bottoms are common in the Canadian arctic. A merely descriptive analysis without accounting for landform history, Holocene regional climate and fluvial dynamics seems not to be sufficient to explain the actual landscape. The analysis of actual dynamics of sediment displacement is therefore an essential tool for the interpretation of landscape features, despite the limitations of extrapolating from contemporary processes.

Actually, in this context only a few studies have been published. Since the statement of CLARK (1988) on the lack of investigations on fluvial geomorphic impact in periglacial areas, a small number of studies have been compiled. An overview is given by LEWKOWICZ & WOLFE (1994) and by FRENCH (1996). Moreover, interpretation of landforms created by fluvial sediment displacement based on process analysis is rare. A related discussion covering the main processes of sediment transports in the fluvial environment, i.e. break-up by slushflows and runoff events, therefore remains preliminary due to limited knowledge.

3.1 *Erosion, transport and deposition by slushflows*

Slushflows have to face the same resistance against erosion in the channel as runoff, but they can perform a much higher shear stress on the channel and the adjacent slopes. The viscosity of slushflows is assumed to be significantly higher than runoff due to the high snow content. In combination with the supercritical velocity of high-energy slushflows (slush torrents), they account for extraordinary shear stress on ground surface and bare rock. Slushflows have to plough their way through the snow cover of the valley bottom and therefore, they frequently flow off the channel course where these are only slightly incised. The course of slushflows more likely is controlled by the valley slopes than by streambanks. This effects a scouring on the slopes which is most efficient with high energy slush torrents revealing an enormous erosional impact (Fig. 4) (cf. e. g. NYBERG 1985, BARSCH et al. 1994a). But ground ice in streambanks and channels, which is a frequent feature in early summer, counteracts the erosion capacity of the events (CARTER et al. 1987). The sediment acquisition of slushflows consequently is extended over a much larger area compared to runoff events.

The extraordinary erosional impact of slush torrents is revealed by sediment yields displaced by single events. RAPP (1960) describes a single boulder of approx. 75 t being moved more than 100 m by a slush avalanche (slush torrent) on an alluvial fan section with 5° inclination. It should be stressed that the total length of the stream draining a lake is only 1,200 m. The overview compiled by NYBERG (1985) on slush avalanches in N-Sweden quotes typical sediment transports of some hundreds of tons per slush torrent event. Slush torrent investigations carried out in Liefdefjorden (NW-Spitsbergen) revealed a sediment yield of at least 6,000 t transported by one event from a catchment of 5 km^2 (BARSCH et al. 1994a). Several boulders moved by the events in Liefdefjorden were approx. 2 m in length. Similar amounts of 3,500 to 18,000 t are measured by ANDRÉ (1995) in Austfjord (central Spitsbergen) based on lichen analysis of deposits transported by single slush torrents. A comparison of sediment transport parameters in Liefdefjorden stresses the importance of slushflows (Table 2) (BARSCH et al. 1994a).

The extension of the erosion impact on slope sections implies also a modification in the characteristics of the mobilized material. Different types of sediments can be well distinguished, since material from the channel bed is at least partly rounded, whereas slope material frequent-

Fig. 4. Section of Kvikkåa canyon with erosion scars on slopes (Liefdefjorden, NW-Spitsbergen).

Table. 2. Magnitudes (total sediment transport), intensities (peak sediment fluxes) and sediment concentrations of fluvial events in Kvikkåa and Beinbekken in 1990/91/92 (Liefdefjorden, NW-Spitsbergen).

	Runoff events	**Minor slushflows**	**Slush torrents**
Magnitude (t)	50	50	6,000
Intensity ($kg \cdot s^{-1} \cdot km^{-2}$)	0.3–0.7	10–20	30,000–120,000
Concentration ($g \cdot l^{-1}$)	1.0–1.5	0.1–5.0	50–100

ly consists of stone fragments without rounded features. A high content of the latter sediment type is documented in a number of slushflow deposits (e.g. Rapp 1960, Nyberg 1985, Clark 1988, Barsch et al. 1994a).

Typical deposition features are comprehensively summarized by Nyberg (1985). Due to the spatial overlap of fluvial event and slushflow deposition, a variety of interference forms occurs. These multi-process fans are typically situated at mouths of incised channel sections, where slushflows can spread out. On the other hand, single-process slushflow fans or tongues can be recognized by their wale-back shape and the high content of coarse and angular debris. Caused by high sedimentation rates compared to fluvial accumulation, these fans are typically situated on higher elevations than the surrounding flood plain or terrace (Fig. 5). Beside accumulation forms, slushflows can effect impact pools or pits, where slushflows shooting down steep channel sections release their kinetic energy on a flat area at the mouth of the valley.

Fig. 5. Kvikkåa alluvial fan adjacent to slushflow fan (Liefdefjorden, NW-Spitsbergen).

3.2 *Erosion, transport and deposition by runoff events*

Sediments, i.e. solutes, suspended and bed load material, are acquired directly in the channel, or are delivered to the channel by slope processes. Since in periglacial environments frost scattering and slope processes are widespread and intensive, river channels are assumed to be filled with huge amounts of rock fragments (BÜDEL 1982, WASHBURN 1980, CLARK 1988, FRENCH 1996). Additionally, streambank erosion provides significant amounts of sediment, and erosion is distinctly facilitated in loose sediments remaining from former glaciation.

Despite this comprehensive supply of sediments for fluvial transport, attention should be given to the timing of suitable sediment delivery to the river. A set of factors that are critical concerning event dynamics can be identified as influencing the sediment availability for the fluvial system: existence of ground and surface ice, vegetation cover and the degree of glaciation in the basin. Especially the latter implies a high variability in sediment transport. Comparison of periglacial and glacial fluvial transports stresses the restrictions in the periglacial transport system. Whereas proglacial rivers display high concentrations of suspension material even in low water stages, periglacial rivers lack suspension material almost completely under similar hydraulic conditions. This behaviour is well documented in a number of rating curves and sediment discharge hydrographs which is exemplified in Fig. 6 (cf. CLARK 1988, BARSCH et al. 1994a).

Low transport yields are also reported concerning bed load. These facts are indicative for a supply-restricted system, which is in contradiction to the statement of ubiquitous supply described above. As known from rivers in mid-latitudes, sediment transport typically starts when a threshold of shear stress on the channel bed surface associated with a certain discharge level is exceeded, and the imbricated pavement of the channel gravels is destroyed (cf. ERGENZINGER 1988). In high latitudes, transport of solids in rivers is equally controlled by hydraulics and therefore restricted to high discharge. This has remarkable consequences for fluvial geomorphology in these areas. The general statement of more pronounced valley forming by fluvial activity in the arctic than in lower latitudes (cf. BÜDEL 1982) ignores the fact that periods of highest discharge normally do not overlap with periods of highest sediment availability. Peak discharge periods are reported to occur typically in the very beginning of open water flow in the channels (break-up events and first nival floods), but at that time, channel floor sediments normally are still fixed by interstitial ice or aufeis (e.g. CARTER et al. 1987, CLARK 1988, FRENCH 1996). Furthermore, meltwater reaching the channel displays a temperature only slightly above 0°C. Consequently, thermal erosion is assumed to be weak during this period.

A number of studies have focused on suspension transport by high-latitude rivers, but only few data are published concerning bed load transport. A comparison of sediment yields is compiled by CLARK (1988) and more recently (but concentrated on Canadian rivers) by LEWKOWICZ & WOLFE (1994). Typical peak concentrations for suspended sediments in connection with runoff events range at 1,000–8,000 $mg{\cdot}l^{-1}$. This coincides with detailed measurements in small glaciated and non glaciated rivers in NW-Spitsbergen, where peak concentrations in the latter catchments slightly exceeded 1 $mg{\cdot}l^{-1}$ (BARSCH et al. 1994b). Confusion arises when comparing total sediment transports (dissolved, suspended, and bed load) to the bed load yield. Whereas in a basin on Baffin Island, bed load transport was found to be responsible for 80–95% of total sediment transport, bed load comprised only 3–15% of the total sediments in a basin on Cornwallis Island (LEWKOWICZ & WOLFE 1994) and BARSCH et al. (1994a) quote a component of 1–5% deriving from bed load transport. It has to be stressed that the data of

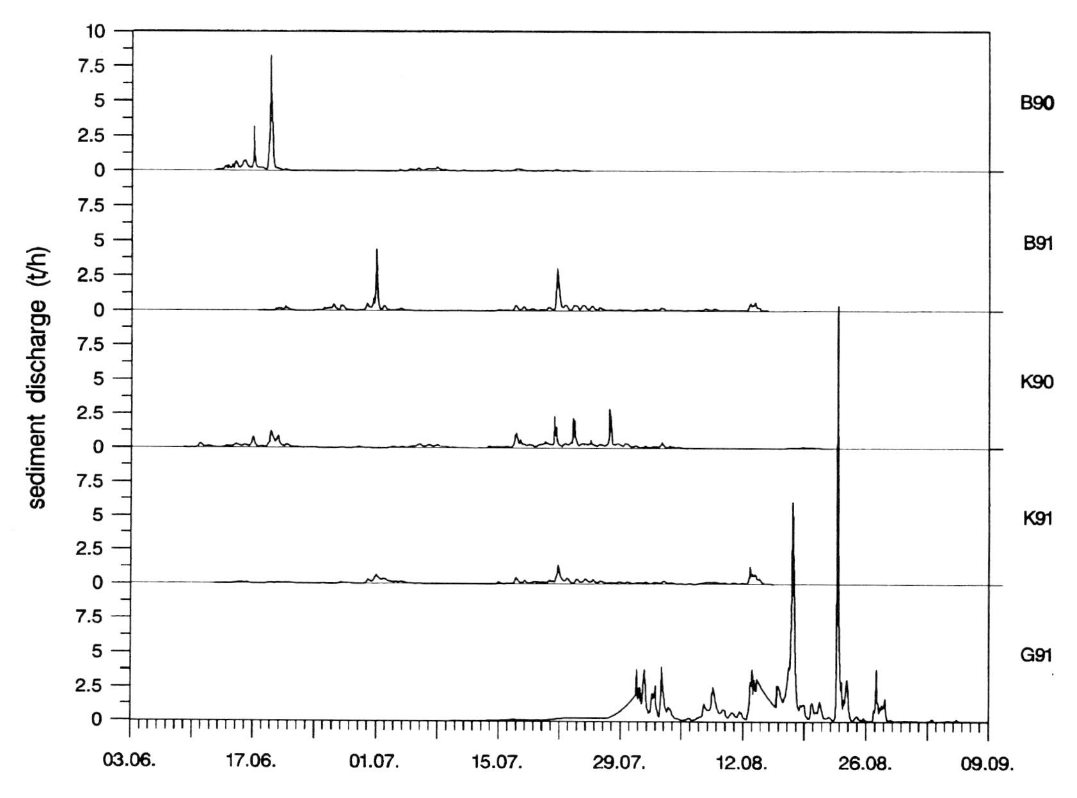

Fig. 6. Sediment hydrographs for the years 1990 and 1991 of three basins with similar geomorphic setting but different glacier coverage. B: Beinbekken, no glaciers; K: Kvikkåa, 30% glacier coverage; G: Glopbreen basin, 50% glacier coverage (Liefdefjorden, NW-Spitsbergen) (after: BARSCH et al. 1994a).

CHURCH (1972) are based on calculations of potential sediment transport, whereas the latter two are based on direct measurements of transports in the rivers. Nevertheless, the contradiction is obvious, and further measurements are needed since the simple bed load equation seems not suitable due to the influence of ice and snow in the channel (cf. CARTER et al. 1987).

4 *Magnitude-frequency in the context of geomorphodynamics*

4.1 *Atmospheric processes*

The main control for the triggering of fluvial events by snowmelt is the duration of a weather situation with high influx of energy by net radiation, by sensible heat or by a combination of both. The typical magnitude-frequency relation of snowmelt events in periglacial environments, especially at high latitudes, reflects this fact, since the mean magnitude of the diurnal cycle is much weaker than that of the yearly cycle.

Beside the two external cycles (earth rotation and revolution), there are two further frequency ranges of relevance. Significant magnitudes are typically observed at intermediate frequencies between 1 d^{-1} and 1 a^{-1} resulting from non-periodic but regular meteorological situations. The highest magnitudes are connected to the lowest frequencies as a result of rare extreme events, where a combination of high mean energy fluxes by net radiation and sensible heat are maintained over a period of several days or longer. In periglacial environments at high latitudes, where seasonal snow covers are removed during one pronounced snowmelt period in later spring, such strong events are always occurring during the main snowmelt period. At this time, net radiation is able to stay positive also during 'night-time' preventing the snow cover to stabilize itself by partial refreezing, which regularly occurs in alpine periglacial regions at mid- and low latitudes.

Strong rainfall events are not restricted to certain periods in the year, although almost every region shows a characteristic seasonality in the yearly distribution of rain precipitation. For example, a rain event of 272 mm occurred in November 1993 in Ny Ålesund/Svalbard. The extreme magnitude of this event can be assessed by the mean monthly total of 32 mm for November, and by the fact that nearly 100% of the precipitation in November is usually falling as snow.

Rainfall in regions with periglacial conditions is strongly coupled to cyclonic activity, which is responsible for the advection of moist air. In consequence, the duration of a rainfall event is more important for the magnitude than the mean intensity. As long as an extended snow cover is present, convective rainfall is negligible. With respect to high-magnitude fluvial events in snow-free areas of periglacial environments, they are not of relevance, too.

Rainfall events are most probable in polar regions during the summer months, although in maritime regions (e.g. in Svalbard) rainfall is regularly observed also in winter. In subpolar maritime regions (e.g. in Norway), high-magnitude rainfall events are concentrated in the autumn and winter season, while in continental regions high-magnitude rainfall events are rare, in general.

The magnitude-frequency relation of rainfall events shows a higher spatial variability than snowmelt events. In addition, the periodicity of rainfall events is smaller than that of snowmelt events. Both relations show the common trend (realized in most magnitude-frequency relations) that rare events, i.e. events of low frequencies show the highest magnitudes.

4.2 *Fluvial events*

Analyses of magnitude and frequency of slushflows were compiled for N-Sweden (NYBERG 1985, RAPP 1995) and Spitsbergen (ANDRÉ 1995, BARSCH et al. 1994a, SCHERER 1994) based on historic, stratigraphic and lichenometric data. NYBERG (1985) presented the most comprehensive inventory on slushflow activity, site characteristics, and release conditions. NYBERG (1985) stresses that the occurrence displays a great variability from site to site, but under certain meteorological conditions, slushflows occur in many suitable sites in a larger region, a fact supporting the control of the release by meteorological factors (cf. ONESTI 1985, HESTNES 1985, SCHERER 1994, SCHERER et al. 1998). Slushflows with geomorphic significance span a frequency range of few years up to decades in northern Sweden, according to NYBERG (1985). Based on historic data and scientific reports, RAPP (1995) compiled a list of slushflows and slush torrents that occurred in European northern areas, and demonstrated the risk connected with these processes. Records of the railway company provided a complete list of events that affected the railway in a channel at Kaisepakte (Abisko mountains, N-Sweden): six events reached the magnitude to cause damages during 90 years of record. A similar frequency of larger events can be summarized on basis of the intense field work in Kärkevagge in the same area (RAPP 1995). BULL (1995) deduced for events with significant debris transport in Kärkerieppe site (Kärkevagge) a frequency in the same range by analyzing lichen diameters on the slushflow fan. On Spitsbergen, ANDRÉ (1995) analyzed lichen diameters on slushflow deposits and recovered a recurrence interval as low as 500 a for major events (slush torrents). But it is also stressed that minor events (small slushflows) occur in this area at annual or decennial recurrence intervals.

In summarizing these investigation results, it seems not possible to deduce a consistent magnitude-frequency relation for slushflows and slush torrents. Nevertheless, geomorphic evidence and process analysis supports the assumption of two different processes. Minor slushflows and slush torrents should be separated in this context, which allows a more differentiated view of break-up event magnitudes and frequencies (Table 3).

Although valley evolution in high latitudes is intensively discussed in relation to fluvial action, high magnitude/low frequency events seem not sufficiently considered as essential geomorphologic agents. As stressed above, related studies concentrated on discharge processes, while slushflows were omitted as part of the fluvial system. This is even more amazing in view of the original definition of slushflows as a "flowage of water-saturated snow along stream courses" (WASHBURN & GOLDTHWAIT 1958). In general, very few process studies offer data on the significance of fluvial geomorpho-dynamics and the effects of high magnitude/low frequency events in the fluvial system, while the database is even more sparse.

In the context of this limited knowledge, one is restricted to only basic assumptions on how magnitude and frequency of fluvial events in high-latitude regions affect landscape evolution:

Table 3. Reports of recurrence intervals of slushflows and slush torrents in different areas.

Location	Process	Recurrence interval (a)	References
Kärkevagge (N-Sweden)	Slush torrents	3–30	RAPP (1995), NYBERG (1985), BULL (1995)
Kaisepakte (N-Sweden)	Presumably slush torrents	100–300	RAPP (1995)
Liefdefjorden (Spitsbergen)	Slushflows and slush torrents	Slushflows: 1–2 Slush torrents: 50–150	BARSCH et al. (1994)
Longyearbyen (Spitsbergen)	Presumably slush torrents	10 (prior to prevention measures in the 50th)	HESTNES (1994)
Spitsbergen (different sites)	Slushflows and slush torrents	Slushflows: 1–20 Slush torrents: 100–500	ANDRÉ (1995)
Brooks Range (Alaska)	Slushflows (pres. partly slush torrents)	1–2	D. KANE (pers. commun.)
Different sites in Siberia	Slushflows/slush torrents	2–14	PEROV (1998)

- Fluvial events are triggered, as in mid-latitudes, by meteorological conditions, but the widespread lack of convective storms in high latitudes limits the magnitude of rainfall runoff events. On the other hand, snowmelt provides an effective water supply process that may reach the intensity of mid-latitude rainstorms.
- Except for proglacial rivers, sediment transport is significantly dependent on single events. The dynamics of coarse sediment transport especially stresses this fact, since the channel bed will withstand shear stress until a certain threshold during an event is exceeded.
- The erosion of high magnitude events is significantly affected by the occurrence of ground ice in streambanks and channel. Nevertheless, these events contribute remarkably to total fluvial sediment yields.
- Geomorphic features testifying to the occurrence of events with a significantly higher magnitude than normal runoff events are reported from numerous sites in high latitudes. Similar processes in many cases, especially in N-Sweden, are identified to be slushflows.
- Although slushflows are typically confined to channels and adjacent areas, very little work has been done until now on comparative evaluations of the geomorphic significance of the event sequences including runoff events and slushflows in fluvial environments.

To assess the geomorphic impact, an analysis of sediment displacement by fluvial events in high-latitude regions can either be related to mean denudation rates in the basin or to the real erosion and deposition features left by single events. The former approach faces the problem of obscuring the internal sediment dynamics and displacements within the catchment. Nevertheless, it offers valuable data as the calculations of RAPP (1960) and BARSCH et al. (1994a) were able to demonstrate (cf. Table 4).

Calculation of sediment mass transfer rates provides a more detailed comparison. Data for NW-Spitsbergen and for Kärkevagge (RAPP 1960) are summarized in Table 4. Horizontal and vertical components of the mass transfer rates between the two sites differ due to the steeper flow tracks in Kärkevagge, but both data sets coincide surprisingly, and this holds also for the denudation rates. A comparison with transports by discharge in Kvikkåa and Beinbekken shows that slushflow transport ranges in the same order of magnitude, but nevertheless they are significantly lower. Actually, the data demonstrate the importance of slushflows for sediment

Table 4. Sediment mass transfer rates performed by runoff events and slushflows (minor slushflows and slush torrents) considering their magnitude-frequency-distribution.

Location/Reference	Transport Process	Horizontal ton-meters ($m \cdot t \cdot km^{-2} \cdot a^{-1}$)	Vertical ton-meters ($m \cdot t \cdot km^{-2} \cdot a^{-1}$)	Denudation ($t \cdot km^{-2} \cdot a^{-1}$)
N-Sweden (Kärkevagge) RAPP (1960)	Slushflows	2,500	1,400	14
NW-Spitsbergen	Slushflows	6,000	480	20
(Kvikkåa/Beinbekken)	Runoff events:			
BARSCH et al. (1994), GUDE (1998)	dissolves	16,000	1,300	15
	suspension	17,000	1,400	35
	bed load	40	3	1

transport in the basins, but the information required to make general statements is too weak and further investigations are needed.

One of the main problems to be considered in this context emerges from the evaluation of event sequences in the fluvial environment and the definition of magnitude and frequency of effective or formative events in the meaning of BRUNSDEN (1996). It seems obvious that in fluvial systems with slushflow occurrence, the latter dominates the landforming process complex. ANDRÉ (1995) demonstrates a life-expectancy for slush torrent depositional forms of at least 2000 years. Lifetimes in the same order can also be expected for the forms described from N-Sweden, although the forms described e. g. by RAPP (1960) and NYBERG (1985) typically result from several events rather than from a single one. These landforms range in the order of mesorelief, whereas the microrelief may be predominantly formed by runoff events. In fluvial systems lacking slushflow and other extreme events (e.g. river ice break-up), runoff events represent the upper end of the event sequence.

Since the events and, indeed, the whole sequences are induced by atmospheric processes, a change in magnitude and frequency of events as a consequence of climatic change has to be expected . The effects are assumed to be most prominent in case of slush torrents, because a shift in magnitude and frequency affects both total sediment yields and landforms more significantly compared to runoff events. This is mainly due to the fact that the increase of magnitude of slush torrents in many sites is virtually unlimited, whereas flood magnitudes are clearly controlled by storm intensities. Furthermore, an increase of slush torrent frequency is expected

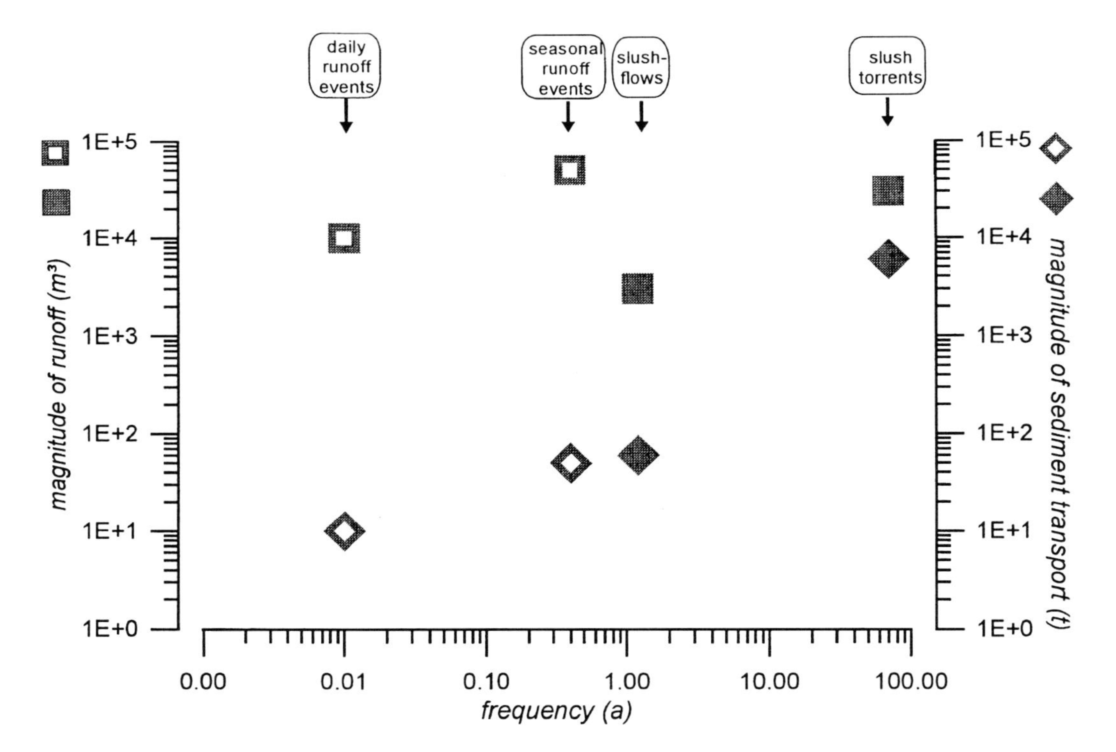

Fig. 7. Magnitude-frequency diagram of the hydrological and sedimentological significance of slushflows and runoff events in the Kvikkåa creek (Liefdefjorden, NW-Spitsbergen) in 1990/91/92.

due to the predicted pronounced warming in high-latitude regions (SCHERER & PARLOW 1996). The magnitude-frequency diagram (Fig. 7) illustrates the importance of slush torrents especially for sediment transfer yields. But actually, available field data sets as well as modeling approaches do not offer a sufficient base to derive general statements, since prognostics concerning GCM and consequences of climate change on high-latitude hydrology are still uncertain (cf. e.g. WOO 1996).

Acknowledgements

The research in NW-Spitsbergen and in N-Sweden was supported by the German and the Swiss Science Foundations (DFG and SNF) and was significantly supported by the Swedish collegues Anders Rapp and Christer Jonasson as well as by the Abisko Scientific Research Station. Fieldwork and the paper was greatly encouraged by Dietrich Barsch, Eberhard Parlow and Roland Mäusbacher. Comments on an earlier draft of this paper where given by Mike Crozier.

References

ANDRÉ, M.-F. (1995): Holocene climate fluctuations and geomorphic impact of extreme events in Svalbard. – Geogr. Ann. 77A (4): 241–250.

BARSCH, D., M. GUDE, R. MÄUSBACHER, G. SCHUKRAFT, A. SCHULTE & D. STRAUCH (1993): Slush stream phenomena – process and geomorphic impact. – Z. Geomorph. N.F., Suppl.-Bd. **92**: 39–53.

BARSCH, D., M. GUDE, G. SCHUKRAFT & A. SCHULTE (1994a): Recent fluvial sediment budgets in glacial and periglacial environments, NW-Spitsbergen. – Z. Geomorph. N.F., Suppl.-Bd. **97**: 111–122.

BARSCH, D., M. GUDE, G. SCHUKRAFT & A. SCHULTE (1994b): Geomorphic impact of slush streams in a changing climate. – In: K. SAND & Å. KILLINGTVEIT (ed.): Proc. 10th Int. Northern Research Basins Symposium and Workshop, Spitsbergen 1994: 359–367.

BLÖSCHL, G., D. GUTKNECHT & R. KIRNBAUER (1988): Berechnung des Wärmeeintrages in eine Schneedecke – Analyse des Einflusses unterschiedlicher meteorologischer Bedingungen. – Deutsche Gewässerkundl. Mitt. **32**/1–2: 34–39.

BRAUN, L.N., E. BRUN, Y. DURAND, E. MARTIN & P. TOURASSE (1994): Simulation of discharge using meteorological data distribution, basin discretization and snow modelling. – Nordic Hydrol. **25**: 129–144.

BRUNSDEN, D. (1996): Geomorphological events and landforms change. – Z. Geomorph. N.F. **40** (3): 273–288.

BÜDEL, J. (1982): Climatic Geomorphology. – Princeton University Press, Princeton, New Jersey.

BULL, W.B., P. SCHLYTER & S. BROGAARD (1995): Lichenometric analysis of the Kärkerieppe slush-avalanche fan, Kärkevagge, Sweden. – Geogr. Ann. **77**A (4): 231–240.

CARTER, L.D., J.A. HEGGINBOTTOM & M.-K. WOO (1987): Arctic Lowlands. – In: GRAF, W.L. (ed.): Geomorphic systems of North America. – Geol. Soc. Amer., Centennial Spec. Vol. 2.

CHURCH, M. (1972): Baffin Island sandurs: a study of arctic fluvial processes. – Geol. Surv. Canada, Bull. **216**, 208 pp.

CHURCH, M. (1988): Floods in Cold Climates. – In: BAKER, V.R., R.C. KOCHEL & P.C. PATTEN (eds.): Flood Geomorphology: 205–229.

CLARK, M.J. (1988): Periglacial Hydrology. – In: CLARK, M.J. (ed.): Advances in Periglacial Geomorphology: 415–462.

CLARK, M.J. & M. SEPPÄLÄ (1988): Slushflows in a subarctic Environment, Kilpisjärvi, Finnish Lapland. – Arctic Alpine Res. **20**, 1: 97–105.

COLBECK, S.C. (1975): A Theory for Water Flow Through a Layered Snowpack. – Water resources Res. **11**, 2: 261–266.

ERGENZINGER, P. (1988): The nature of coarse material bed load transport. – IAHS Publ. **174**: 207–216.
FRENCH, H.M. (1996): The periglacial environment. – London, Longman Group Limited, 2nd ed., 341 pp.
GLENN, S.M. & M.-K. WOO (1997): Spring and summer hydrology of a valley bottom wetland, Ellesmere Island, Northwest Territories, Canada. – Wetlands **17**, 2: 321–329.
GUDE, M. (1998, in press): Ereignissequenzen und Sedimenttransporte im fluvialen Milieu kleiner Einzugsgebiete auf Spitzbergen. – Heidelberger Geogr. Arb.
GUDE, M. & D. SCHERER (1995): Snowmelt and slush torrents – Preliminary report from a field campaign in Kärkevagge, Swedish Lappland. – Geogr. Ann. **77A** (4): 199–206.
GUDE, M. & D. SCHERER (1998): Snowmelt and water movement in saturated snow layers of high latitude areas. – Proc. 11th Int. Symp. and Workshop Northern Research Basins, Fairbanks 1997.
HESTNES, E. (1985): A Contribution to the Prediction of Slush Avalanches. – Ann. Glaciol. **6**: 1–4.
HESTNES, E. (1994): Impact of rapid mass movement and drifting snow on the infrastructure and development of Longyearbyen. – In: SAND, K. & A. KILLINGTVEIT (eds.): Proc. 10th Int. Northern Research Basins Symposium and Workshop, Spitsbergen 1994: 23–46.
KANE, D.L., L.D. HINZMAN, M.-K. WOO & K.R. EVERETT (1992): Arctic hydrology and climate change. – In: CHAPIN, F.S., R.L. JEFFERIES, J.F. REYNOLDS, G.R. SHAVER & J. SVOBODA (eds.): Arctic ecosystems in a changing climate. An ecophysiological perspective. – San Diego a.o.: 469.
KIRNBAUER, R., G. BLÖSCHL & D. GUTKNECHT (1994): Entering the Era of Distributed Snow Models. – Nordic Hydrol. **25**: 1–24.
LEWKOWICZ, A.G. & P.M. WOLFE (1994): Sediment transport in Hot Weather Creek, N. W. T., Canada, 1990–1991. – Arctic Alpine Res. **26**, 3: 213–226.
MALE, D.H. & R.J. GRANGER (1981): Snow surface energy exchange. – Water Resour. Res. **17**/3: 609–627.
MARKS, D. & J. DOZIER (1992): Climate and energy exchange at the snow surface in the alpine region of the Sierra Nevada. Part 2: Snow cover energy balance. – Water Resour. Res. **28**/11: 3043–3054.
MARSH, P. & M.-K. WOO (1981): Snowmelt, glacier melt, and high arctic streamflow regimes. – Canad. Journ. Earth Sci. **18**: 1380–4.
MCCANN, S.B. & J.G. COGLEY (1973): The geomorphic Significance of fluvial activity at high latitudes. – In: FAHEY, B.D. & R.D. THOMPSON (Eds): Research in Polar and Alpine Geomorphology. – Proc. 3rd Guelph Symposium on Geomorphology: Norwich, England: 118–135.
NOBLES, L.-H. (1966): Slush Avalanches in Northern Greenland and the Classification of Rapid Mass Movements. – Ass. Internat. Hydrol. Sci. Publ. **69**: 267–272.
NYBERG, R. (1985): Debris Flows and Slush Avalanches in Northern Swedish Lappland. – Medd. Lunds Universit. Geograf. Inst., Avhandli. **XCVII**: 222.
NYBERG, R. (1989): Observations of slushflows and their geomorphological effects in the Swedish mountain area. – Geogr. Ann. **71A** (3–4): 185–198.
OHMURA, A. (1981): Climate and energy balance on arctic tundra. – Züricher Geograph. Schr. **3**, 448 pp.
ONESTI, L.J. (1985): Meteorological Conditions that Initiate Slushflows in the Central Brooks Range, Alaska. – Ann. Glaciol. **6**: 23–25.
ONESTI, L.J. & E. HESTNES (1989): Slush-flow Questionnaire. – Ann. Glaciol. **13**: 226–230.
PEROV, V.F. (1998): Slushflows: basic properties and spreading. – In: HESTNES, E. (ed.): Proc. Conf: 25 years of snow avalanche research, NGI Publ. 203, Oslo: 203–209.
PLÜSS, C. & R. MAZZONI (1994): The role of turbulent heat fluxes in the energy balance of high alpine snow cover. – Nordic Hydrol. **25**: 25–38.
RAPP, A. (1960): Recent Development of Mountain Slopes in Kärkevagge and Surroundings, Northern Scandinavia. – Geogr. Ann. **17** (2–3): 71–200.
RAPP, A. (1995): Case studies of geoprocesses and environmental change in mountains of northern Sweden. – Geogr. Ann. **77A** (4): 189–198.
RUDBERG, S. (1988): High arctic landscapes: comparison and reflexions. – Norsk Geogr. Tidskr. **42**: 255–264.
SCHERER, D. (1994) Slush stream initiation in a High Arctic drainage basin in NW-Spitsbergen. – Stratus 1, Basel, 95 pp.
SCHERER, D. & E. PARLOW (1994) Terrain as an important controlling factor for climatological, meteorological and hydrological processes in NW-Spitsbergen. – Z. Geomorph. N.F., Suppl.-Bd. **97**: 175–193.

SCHERER, D. & E. PARLOW (1996) Meteorology and hydrology of slush stream initiation in a changing climate. – In: K. SAND & Å. KILLINGTVEIT (eds.): Proc. 10th Int. Northern Research Basins Symposium and Workshop, Spitsbergen 1994: 346–358.

SCHERER, D., M. GUDE, M. GEMPELER & E. PARLOW (1998): Atmospheric and hydrological boundary conditions for slushflow initiation due to snowmelt. – Ann. Glaciol. 26.

THORN, C.E. (1992): Periglacial geomorphology: what, where, when? – In: DIXON, J.C. & A.D. ABRAHAMS (ed.): Periglacial Geomorphology. Proc. of the 22nd Annual Binghamton Symposium in Geomorphology; Wiley, Chichester: 3–31.

WASHBURN, A.L. (1979): Geocryology: A survey of periglacial processes and environments. – London, Edward Arnold, 406 pp.

WASHBURN, A.L. & R.P. GOLDTHWAIT (1958): Slushflows. – Geol. Soc. Amer. Bull. **69**: 1657–1658.

WOO, M.-K. (1996): Hydrology of northern north America under global warming. – In: JONES, J.A. A., C. LIU, M.-K. WOO & H.-T. KUNG (eds.): Regional hydrological response to climate change. – Kluwer Academic: 73–86.

WOO, M.-K. & J. SAURIOL (1980): Channel Development in Snow-Filled Valleys, Resolute, N.W.T., Canada. – Geogr. Ann. **62**A (1–2): 37–56.

WOO, M.-K. & R. HERON (1981): Occurrence of Ice Layers at the Base of High Arctic Snowpacks. – Arct. Alp. Res. **13** (2): 225–230.

Addresses of the authors: MARTIN GUDE, Department of Geography, University of Jena, Löbdergraben 32, D-07743 Jena, Germany. DIETER SCHERER, MCR Lab., Dept. of Geography, University of Basel, Spalenring 145, CH-4055 Basel, Switzerland.

Z. Geomorph. N.F.	Suppl.-Bd. 115	113–123	Berlin · Stuttgart	Juli 1999

Extreme storms and rainfall erosivity factor in Évora (Portugal)

Cláudia Brandão and Marcelo Fragoso, Lisboa

with 7 figures and 3 tables

Summary. In Alentejo region, situated in the south of Portugal, with a Mediterranean type of climate, rainfall has a strong seasonality, high interannual and spatial variability, occurring in a small number of events, some of them with high intensities. Alentejo has traditionally an extensive agricultural land use of winter cereals and pasture and so presents each year an extensive area of bare soils potentially subject to severe precipitation erosion. The aim of research is to determinate the relationships between hydrologic characteristics, total precipitation, precipitation intensity, rainfall erosivity factor and precipitation time distribution and the meteorological conditions of extreme storms, at a synoptic scale, so as to be able to forecast the magnitude and time frequency of extreme storm erosivity in Alentejo region.

Résumé. *Évenements de fortes précipitations et le facteur "R" des pluies à Évora (Portugal).* – Située au Sud du Portugal et caractérisée par un climat de type méditerranéen, la région du Alentejo est marquée par une forte variabilité spatiale et interannuelle des pluies et encore par un régime pluviométrique irrégulier. Celui-ci comporte des événements pluvieux peu fréquents mais qui peuvent se manifester avec une forte intensitée. Traditionnellement, l'exploration agricole des sols du Alentejo est extensive et se fait avec des céréales d'hiver et des pâturages, c'est pourquoi, toutes les années, ces champs découverts sont exposés aux sévères précipitations érosives. Ainsi, l'object de cette recherche consiste à déterminer les rapports existants entre les caracteristiques hydrologiques des événements (la précipitation totale, l'intensité des pluies, le facteur d'érosivité des pluies et distribution temporelle des précipitations) et ses conditions météorologiques à l'échelle synoptique, de façon à évaluer la magnitude et la fréquence de l'érosivité provoquée par les fortes pluies.

Zusammenfassung. *Die gelegentlich starken Niederschläge und der Erosionsfaktor des Niederschlags in Évora (Portugal).* – Die im Süden Portugals gelegene und durch ein mediterranes Klima gekennzeichnete Region des Alentejo weist im Jahresverlauf räumlich erhebliche Schwankungen in Form und Menge des Niederschlags auf. Trockene und feuchte Perioden wechseln einander ab, wobei letztere zum Teil Regenfälle umfassen, die – wenn auch selten – sehr intensiv sein können. Die Böden des Alentejo werden traditionell extensiv bewirtschaftet. Dabei überwiegen Wintergetreide und Weiden, was Ursache dafür ist, daß die nur durch eine schwache bzw. geringe Gründecke geschützten Felder jedes Jahr von der erosiven Wirkung des Regens betroffen sind, vor allem bei Regenfällen stärkerer Intensität. In dieser Arbeit soll der Zusammenhang zwischen den hydrologischen Merkmalen der Episoden starken Niederschlags (gesamte Niederschlagsmenge, Intensität des Niederschlags, Erosionsfaktor und zeitliche Verteilung des Niederschlags) und den hierfür verantwortlichen meteorologischen Bedingungen untersucht werden. Dies soll es ermöglichen, die Größe und Häufigkeit der Episoden vorherzusehen, die potentiell erosive Auswirkungen auf die Böden des Alentejo haben können.

Introduction

In Alentejo region, situated in the south of Portugal, the precipitation has a large spatial and annual variability (Ventura 1994). The Alentejo region, with traditional agriculture of winter cereals and pasture, is potentiality subject to severe soil erosion due to precipitation. Many Portuguese studies have been made in order to predict the soil loss due to precipitation (hydric soil erosion)

0044-2798/99/0115-0113 $ 2.75

in agricultural areas (FERREIRA et al. 1985, ROXO et al. 1993, 1994, ROXO 1994) as in forests (ROCHA et al. 1986): but the results show that is difficult to find a unique rainfall index. Besides, the field determination of soil erodibility is not yet available at regional scale. Thus, it has not been easy to identify the threshold of precipitation from which the significative hydric soil erosion occurs.

The study concerns the characteristics of the erosive precipitation in Alentejo and presents the preliminary results. The overall objective of the research is to determine the relationships between hydrologic characteristics, total precipitation, precipitation intensity, rainfall erosivity factor and precipitation time distribution and the meteorological conditions of extreme storms, at a synoptic scale, in order to forecast the magnitude and time frequency of extreme storm erosivity at Évora-Cemitério raingauge, in Alentejo region, based on the meteorological conditions.

Data Acquisition

Hydrologic data

The hydrologic data acquisition was made in four steps:

1. Digitization of daily rainfall charts from 1941 to 1992 of Évora-Cemitério recording raingauge (with a total of 18994 charts).
2. Based on the digitized record, storms were listed on the basis of: date and hour of beginning, total precipitation, average precipitation intensity and total duration of storms. According to many authors the total storm duration is the time period with precipitation, possibly intermittent, preceded and followed at least by 6 h of precipitation absence. This study assumes that a 6 h interval assures the existence of independent storms.
3. Selection, from the list of storms, of the extreme storms, which are the storms with produced at least 25.4 mm (1 inch) of precipitation.
4. Using time intervals in order of magnitude of the minute extreme storms were analysed for: total precipitation, average precipitation intensity, total storm duration, hyetogram with one minute of time increment chronological graph of rainfall phenomena (expressed in percentage) and localisation of maximum cumulative precipitation between the four quarters the total duration of storm.

Meteorological data

In order to study the main synoptic conditions of the extreme storms, the synoptic charts (surface and 500 hPa charts) and meteorological data tables from the Portuguese daily weather bulletin[1] were analysed and sixteen meteorological parameters were collected. The analysis was made for 1950 to 1992 period, sufficient to permit a statistical study.

Methodology and results

The distribution of extreme storms were studied by decade, months and quarter time period of total precipitation (Table 1) and to precipitation intensity in 30 min (Table 2).

[1] Boletim Meteorológico Diário, Instituto de Meteorologia, Lisboa.

Table 1. Extreme Storms characterization. Évora-Cemitério recording raingauge.

	Extreme Storms Number	
Decade	P>= 25.4 mm	P>= 50.8 mm
1940–1949	36	9
1950–1959	45	11
1960–1969	82	12
1970–1979	44	8
1980–1989	61	12
1990–1992	8	0
Total	276	52

	Extreme Storms Number	
Month	P>= 25.4 mm	P>= 50.8 mm
October	28	5
November	46	9
December	34	10
January	42	8
February	38	7
March	33	6
April	19	2
May	12	1
June	8	1
July	4	1
August	1	0
September	11	2
Total	276	52

	Extreme Storms Number	
Quarter	P>= 25.4 mm	P>= 50.8 mm
Quarter1	85	15
Quarter2	74	14
Quarter3	68	12
Quarter4	47	12
Total	272	53

The extreme storms erosivity factors, R (WISCHMEIER 1978) were determinated according to LENCASTRE & FRANCO (1984) by

$$R = \frac{\sum_{i=1}^{n} E_i I_{30}}{1735} \qquad I_{30}\ (\text{mm/h});\ R\ (\text{t.m.pé/acre}) \qquad (1)$$

and

Table 2. Extreme Storms precipitation intensity in 30 min. Évora-Cemitério recording raingauge.

	Extreme Storms Number	
Decade	I(mm/h)< T= 50 years	T= 50 years <I(mm/h)< T= 100 years
1940–1949	34	2
1950–1959	45	0
1960–1969	76	6
1970–1979	43	1
1980–1989	58	3
1990–1992	8	0
Total	264	12

	Extreme Storms Number	
Month	I(mm/h)< T= 50 years	T= 50 years <I(mm/h)< T= 100 years
October	26	2
November	45	1
December	34	0
January	42	0
February	37	1
March	32	1
April	18	1
May	12	0
June	5	3
July	2	2
August	0	1
September	11	0
Total	264	12

	Extreme Storms Number	
Quarter	I(mm/h)< T= 50 years	T= 50 years <I(mm/h)< T= 100 years
Quarter1	77	6
Quarter2	71	3
Quarter3	67	1
Quarter4	45	2
Total	260	12

I(T=50 years)=39.1 mm/h; I(T=100 years)=67.1 mm/h (Brandão, 1996)

$$E_i = (12.13 + 8.9 \, \mathrm{Log} I_i) h_i \qquad I_i \text{ (mm/h)}; h_i \text{ (mm)}; E_i \text{ (t.m/ha)} \qquad (2)$$

using four methodologies:

Method 1 – The discretization of each extreme storm into time-step increments is not considered. E_i is calculated using the total precipitation (h) and average precipitation intensity (I) of the extreme storm. R is calculated using the E of the extreme storm and the maximum precipitation intensity in 30 min (I_{30}).

Method 2 – Each extreme storm duration is sub-divided in 30 min time-steps. E_i is calculated using the precipitation (h_i) and the precipitation intensity (I_i) in 30 min time-steps. R is calculated using extreme storm E_i and the greater precipitation intensity occurred in the 30 min time-steps previously consider (I_{30}).

Method 3 – Each extreme storm duration is sub-divided in 30 min time-steps. E_i is calculated using the precipitation (h_i) and the precipitation intensity (I_i) in 30 min time-steps. R is calculated using extreme storm E_i and the maximum precipitation intensity in 30 min (I_{30}).

Method 4 – Each extreme storm duration is sub-divided in 1 min time-steps. E_i is calculated using the precipitation (h_i) and the precipitation intensity (I_i) in 1 min time-steps. R is calculated using extreme storm E_i and the maximum precipitation intensity in 30 min (I_{30}).

The application of these four methodologies allow to calculate four erosivity factors, R, for each extreme storms selected (Fig. 1).

The four group of R were organize on increase way. These incresing ordination show the methodology influence on the R magnitude (Fig. 2).

The time periods between occurrence of extreme storms with erosivity factor greater than 10 and its magnitude values were estimated by five probability distribution function: exponential, gumbel, gamma, log-normal, weibull and pearson III.

The exponential function was selected after the calculation of tree fitting tests, Qui², Kolmogorov-Smirnov and Cramer-Von-Mises.

The exponential function (3) with b equal 9.916 estimates the value magnitude and with b equal 8.803 estimates the time periods between the occurrence of extreme storms with erosivity factor greater than 10.

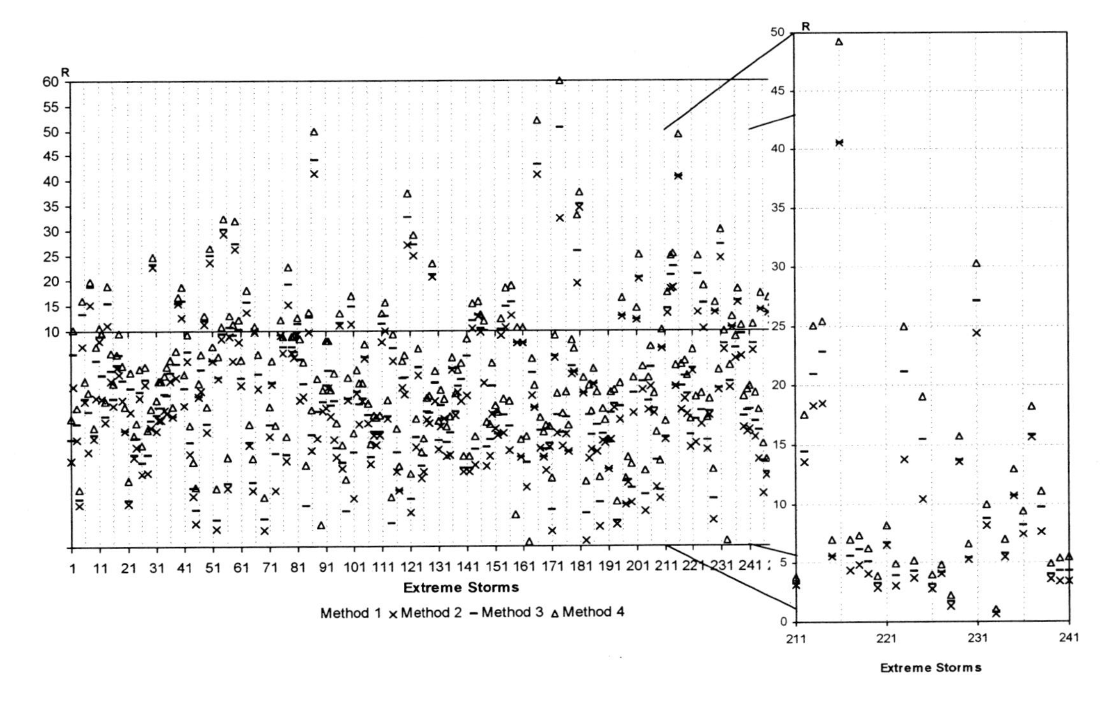

Fig. 1. Extreme storms erosivity factor in chronological order (WISCHMEIER 1978).

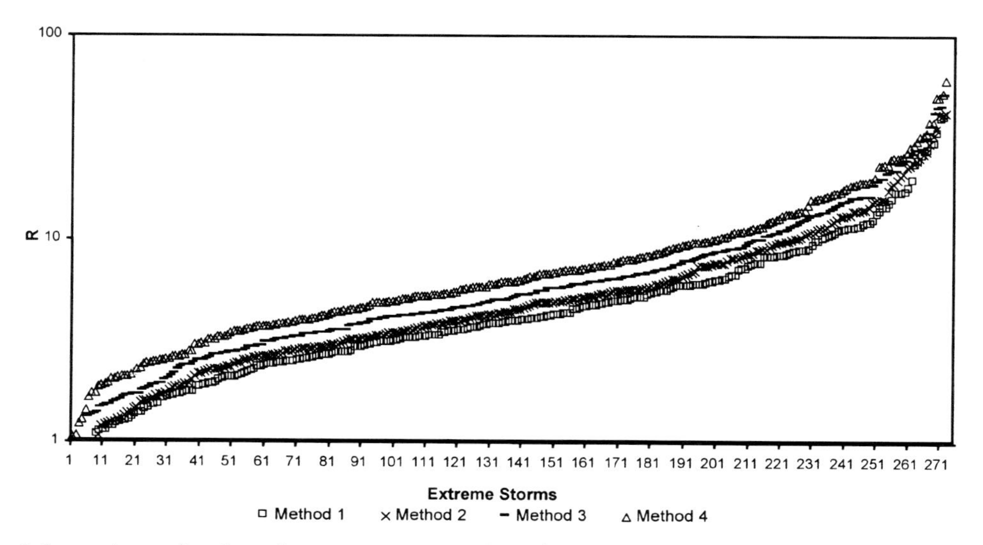

Fig. 2. Increasing ordination of extreme storms erosivity factor (WISCHMEIER 1978).

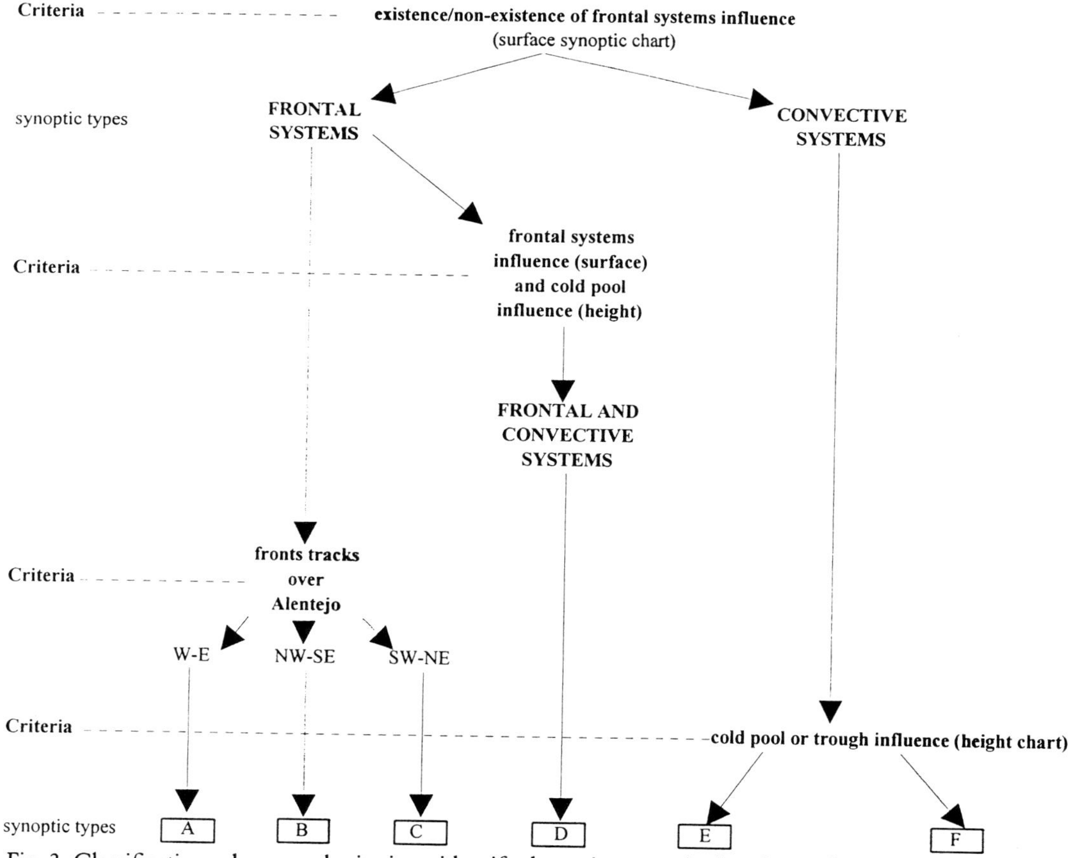

Fig. 3. Classification scheme and criteria to identify the main synoptic situations of extreme storms.

Table 3. Six synoptic situations (A to F).

SYNOPTIC SYSTEMS	Type	Fronts tracks over Alentejo	Predominant position of the low pressure center (surface)	Upper level (500 hPa) circulation
FRONTAL	A	W – E	– the low is southwest of Ireland – the low is northwest of Galice or over Galice	– Zonal – Trough (eastern part or axis)
	B	NW – SE	– the low is southwest of Ireland	– Zonal – Trough (eastern part)
	C	SW – NE	– the low is west of Iberian Peninsula or over western Iberian Peninsula	– Zonal – Trough (eastern part)
FRONTAL AND CONVECTIVE	D	W – E or SW – NE	– the low is west of Iberian Peninsula or over western Iberian Peninsula – the low is between Azores and Portugal – the low is south of Algarve (between Madeira and Cádiz Gulf)	– Cold pool is north of Galice or over Galice – Cold pool between Azores and Portugal – Cold pool between Madeira and Algarve or over Cádiz Gulf
CONVECTIVE	E	–	– the low is west of Iberian Peninsula or over western Iberian Peninsula – the low is between Azores and Portugal – the low is south of Algarve (between Madeira and Cádiz Gulf)	– Cold pool is north of Galice or over Galice – Cold pool between Azores and Portugal – Cold pool between Madeira and Algarve or over Cádiz Gulf
	F	–	– variable	– Trough (eastern part)

$$P\left[X_n \leq x\right] = 1 - e^{-x/\beta} \qquad (3)$$

The study of synoptic situation allow to define a typology, based on the joint analysis of atmospheric conditions observed in altitude and surface maps. A restrict group of classification criteria was used. In Fig. 3, the adopted criteria and the resulting classification structure, with six different synoptic types, are presented. It must be pointed out that the typology definition was established taking into consideration the most representative synoptic situations of Ferreira (1985) and Ventura (1994) works. A short characterization of the different synoptic types is presented in Table 3.

Figs. 4, 5 and 6 show the summary statistics of the synoptic analysis of the extreme storms.

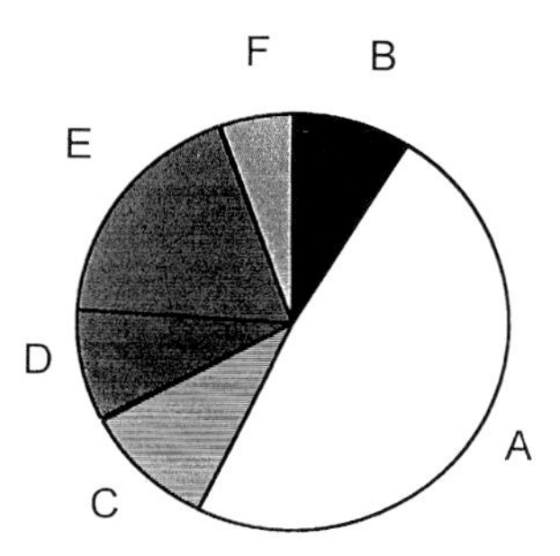

Fig. 4. Global frequency of the eight synoptic types (1950–1992).

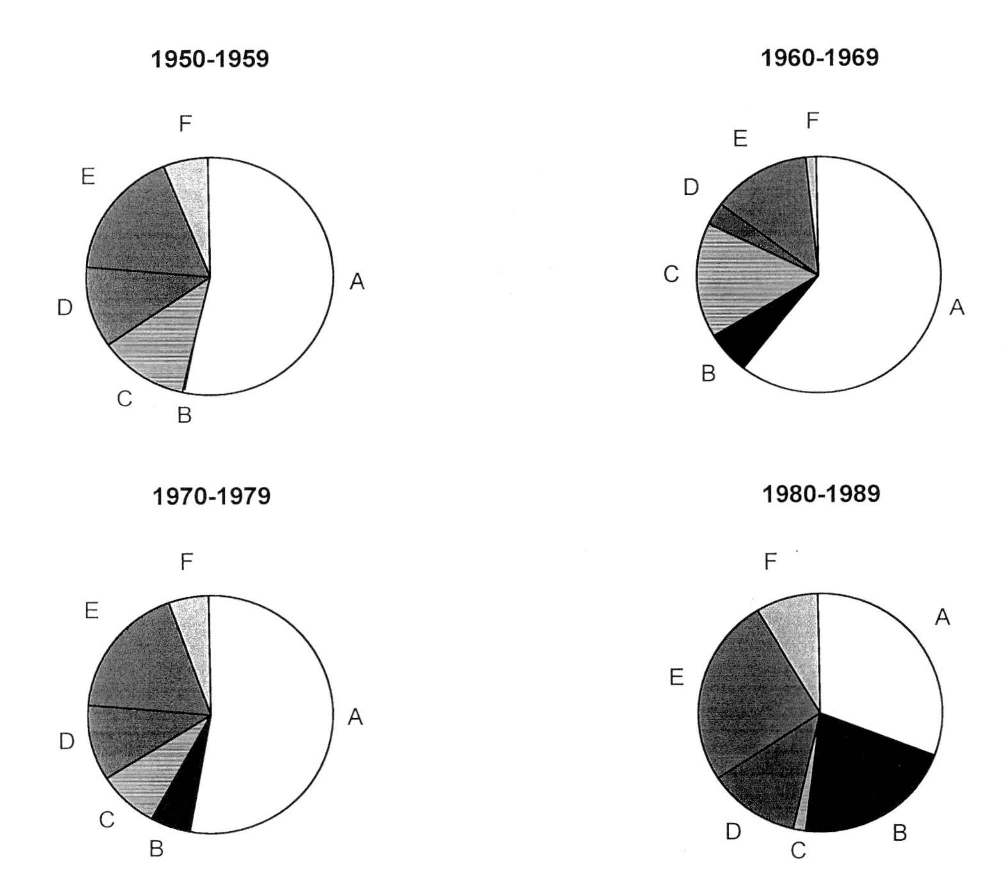

Fig. 5. Decade frequency of the eight synoptic types.

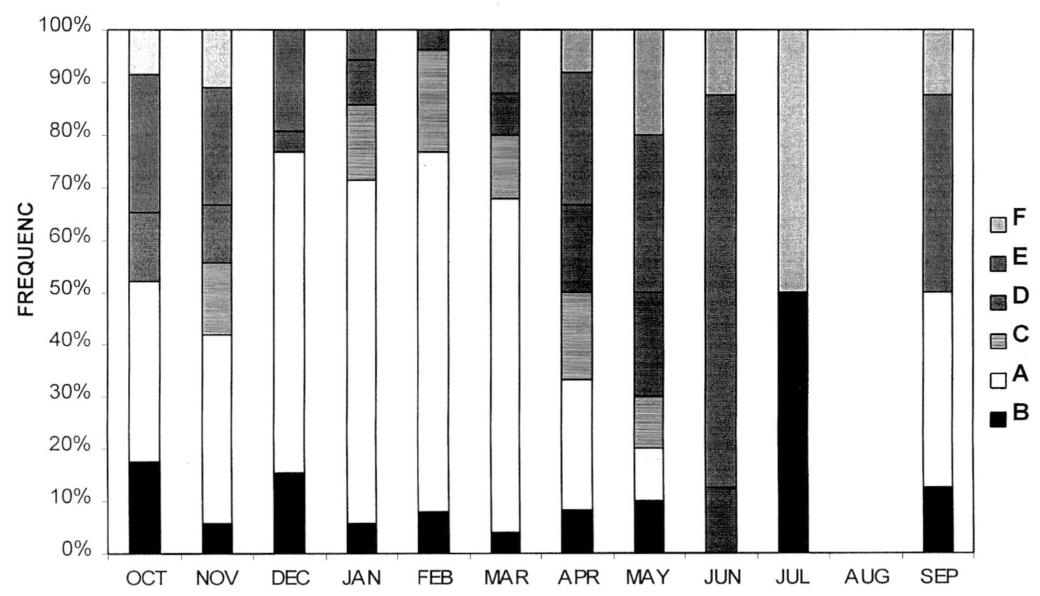

Fig. 6. Monthly frequency of the eight synoptic types (1950–1992).

Conclusions

The results of the analysis demonstrate some hydrologic characteristics and meteorological conditions of extreme storms and also the relationship between them.

The greatest number of extreme storms occurred during the sixties (29,7%).

November is the month with the greatest number of extreme storms (16,7%).

The highest values of erosivity Factor R were obtained applying method 4. They are close to those obtained by method 3.

The probability of the existence of R values greater than 10 as well as the estimation of its values, for a certain frequency, is carried out by exponential probability distribution function.

Comparing the global frenquency of the eight synoptic situations (Fig. 4): it is evident that frontal situations (A, B and C synoptic types) are predominant (67%) over convective situations (E and F synoptic types, with 24%) and mixed type (D synoptic type, with 9%). The A synoptic type is predominant (48%): followed by E synoptic type (19%).

Considering the frontal situations, there is a clear predominance of frontal situations with WE trajectory over Alentejo region, which are responsible for about an half of the 211 extreme storms studied. The majority of the extreme storms observed at Évora-Cemitério raingauge are associated to frontal systems with W trajectory, being the heavy precipitation, generally, connected to cold frontal activity.

The second most representative synoptic type is associated to the cold low (E type) which can be characterizated by strong convective instability. Predominantly, these situations are due to the action of cold pools whose nucleus stabilizes over the Atlantic ocean between Madeira and Algarve, staying Alentejo region under the influence of the cold pools (nonfrontal synoptic systems) orient sectors.

The relative importance of synoptic types is similar in 50's, 60's and 70's decade (Fig. 5). However, the dominance of frontal types (A, B and C types) is more evident during the sixties. In contrast the synoptic distribution in 80's decade when there was a significant decrease on the frequency of W frontal situations (A type) which led to an increase of convective situations, namely the E synoptic type.

The synoptic types characterised by frontal situations present maximum absolute frequency between December and March while synoptic types characterised by convective situations are dominant, in relative frequency, between May and June (Fig. 6). Between September and November, frontal and convective situations present equal frequencies. In these months, there were much more extreme storms than in Spring.

The extreme storms are associated to different causes during the year and the month frequency can be linked to an annual rhythm. In Winter, frontal situation with W-E trajectory (A type) has the highest frequency. In Spring, the frontal situations have an accentuated fall and the convective situations rise.

The precipitation group 25.4–38.1 mm is predominant (60%): belonging 29% to A synoptic type. The quartil 1 group is predominant (28.6%). However, the association of quartil 2 and A synoptic type is dominant with 14.1%. The maximum precipitation intensity group in 30 min 2.8–18.6 mm/h is predominant (74.4%): in which 37.8% belongs to A synoptic type. The

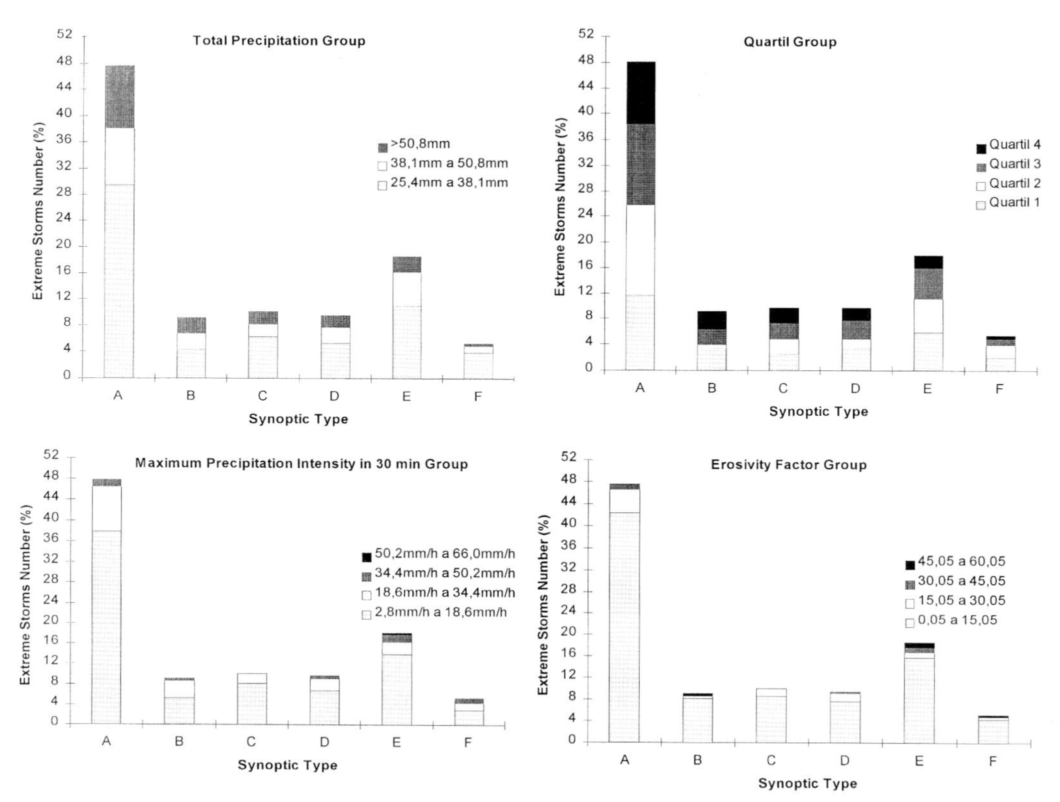

Fig. 7. Synoptic type and precipitation, quartil, maximum precipitation intensity in 30 min and erosivity factor groups.

R group 0.05–15.05 is predominant (86.7%): and 42.4% of this group belongs to A synoptic type (Fig. 7).

In the future the determination of erosivity factor (R) and the characterisation of synotic type of extreme storms will be made for the highest region of Portugal (Serra da Estrela region). This analyse will be based principally on two daily raingauges, Penhas Douradas and Covilhã. These raingauges are located at 1380 m (Penhas Douradas) and 745 m (Covilhã). Due to topographic characteristics of this region, it is expect a change on the predominant synoptic type and on the probability distribution function for the erosivity factor of extreme storms.

Acknowledgements

Marcelo Fragoso research is founded by FEDER and PRAXIS XXI, 2/2.1/CSH/864/95.

References

BRANDÃO, C. (1995): Análise de precipitações intensas. – Dissertação para obtenção do grau de mestre em hidráulica e Recursos Hídricos. IST. Lisboa.

FERREIRA, D.B. (1985): Les depressions convectives du bassin atlantique nor subtropical oriental. – Finisterra, XX, 39: 25–45.

FERREIRA, I., A. FERREIRA & O. SIMS (1985): Projecto de drenagem e conservação do solo no Alentejo. – Direcção de Hidráulica e Engenharia Agrícola, Lisboa, 33 p.

HAHN, G.J. & S.S. SHAPIRO (1967): Statistical models in engineering. – John Wiley & Sons, p. 82–91, p. 302–307.

KIRKBY, M.J. & R.P.C. MORGAN (1980): Soil erosion. – John Wiley & Sons, p. 23–31.

LENCASTRE, A. & F.M. FRANCO(1984): Lições de hidrologia. – Universidade Nova de Lisboa, FCT, p. 342–346.

ROCHA, J., O. BOTELHO,T. MALÓ, N. NUNES & C. MONTEIRO (1986): A floresta e a erosão do solo. – 1° Congresso florestal nacional, Lisboa, p. 2–6.

ROXO, M. J., P.C. CASIMIRO & J.M. MOURÃO (1993): First annual report covering the period 1 January to 31 December 1993.

ROXO, M. J., P.C. CASIMIRO & J.M. MOURÃO (1994): Second annual report covering the period 1 January to 31 December 1994.

ROXO, M.J. (1994): Human impact on soil degradation processes – Serpa and Mértola hills. – Doctoral dissertation.

VENTURA, J.E. (1994): As precipitações no Sul de Portugal (ritmo e distribuição espacial). – Dissertação apresentada à Universidade Nova de Lisboa para a obtenção do grau de Doutor, Lisboa.

WISCHMEIER, W.H. (1978): Predicting rainfall erosion losses-a guide to conservation planning. – U.S. Department of Agriculture, Handbook n° 537.

Addresses of the authors: CLAUDIA BRANDÃO, Instituto da Água, Av. Almirante Gago Coutinho, 30, 1000 Lisboa, Portugal, and MARCELO FRAGOSO, Centro de Estudos Geográvicos, Cidade Universitária, 1699 Lisboa, Portugal.

Z. Geomorph. N.F.	Suppl.-Bd. 115	125–139	Berlin · Stuttgart	Juli 1999

Head scarps and toe heaves

M.-L. Ibsen and E.N. Bromhead, Kingston upon Thames, UK

with 10 figures and 1 table

Summary. Where the historical archival record is poor, evidence of the temporal occurrence and magnitude of mass movement must be derived from the identification and investigation of the landslides themselves. Inevitably, the geomorphologist must learn to recognise the significance of subtle elements of old landslides, in order to correctly anticipate the necessary scope and scale of sub-surface geotechnical investigations without which there can be no confidence in the design of remedial measures.

Features described loosely as head scarps and toe heaves, relating to a variety of landslide types, are described in this paper, with examples drawn from small and medium size landslides. Accurate recognition of these features permits preliminary assessments of slip surface position and shape to be made using general guidelines. Related problems associated with graben geometry are also discussed, and the observations of Cruden et al. 1991, are extended.

It is often the case that the head scarp of a developing landslide is easier to distinguish than the toe heave. The reasons for this are discussed. Movements at both the head and toe of a landslide failure often have virtually the same dimensions, indicating that the material has merely been displaced downslope, yet the resulting features are different in shape, and the head scarp is significantly easier to detect.

Some of the results of this study explain the development of the composite landslide type termed a "slump earthflow". During the production of 'Landslide Recognition' (Dikau et al. 1996), the authors came across the problem of defining the characteristics of a slump-earthflow. It was considered essentially a complex failure comprising an upper rotational section which extends into a mudslide. Clear evidence of this comes from the recognition in the 'slump earthflow' of the toe and head scarp of the constituent slides.

The paper concludes with a number of observations on the degradation with time of head scarps and toe heaves, and the problems of assessing the frequency and magnitude of movement from such geomorphological features.

Zusammenfassung. Wo nur spärliche Archiv-Informationen über Erdbewegungen existieren, müssen die Existenz von Häufigkeit und das Ausmaß an Erdbewegungen aus der Untersuchung von Erdrutschen selbst nachgewiesen werden. Der Geomorphologe muß dabei die Signifikanz von subtilen Spuren alter Erdrutsche erkennen lernen, um den Umfang unterirdischer geotechnischer Untersuchungen korrekt abschätzen zu können, ohne die Präventivmaßnahmen nicht überzeugend geplant werden können.

In dem vorliegenden Beitrag werden die Charakteristika Kopfsekundärabriß und Fußspitze im Zusammenhang mit verschiedenen Erdrutschen und anhand von Beispielen kleiner und mittlerer Erdrutsche beschrieben. Ein genaues Erkennen dieser Charakteristika ermöglicht ein vorläufiges Determinieren der Position von Rutschoberflächen und deren Form. Probleme, die mit der Grabengeometrie zusammenhängen, werden ebenfalls erläutert, und die Beobachtungen von Cruden et al. 1991 werden weitergeführt.

Oftmals ist der Kopfüberhang eines sich entwickelnden Erdrutsches leichter zu erkennen als die Fußspitze. Die Gründe hierfür werden erläutert. Bewegungen an Kopfsekundärabriß und Fußspitze eines Erdrutsches haben häufig die gleichen Dimensionen, was darauf hindeutet, daß das Material sich lediglich hangabwärts bewegt hat. Allerdings sind die dabei entstehenden Formen unterschiedlich, wobei der Kopfüberhang wesentlich leichter auszumachen ist.

Einige Ergebnisse dieser Untersuchung erklären die Entwicklung des zusammengesetzten Erdrutsches, der als "slump earthflow" bezeichnet wird.

0044-2798/99/0115-0125 $ 3.75

Während der Arbeiten zu "Landslide Recognition" (Dikau et al. 1996) stießen die Autoren auf das Problem, die Charakteristika eines "slump earthflow" zu definieren. Es wurde grundsätzlich als komplexe Ruptur betrachtet, mit einer oberen rotierenden Sektion, die sich zu einem Schlammrutsch entwickelt. Ein klarer Beleg dafür ist das Erkennen eines "slump earthflow" in der Fußspitze und dem Kopfsekundärabriß der jeweiligen Erdrutsche.

Der Beitrag schließt mit einigen Observationen zum Abflachen von Kopfüberhangen und Fußspitze Über eine Zeitspanne sowie mit den Problemen, die sich beim Determinieren von Häufigkeit und Ausmaß von Erdbewegungen aus solchen geomorphologischen Charakteristika ergeben

Resumé. La où les archives historiques sont pauvres, la preuve de l'existence d'un mouvement de masse et son amplitude doit être déduite de l'identification et de la recherche des glissements de terrain eux même. Il va de soi que le géomorholoque doit apprendre à reconnaître la signification des subtils élèments des anciens glissements de terrain afin de prévoir avec justesse l'étendue et l'importance des recherches géotechniques souterraines sans lesquelles il ne peut y avoir d'assurances quant à l'étude des mesures correctives.

Les caractéristiques vaquement décrites comme les têtes d'escarpement et les fronts de soulèvement, qui font références à différents type de glissemnt de terrain, seront décrites dans cet article avec des exemples tirés de glissements de petite et moyenne taille. Une reconnaissance exacte de ces caractéristiques permet en utilisant des règles générales une évaluation préparatoire de la surface, de la position, et de la forme du glissement. Les problèmes apparentés, liés à la 'géometrie graben' seront aussi abordés et nous étendrons les observations de Cruden et al. 1991.

C'est souvent le cas que la tête d'escarpement d'un glissement en préparation est plus facile à distinguer que son 'front de soulèvement' est fréquent, et nous en expliqueront les raisons. Les mouvements au sommet et au pied du glissement ont souvent des dimensions semblables, indiquant que les matériaux se sont simplement déplacés vers le bas de la pente, toutefois les caractéristiques résultants possèdent des formes différentes et la tête d'escarpement est sensiblement plus facile à détecter.

Quelques résultats de cette étude expliquent le développement de glissements composite étiquetès sous le nom 'slump earth flow'. Pendant la production de 'Landslide Recognition' (Dikau et al. 1996), les auteurs ont rencontré le problème de la définition des caractéristiques d'un 'slump earth flow'. Cela était considéré comme une 'défaillance' complexe comprenant une parie supérieure en rotation se prolongeant en une coulée de boue. Des preuves nettes de ceci proviennent de la reconnaissance dans un 'slump earth flow' du front et de la tête du glissement le constituant.

Cet article se termine par plusieurs observations sur la dégradation au cours du temps des têtes d'escarpement et des fronts de soulèvement et sur la difficulté d'estimer la fréquence et l'ampleur des mouvements de tels traits géomorphologique.

Introduction

The study of historical records for landslide activity cannot be completely relied upon to give the entire historical context of a particular phenomenon so that future movement can be predicted with confidence. There are frequently gaps in the archive, misunderstandings and inconsistent records (Brunsden et al. 1995, Ibsen & Brunsden 1996). It requires geomorphological and engineering studies to be undertaken to establish surface and subsurface details of landslide problems.

This paper considers a small number of linked topics associated with the identification of elements of landslide morphology. Firstly, those features described loosely as head scarps and toe heaves are outlined. These features permit us not only to delineate the maximum extent of moving zones, but also by appropriate methods allow preliminary assessments of slip surface position and shape. The relative size of head scarps to toe heaves is also covered and a suggestion given as to why sometimes only one or other feature is obvious. Graben geometry, which is a common element of a number of landslide types, is equally indicative of slip surface position and has therefore been given a substantial proportion of this paper.

It is concluded that some of the results of this study explain the development of the composite landslide type termed a "slump earthflow", explaining why it is to some writers a fundamental class of landslide, whereas to others it does not merit separation (DIKAU et al., 1996). Finally, the possibilities and limitations of morphological analysis to the magnitude and frequency of these events is given some discussion.

Features at the head and toe of a landslide

Head scarps

A head scarp is the detachment scar upslope of a landslide. It is usually concave towards the slide, very steep and (particularly where the slide is fresh) bare. Often the scar is in natural, or unslipped, ground (Fig. 1). This is always the case in a "first-time" failure, but where the landslide is a reactivation, the head scarp may be developed in landslide debris from earlier slide events. The overall height of a head scarp is directly proportional to the cumulative displacement of the landslide, and it may be many metres high, or it may be little more than a tension crack. Each episode of movement in a landslide which moves intermittently will add to the height of the head scarp, but since head scarps degrade comparatively rapidly, clear evidence of the head scarp position may be confined to the last event or few events. It is rarely the case that the slickensided slip surface of a landslide extends to the ground surface in this part of a slide and the uppermost part of the detachment scar usually signifies tension rather than shear.

At the head of a landslide, the relative orientation of the slip surface to the ground surface may conceivably be (Fig. 2):

Fig. 1. Isle of Sheppey rotational slide, head scarp showing upper scar in natural ground and tension cracks.

Slip surface steeper than ground profile:
throw is bigger than gap

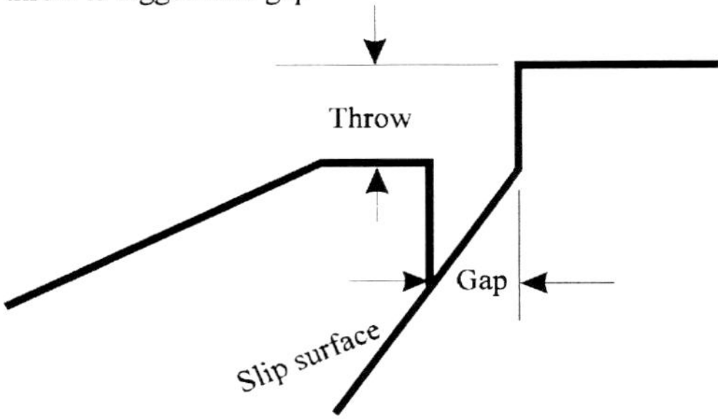

Slip surface sub-parallel to ground profile:
gap is bigger than throw

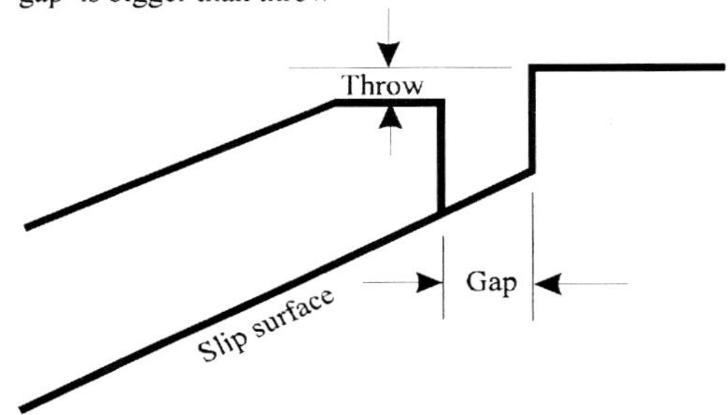

Slip surface flatter than ground profile -
normally improbable:
(except for counterscarp,
where throw direction
is reversed

Gap
Throw
Slip surface
(counterscarp
or antithetic
scarp)

Fig. 2. Head scarps case a, b and c showing head morphometry.

(a) steeper than the ground slope – which is the condition normally expected in first-time failures;
(b) equivalent to the ground slope – which is the common situation where sliding takes place approximately parallel to the ground surface;
(c) flatter than the ground slope – which contradicts the normal expectations of conditions at the head of a slide, and if ever encountered, must be a rarity.

Ground strains, as the slide is initiated, cause tension in the head zone of the slide resulting in cracks. It takes appreciable movement to create a noticeable level difference between the formerly contiguous sides of the gap at ground level. This level difference is the key feature in showing that a tension crack has become a head scarp in the process of development.

Tension cracks may occur in the ground as a result of drying shrinkage or in a moving landslide mass where the particular stress conditions in the soil mass indicate tension. This latter case is common where a landslide passes over a convex-upwards segment of slip surface in moving down a cascade of landslides. Tension cracks may also develop in the intact ground above the head scarp of a developing landslide (Fig. 1).

Toe heaves

A toe is the recognisable morphological feature at the downslope extremity of a landslide. It is generally the accumulation of material which can take on the form of a pile of debris, have a characteristic lobate appearance or simply be a flow of material. At the toe of a slide, the slip surface may break out at a variety of positions and inclinations. Often in sedimentary deposits, landslide slip surfaces follow the outcrop of particular beds, and sometimes, therefore, the slip surface emerges in such a position and orientation that it can be followed with certainty by eye (Fig. 3). These are specifically termed perched toes which frequently emerge high on the sea or river cliff where continued falls of debris may keep the appearance fresh. Other positions and inclinations at lower elevations give rise to an accumulation of debris which obscures the toe breakout position. Occasionally where there are cascades of landslides both are evident (BARTON & COLES 1984).

Toe features are often undetectable because they are resubmerged or have runout (Fig. 3). Where they have been resubmerged or overridden, a number of layers of overthrust and repeatedly sheared accumulated debris may be discovered. In the case of a landslide where the toe of the slide coincides with the toe of the slope, for example, on to a wave-cut platform or base of an excavation, the toe advances maintaining its shape and relationship to the platform and its original breakout position may therefore not be readily detectable.

Often associated with deep seated landslides (and thus by implication, with graben features in the upper part of the slide) is a toe area which involves the upheaval of debris. Fig. 4 reveals the toe feature of a coastal landslide which has been extruded onto the beach (MACQUAIRE 1993). This generally occurs when the slip surface breaks out a long distance in front of the toe of the slope (HUTCHINSON et al. 1980).

Relationship between head scarps and toe heaves

Intriguingly, the slide-head tension cracks and corresponding developing head scarps are often detected long before toe heaves are seen. A landslide phenomena is a displacement of material downslope and the magnitude of the toe heave should be a function of that of the head scarp. This can be shown by using a purely mechanistic approach (Fig. 5). A circular arc slip with rigid body movement produces a head scarp which is a higher than the toe heave despite total displacement in both locations being the same. The height difference is due to the inclination of the slip surface at the head being steeper than at the toe. The collapse of the toe breakout would further reduce the height of the toe heave, also making it a less obvious feature. Real landslides usually exhibit some degree of non-circularity and are rarely rigid-body motions

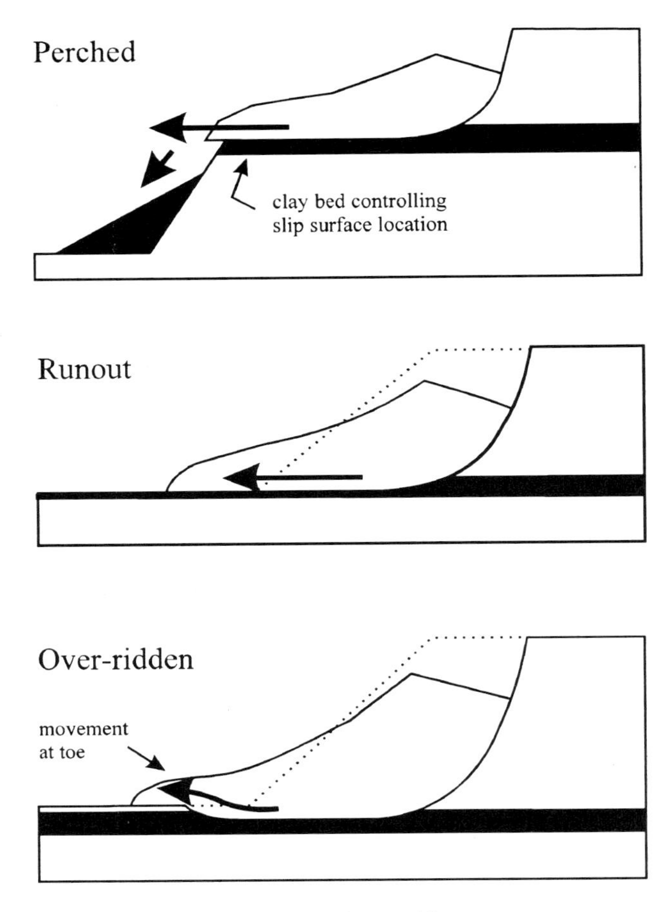

Fig. 3. Toe morphometry types: perched, runout and overidden.

which further increases the difference in height between the head and the toe. Additionally, the head scarp is seen as a sharp feature which is in contrast to the toe heave which is often a diffuse rounded feature.

The size of a head scarp in a first time failure is a function of the brittleness index or the percentage decrease from peak to residual shear strength. Prior to slope failure the material is in equilibrium with the peak strength. Once failure has occurred, the slide mass cannot return to equilibrium except with residual strength acting along the slide surface. Large changes of geometry may occur if the loss in strength or brittleness is appreciable, and correspondingly large slide displacements are experienced, which in turn give rise to large head scarps. For reactivations of a landslide, however, the size of the head scarp may be much smaller, since it is often (but by no means always) the case that total displacement in an individual slide event is small, and the corresponding head scarp features may be on a commensurate scale.

A head scarp is also related to the type of failure. In a rotational landslide as the material moves down the slip surface a backward tilt in the ground surface is created exposing the head

Fig. 4. Active block slide, Northern France, reveals a toe heave where the upheaval of debris has been extruded onto the beach (Photograph courtesy of O. MACQUAIRE).

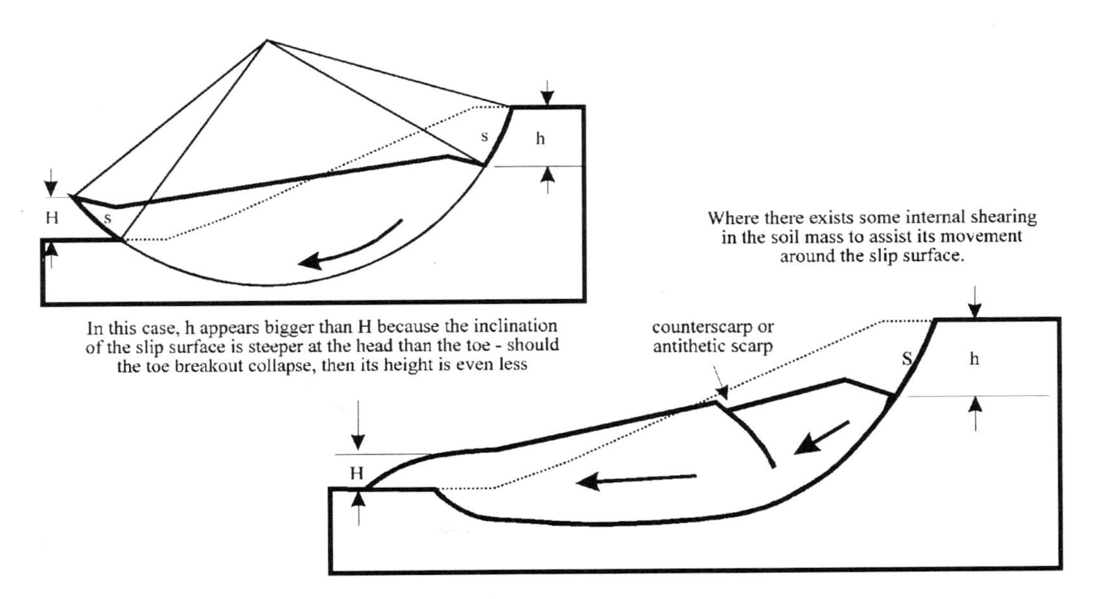

Fig. 5. Head and toe magnitude, in this case h is greater than H due to the inclination of the slip surface at the head being steeper than at the toe. The difference in magnitude is further increased by a non-circular slip surface and/or the collapse of the toe breakout.

scarp (Fig. 6). In a text-book rotational slide (never ever found in practice!) no gap appears between the top of the slide mass and the head scarp. For a translational movement, however, the material which has slipped detaches itself forming a gap and the actual head scarp is the difference in level between the former ground surface and the present surface.

Graben features and morphometry

Graben features

Where significant amounts of ground rupture are present in the vicinity of the head of a landslide, a characteristic feature known as a *graben* is frequently found. A graben, named after the structural geology term, signifies a zone of relative subsidence at the head of a landslide bounded by two rupture surfaces (Fig. 7). This forms a platform or depression which may have a slight backward tilt significant of the curvature of the upper slip surface. It develops in response to the kinematics of the landslide mass moving along its slip surface. Where a slip surface has, for instance, a steeply rising section and a flatter, predominantly bedding-controlled sole (Fig. 8), the fraction of the slip mass which descends the steep part must move downwards relative to the laterally-moving main body of the slide. Normally this produces a distinct

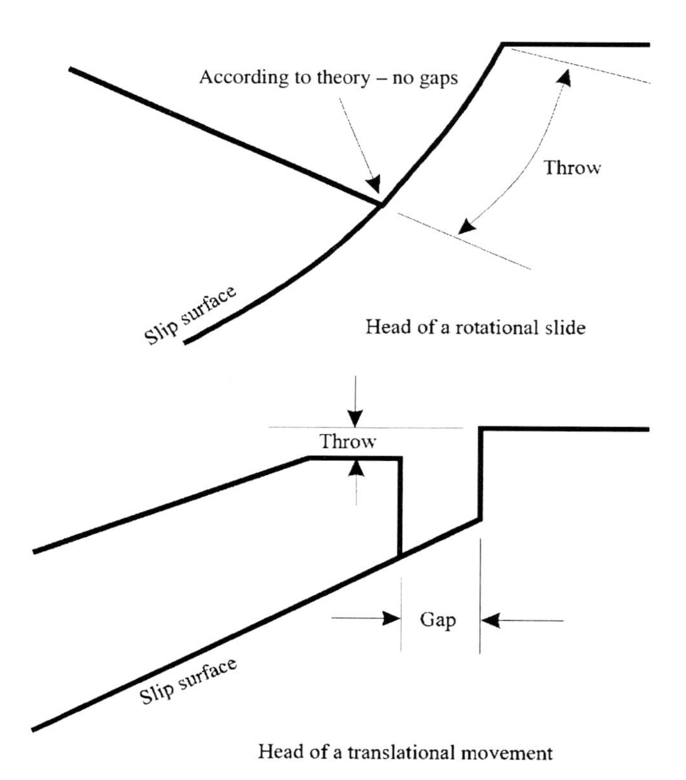

Fig. 6. Head scarps and the type of failure (a) a rotational landslide; (b) a translational movement.

Fig. 7. Active block slide, Northern France, graben feature denoted by the relative subsidence at the head of the landslide and bounded by two rupture surfaces. (Photograph courtesy of O. MACQUAIRE).

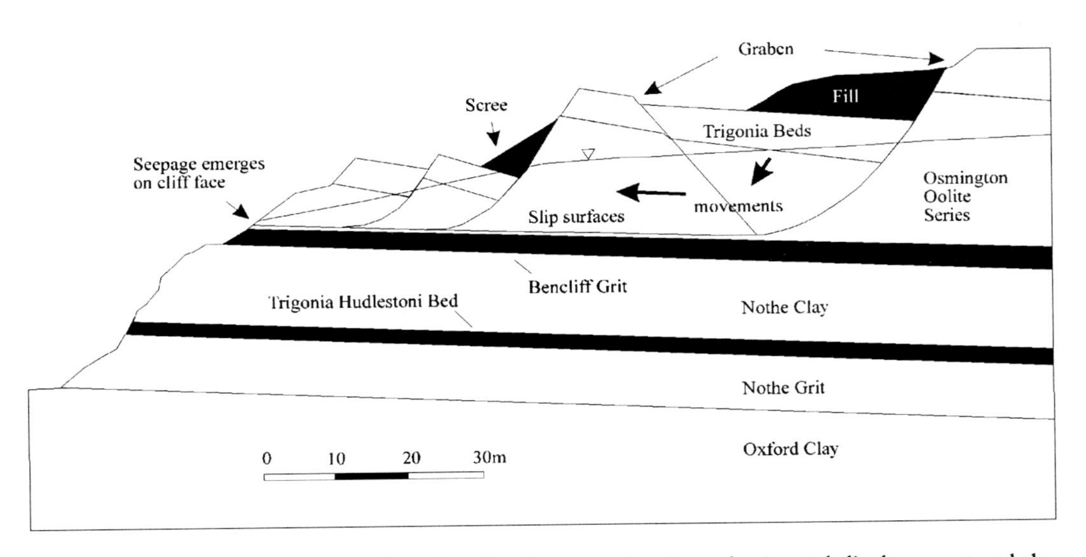

Fig. 8. Red Cliff, East of Weymouth, graben development showing a horizontal displacement and the subsequent collapse.

intact ridge some distance from the rear scarp. The differential vertical movement causes rupture in the slide mass, and the emergence of this rupture surface at ground level provides a shear surface complementary to the developing head scarp sometimes termed a *counterscarp* or an *antithetic scarp* (CRUDEN et al. 1991). Landslide types which may exhibit graben development include block slides, slab slides and lateral spreading (DIKAU et al. 1996). Referring back to Fig. 2, the counterscarp is an instance where the slip surface emerges reversing the normal direction forming a step upwards in the overall movement. Such a feature may also occur in a multiple toppling failure, but this is not being dealt with in this paper.

A clear graben feature cannot always be detected as it is dependent on the curvature of the upper slip surface, the material involved, the magnitude of movement and the precise position in the slope profile where the counterscarp breaks out. The type of translational slide seen at St Catherine's Point (HUTCHINSON et al. 1991), which is equally referred to in the geotechnical literature as a non-circular or flat-soled bedding-controlled landslide, and the failure at the earthworks of the Carsington Dam (SKEMPTON & COATS 1985) show a typical rotational failure in the upper section of the landslide. These rotations are combined with translational movement on the lower part of the failure by evidence of a counterscarp. The upper rotational part of such a landslide is analogous to the graben, although the counterscarp will no longer appear to be a dominant feature. This paper has considered such landslide types as indistinguishable from conventional graben landslides.

Graben development

As noted above, grabens in landslides, take their characteristic morphology from the changes in direction in the slip surface shape, and the abruptness or otherwise of those changes. A model for graben geometry was developed by CRUDEN et al. 1991, which stated that it was possible to estimate the depth of the slip surface, from the original ground surface, using the initial graben width.

Observations of ten landslides in Alberta, Canada, indicated that the depth to the translational slip surface was approximately 1.1 times the original graben width. In order to extend the CRUDEN et al. (1991) correlation a further set of landslides exhibiting graben features, from Britain and on the Northern French coast, have been considered. Each of these cases is fully described in the literature from which key elements have been obtained. Table 1 list the cases, their graben geometry and the source documents. In most cases the graben geometry is fully defined by subsurface investigations. Landslides with more than one graben feature such as Sevenoaks, Kent, in Britain (WEEKS 1970) have been excluded. Reanalysing the data with the additional cases shows that the association between width and depth given by CRUDEN et al. (1991) is a poor fit with the additional data. Using a new regression analysis, the relationship can be explained more accurately (Fig. 9).

The revised correlation has been used to predict the slip surface position of the famous landslide at Bindon on the Devon/Dorset border in the UK. This failure occurred on Christmas Eve in 1839 and incorporated an area known as Goat Island which became detached from the mainland, creating the graben feature known as the Chasm. It is the only known landslide to have a piece of music written about it: Ricardo Linter's 'The Landslip Quadrille', which was used by the local population to celebrate the event. The landslip s notoriety is dependent upon the presence of two famous geologists and some local artists who reported the extraordinary

Table 1. Graben width and slip surface depth for various landslides. Canadian landslides are taken from CRUDEN et al. 1991.

	Location	Depth to slip surface (D) m	Width of graben (W) m	D/W
1	St Catherine's Point, Isle of Wight, UK (HUTCHINSON et al. 1991)	164.7	205.9	0.80
2	Le Bouffay, N. France (DIKAU et al. 1996)	78.0	91.8	0.85
3	Red Cliff, Dorset, UK (BROMHEAD 1992)	25.7	42.9	0.60
4	Miramar, Herne Bay, UK (BROMHEAD 1992)	28.1	42.6	0.66
5	Lesueur, Alberta, Canada	32.0	12.2	2.62
6	Devon, Alberta, Canada	12.8	12.2	1.05
7	Edgerton 1, Alberta, Canada	38.1	51.8	0.74
8	Quesnell, '74, Alberta, Canada	5.2	5.8	0.90
9	Quesnell, '73, Alberta, Canada	6.7	7.0	0.96
10	Hardisty, Alberta, Canada	11.9	7.6	1.57
11	Mill Creek, Alberta, Canada	19.8	15.2	1.30
12	S. Portal, Alberta, Canada	23.0	32.0	0.72
13	Edgerton 2, Alberta, Canada	38.0	21.3	1.78
14	Tolman, Alberta, Canada	8.5	10.1	0.84
15	Bindon, Devon/Dorset, UK (PITTS 1974)	*296.5* (Prediction of depth from width)	368.4	---

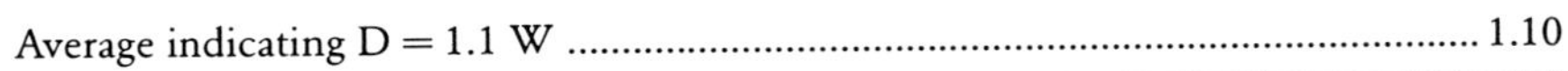

Average indicating D = 1.1 W 1.10

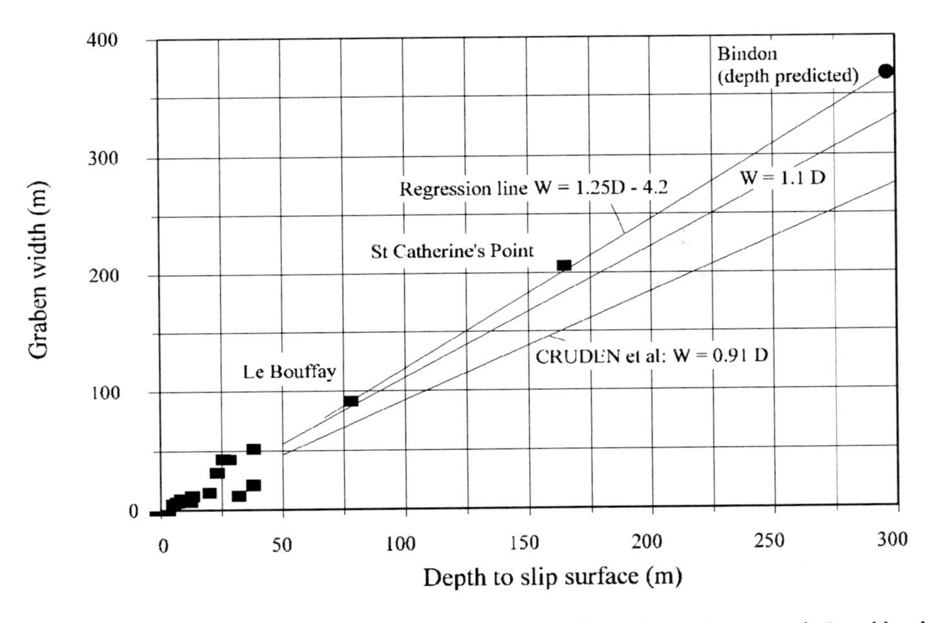

Fig. 9. Relationship between graben width and depth to slip surface using various translational landslides.

event, initiating debate as to its formation (CONYBEARE 1840). Using the regression formula and the knowledge of the width of the graben a slip surface depth of 95 m has been predicted (Table 1). According to the geological section of the landslide this lies on the intersection of the Gault and Upper Greensand which corresponds with the reconstruction given by PITTS & BRUNSDEN (1987).

In view of the usefulness of the improved CRUDEN et al. (1991) model it is clear that there is value in this approach and may have some merit in permitting the engineer or geomorphologist investigating such a landslide to approximate the required depths for boreholes and instrumentation. However, it should be noted that there is some scatter at the lower end of the scale and more information on this morphometric model would be useful.

The slump-earthflow

The term *slump-earthflow* was originally used by ROGERS in 1929 and some ten years later SHARPE (1938) continued to use the term. The landslide type described by the term *slump-earthflow* is clearly a *complex landslide* (CRUDEN & WP/WLI 1993) where two types of movement are exhibited in sequence as the mass moves downslope. The initial failure involves rotational slides, often with a bedding-controlled basal slide surface where the slides occur in low-dip sedimentary rocks, with the toe overthrust breaking down into a mudslide or perhaps a flow (Fig. 10). One can discern the two head scarps of the individual movements, but only one toe zone, since the toe of the initial movement is obscured. The initial rotational slide thus has a form of the *perched* slip surface (Fig. 3). In common with most slides and all flows of debris, the toe zone of the system may be repeatedly overthrust and re-sheared, with many small individual toes detectable. In most cases, the lower slide element is a *slide*, not a *flow*: basal shear surfaces being in evidence if they are sought in trial pits, or being evident to the geomorphologist

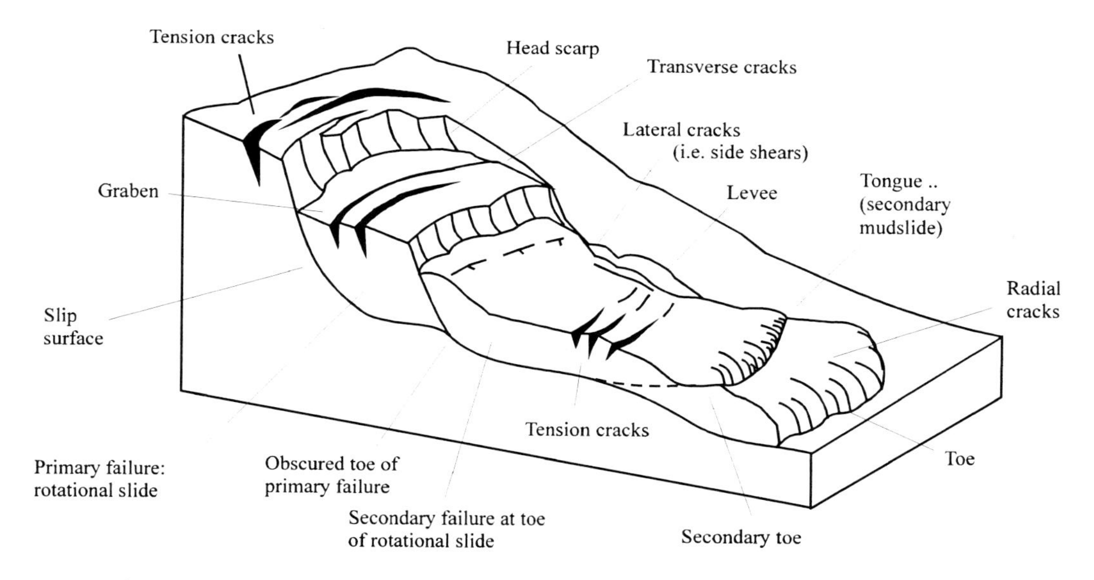

Fig. 10. Slump-earthflow exhibits three distinct morphological sections a head, body and toe.

where they emerge as lateral shears to each side of the slide mass. The term slump on its own is obsolete.

As the landslide system described by the term *slump-earthflow* is initiated, it is clearly *not* a *composite landslide* (CRUDEN & WP/WLI 1993) where two types of movement are exhibited simultaneously as the mass moves downslope. However, once formed, it may be impossible to disentangle which element moves first, and the system may then have to be described as a *complex landslide.*

It is unhelpful to have a landslide type described by two terms, one of which is obsolete, and the other (usually) incorrect. Similarly, the allocation of a specific name to a phenomenon which involves two clearly separable types of landslide mechanism, when other complex and composite landslide systems are described in terms of their separate elements, is illogical. Therefore, the Authors are firmly of the opinion that complex and composite landslides such as the 'slump-earthflow' should be removed from the taxonomy of landslide types.

Frequency and magnitude identification

Determining the frequency and magnitude of individual landslide events is difficult to resolve simply from morphological analysis. For first-time failures the measurement of time and size can be relatively accurate, however with the reactivation of landslides these criteria require the distinction of events. This is usually possible with continuous monitoring and may be determined by morphological analysis by discovering the number of head scarps and/or the number of toe heaves in any given sequence.

In terms of magnitude first-time events are usually larger than the reactivations, since the movement is controlled by the brittleness of the material and its change from peak to residual strength. Reactivations are normally a function of the perturbation of the landslides equilibrium and only when this is of considerable significance will the event be greater than its initial failure. This is usually associated with the rare event, for example, at Folkestone Warren in 1915 (HUTCHINSON et al. 1980) and Gore Cliff in 1978 (BROMHEAD et al. 1991), where the perturbation of equilibrium was larger than the mechanisms acting within the landslide (HUTCHINSON 1987). The magnitude of any given event could be derived from the head scarp, but in composite landslides it would be necessary to integrate across all the head scarps, graben features and toe heaves in order to reflect the true magnitude of the failure.

Frequency can be determined when the magnitude of an event is significant enough to produce identifiable scarps. Continuous or near continuous movements of small magnitude do not produce successive head scarps, since the rate of degradation quickly obscures these events and over time the head scarps gently degrade to lower angles.

Identifiable head scarps degrade naturally through a process of free degradation (HUTCHINSON 1967). The rapid colonisation of certain types of vegetation may be related to quantifying periods of events. Species with shallow roots and used to a dry environment will tend to accumulate on the head scarp. Although it should be noted that the head scarp frequently stays bare for significant periods of time due to its sheer inclination. The toe area, in contrast, will normally include numerous plants which enjoy plenty of water and have deep root systems. Trees or large shrubs are often good indications of the way in which the slip surface is evolving downslope. A backward tilt in the head area would be significant of a curved upper slip surface and a forward tilt at the toe would indicate curvature at the lower end. It should be noted that toe heaves are often destroyed very rapidly in marine or coastal environments and are therefore of significantly lesser use in assessing frequency of occurrence. However, on inland sites, such as Sevenoaks, Kent, in Britain, the long term preservation of toe features in a relatively fresh state may give a misleading impression of frequency of movement.

Translational failures with associated graben features are usually particularly sensitive at the toe and any slight toe erosion can trigger failure involving deep settlement at the head. Different events and their magnitude can be identified by analysing the head scarp and counterscarp changes in material. The degradation of a graben feature usually involves the pinnacles or ridges which form the counterscarp to become rounded with age and gradually diminish in size.

Conclusions

This paper has shown that it is possible to make preliminary assessments of slip surface position and shape from merely observing the head scarp and possible toe feature of a landslide. It has also presented a landslide termed the slump-earthflow which is a composite landslide with two head scarps, but only one toe zone. From the identification of any head scarp and toe heave it is possible to gain some idea of the scope and scale of the sub-surface geotechnical observations which are required for further landslide investigation.

Acknowledgements

The lead author would like to thank the Engineering and Physical Sciences Research Council for their financial support via a generous grant to support a project concerned with assessing the properties of slip surfaces. Both authors would like to thank our numerous colleagues who through discussions generated the ideas for this paper.

References

BARTON, M.E. & B.J. COLES (1984): The characteristics and rates of the various slope degradation processes in the Barton Clay cliffs of Hampshire. – Q. J. Eng. Geol. **17**: 117–136.

BROMHEAD, E.N. (1992): The Stability of Slopes. – 2nd Edition, Blackie, Glasgow.

BROMHEAD, E.N., M.P. CHANDLER & J.N. HUTCHINSON (1991): The recent history and geotechnics of landslides at Gore Cliff, Isle of Wight. – In: CHANDLER, R. J. (Ed.): Slope Stability Engineering. – Applications and Developments, Proceedings of the International Conference for Civil Engineers, Isle of Wight, 15–18 April, Thomas Telford, London, 189–196.

BRUNSDEN, D., M.-L. IBSEN, M. LEE & R. MOORE (1995): The validity of temporal archive records for geomorphological processes. – Quaestiones Geographicae, Special Issue, **4**.

CONEYBEARE, W.D. (1840): Extraordinary landslip and great convulsions of the coast at Culverhole point, near Axmouth. – New Philos. Journal, **29**.

CRUDEN, D.M., S. THOMSON & B.A. HOFFMAN (1991): Observation of graben geometry in landslides. – In: CHANDLER, R.J. (Ed.): Slope Stability Engineering – Applications and Developments. – Proceedings of the International Conference for Civil Engineers, Isle of Wight, 15–18 April, Thomas Telford, London, 33–35.

CRUDEN, D.M. and the Working Party on the World Landslide Inventory (WP/WLI) (1993): Multilingual Landslide Glossary. – BiTech Publishers Ltd., Richmond BC, Canada.

DIKAU, R., D. BRUNSDEN, L. SCHROTT & M.L. IBSEN (Eds., 1996): Landslide Recognition. – Wiley, Chichester.

HUTCHINSON, J.N. (1967): The free degradation of London Clay cliffs. – Proc. Geotech. Conf. on Shear Strength of Natural Soils and Rocks. NGI, Oslo, **1**: 113–118.

HUTCHINSON, J.N. (1987): Mechanisms producing large displacements in landslides on pre-existing shears. – Mem. Geol. Soc. China **9**: 175–200.

HUTCHINSON, J.N., E.N. BROMHEAD & M.P. CHANDLER (1991): Investigation of the coastal landslides at St Catherine s Point, Isle of Wight. – In: CHANDLER, R.J. (Ed.): Slope Stability Engineering – Applications and Developments. – Proceedings of the International Conference for Civil Engineers, Isle of Wight, 15–18 April, Thomas Telford, London, 151–161.

HUTCHINSON, J.N., E.N BROMHEAD & J.F. LUPINI (1980): Additional observations on the Folkestone Warren landslides. – Q. J. Eng. Geol. **13**: 1–31.

IBSEN, M.-L. & D. BRUNSDEN (1996): The nature, use and problems of historical archives for the temporal occurrence of landslides, with specific reference to the south coast of Britain, Ventnor, Isle of Wight. – Geomorphology **15**: 241–258.

MACQUAIRE, O. (1993). The Bessin Cliffs. – In: FLAGEOLLET, J.C. (ed.): Prevention of coastal erosion and submersion risks: knowledge of the risk, impact studies with a view to protection work. – EC CERG publication.

PITTS, J. (1974): The Bindon landslip of 1839. – Proceedings of the Dorset Natural History and Archaeological Society **95**: 18–29.

PITTS, J. & D. BRUNSDEN (1987): A reconsideration of the Bindon landslide of 1839. – Proc. Geol. Assoc. **98**: 1–18.

ROGERS, J.K. (1929): A type of landslide common in clay terraces (Abstract). – Ohio Journ. Sci. **29**: 167; also in the Ohio Acad. Sci. Proc. **8**: 304.

SHARPE, C.F.S. (1938): Landslides and Related Phenomena. A study of mass movements of soil and rock. – New York, Columbia University Press.

SKEMPTON, A.W. & D.J. COATS (1985): Carsington dam failure. – Failure in Earthworks, Thomas Telford, London, 230–220.

WEEKS, A.G. (1970): The stability of the Lower Greensand Escarpment in Kent. – PhD Thesis, Surrey University.

Address of the authors: M.-L. IBSEN and E.N. BROMHEAD, School of Civil Engineering, Kingston University, Penrhyn Road, Kingston upon Thames, Surrey KT1 2EE, U.K.

Z. Geomorph. N.F.	Suppl.-Bd. 115	141–155	Berlin · Stuttgart	Juli 1999

Frequency and magnitude of landsliding: fundamental research issues

M.J. Crozier and T. Glade, Wellington and Bonn

with 5 figures and 1 table

Summary. Some fundamental research issues related to the application of frequency-magnitude analysis to landslides are discussed. It is shown that there are marked differences between the hillslope system and the fluvial system from which many of the original frequency-magnitude concepts were developed.

Parameters used to characterise the episodic behaviour of landsliding are presented and discussed with respect to their value in addressing specific research questions. Frequency-magnitude analysis should distinguish between first-time failures and reactivations of existing landslides and be applicable to both temporal and spatial distributions.

Methodological approaches to the establishment of temporal frequency and magnitude are presented and illustrated. Both empirical and deterministic methods are discussed. In particular, advances in empirical modelling are used to show the extent to which the probability of landslide occurrence can be determined within a regional context. This method requires recognition of a range of landslide-triggering rainfall thresholds and these in turn rely on the establishment of a comprehensive database on both climate and landslide history.

Frequency-magnitude analysis can also be applied to the spatial distribution of landslides. Whereas smaller landslides are more common than larger landslides, geomorphic work and land-forming dominance, in some areas, is achieved by the largest landslides on record.

Zusammenfassung. Einige grundlegende Überlegungen zur Frequenz-Magnituden Analyse von Hangrutschungen werden diskutiert. Es wird gezeigt, daß entscheidende Unterschiede zwischen dem Hangsystem und einem fluvialen System, von dem ursprünglich die Frequenz-Magnituden Analyse entwickelt wurde, bestehen.

Parameter, die das episodische Auftreten von Hangrutschungen charakterisieren, werden dargestellt und in Bezug auf ihre Aussagefähigkeit für bestimmte wissenschaftliche Fragestellungen diskutiert. Frequenz-Magnituden Analysen sollten zwischen Erstauslösung und Reaktivierung bereits existierender Hangrutschungen unterscheiden und sowohl für räumliche als auch zeitliche Fragestellungen anwendbar sein.

Methodologische Ansätze einer zeitlichen Frequenz-Magnituden Analyse werden aufgezeigt und illustriert. Empirische und deterministische Methoden werden diskutiert. Im Besonderen werden neue Erkenntnisse in der empirischen Modellierung benutzt, um die Wahrscheinlichkeit des Auftretens von Hangrutschungen im regionalen Maßstab aufzuzeigen. Diese Methode erfordert die Kenntnis einer Reihe von hangrutschungsauslösenden Niederschlagsschwellenwerten, welche wiederum nur durch die Erstellung einer umfassenden Datenbank, die sowohl klimatische Zeitreihen als auch historische Information über Hangrutschungen erhalten muß, erlangt werden kann.

Eine Frequenz-Magnituden Analyse kann auch auf die räumliche Verteilung von Hangrutschungen angewandt werden.

Obwohl kleinere Hangrutschungen häufiger Auftreten als große Hangrutschungen, wird in einigen Gebieten die geomorphologische Arbeit und die Landformung dominiert von den größten historischen Hangrutschungen.

0044-2798/99/0115-0141 $ 3.75

1 *Introduction*

Over the last few years, frequency-magnitude analysis, originally developed in the context of fluvial geomorphology, has been applied to a wide spectrum of geomorphic processes. The episodic behaviour of landsliding lends itself well to this approach. For example, knowledge of the frequency and magnitude behaviour of landsliding provides the basis for:

- Characterising landslide hazard (CROZIER 1996)
- Assessing rates of geomorphic work (TRUSTRUM et al., this volume)
- Identifying significant temporal change in the environmental factors which affect landsliding, such as, climate, seismic activity, vegetation, and land use. (CROZIER 1997)

To carry out these activities successfully a good quality database is required. In this respect, landslide studies are at a distinct disadvantage compared with more data-rich fields such as fluvial geomorphology. Most studies of landslide frequency and magnitude have had to develop their own primary data and, because of the effort required, are often limited in areal and temporal extent. There are few standardised systems in place for recording landslide activity and even fewer long-term records available. Current discussion on suitable recording protocols (GLADE & CROZIER 1996) has focussed attention on which landslide parameters are of value for specific management purposes and how databases should be structured and maintained.

Traditional approaches to determining frequency and magnitude have centred on fluvial processes (WOLMAN & MILLER 1960) and have dealt with 'frequency' in terms of discrete hydrological events and 'magnitude' by measures of volume or mass of water and sediment associated with those events. They assume a direct relationship between the hydrological processes and the geomorphic response. In contrast, the relationship between a landslide and its initiating process (such as a rainstorm) is further complicated by the intervening behaviour and properties of the hillslope system.

The purpose of this paper is to raise some of the fundamental issues relating to the episodic behaviour of landsliding. Concepts and parameters are defined with respect to the anticipated end-use for the information. Different methodologies for obtaining and analysing information on landslide activity are reviewed. Finally, examples are presented which indicate the extent of our knowledge on the frequency-magnitude behaviour of landslides.

2 *Concepts and parameters*

As argued elsewhere (CROZIER, this volume), frequency and magnitude are just two of the many parameters which are required to characterise adequately the behaviour of an episodic geomorphic process. Bearing in mind the main purposes for studying episodic behaviour of landslides (assessment of hazard, geomorphic work, and environmental change), a comprehensive and purpose-oriented characterisation of episodic behaviour is required. Ideally, this should include information on starting and terminating thresholds, duration, and areal extent of occurrence as well as an appropriate characterisation of frequency and magnitude.

2.1 *Events, frequency, and duration*

'Frequency' of landsliding might initially appear to be a straightforward concept. However, landslides can occur either as 'first-time' failures or as subsequent reactivations. Whereas the effects of landslide events are important irrespective of their origin, the physical conditions and starting thresholds are quite different. First-time failures, in a given material, need to overcome higher material strength values than in the case of reactivations, as represented by differences between peak and residual strength respectively. Determining frequency from the behaviour of the triggering agent (a common approach) therefore needs to take these different strength-related thresholds into account. Reactivation of deep-seated landslides appears to be more frequent than the occurrence of first-time failures. REIMER (1995), for example, has noted that two-thirds of the landslides entering reservoirs are reactivations of existing slides. Because of this difference in the frequency of movement, it is common engineering practice to be much more concerned about the hazard presented by existing landslides than the probability of occurrence of first-time failures. By contrast, situations have been described involving shallow, regolith landslides where first-time failures are more common than reactivations. In these situations, special geomorphological conditions usually apply such as widespread destabilisation of the regolith by recent deforestation. The landslide deposits are either removed from the hillslope system or deposited at depths and angles which preclude further regolith failure (CROZIER & PRESTON 1998).

Thus there are two classes of landslide event to which frequency analysis can be applied; first-time failures and reactivations. Inventories of landslide activity should distinguish between these two types of movement, because of the fundamental geotechnical differences which apply.

Correspondingly, other parameters, such as duration, may refer to the first-time failure or to the reactivation. Differences in duration between the first-time failure and reactivations may be considerable. For example, MCSAVENEY et al. (1992) have shown that, in New Zealand schist terrain, certain existing deep-seated landslides which produce periodic seasonal reactivations may have initially failed around 250,000 years BP. No evidence was found for similar landslides occurring within the last 15,000 years. Thus, the frequency and duration of the initial first-time failure can be markedly different from that of the reactivations. Landslides clearly have distinctive lifetimes. Shallow regolith failures are often instantaneous, with a lifetime measured in seconds or minutes. Other landslides, after initial failure, may be subject to continual movement for hundreds of years before they cease movement altogether (Fig. 1).

Whereas the occurrence of a single landslide may be considered an event, the term 'landslide event' can have a broader connotation. For example, a single triggering event, such as an intense rainstorm, may produce up to hundreds or even thousands of individual landslides over a wide area (CROZIER et al. 1980, GLADE 1997) (Fig. 2). Consequently, frequency and magnitude analysis can also be applied to the spatial distribution of landslides occurring in a single triggering event as well as to temporal distributions. Frequency-magnitude relationships derived from spatial analysis may also be analysed in terms of the amount of geomorphic work achieved (INNES 1985).

Fig. 1. The deep-seated landslide (foreground) is subject to periodic reactivations, particularly during wet conditions of winter. (Photo: National Water and Soil Conservation Authority, New Zealand).

2.2 *Magnitude*

The term 'magnitude' is a measure of scale which, in the strict sense, refers to the mass or volume of material moved in an event. However, the scale of an event may be more appropriately represented by some other measure of magnitude, depending on the overall purpose of the study. For example, in conventional hazard and risk analysis, hazard is defined as 'the probability of occurrence of a given magnitude of event' (VARNES 1984). In this instance, where the intention is to measure danger or threat, it may be better to represent magnitude by some parameter that conveys the scale of danger more accurately than volume. Landslides with similar volumes may be quite different in terms of their hazard potential. For example, two landslides might both have a volume of 5,000 m^3 but one is 10,000 m^2 in area and only 50 cm in depth and the other 1000 m^2 in area and 5 m in depth. Clearly the deeper one is a much more serious threat to buildings and infrastructure than the shallow one, which may, on the other hand, be a greater problem in terms of soil degradation. The parameter used to represent 'magnitude' in the 'frequency-magnitude' calculation of hazard should therefore be one matched with the overall objective of the analysis. Parameters representing hazard potential ('impact

Fig. 2. Shallow first-time landslides triggered during the one rainfall event, North Island, New Zealand. (Photo: Garth Eyles).

characteristics' (CROZIER 1996), besides depth and volume, include areal extent of scar, distance of runout, velocity, duration, and degree of disruption. The frequency distribution in time of any of these parameters is largely unknown and constitutes an important area of future landslide hazard research. The lack of knowledge of frequency-magnitude behaviour of landslides has meant that, to date, most so-called landslide hazard maps simply rank terrain in terms of its susceptibility (CROZIER 1995).

Landslide events producing multiple landslide occurrences (clusters or swarms) are an important phenomenon in many parts of the world. In a given region, multiple occurrence events appear to be much less frequent than single events. In the mountains around the Japanese city of Kobe, for example, individual shallow landslides are estimated to have a return period of about 10 years whereas swarms occur with a return period of 30 years (OKUNISHI et al., in press).

To assess the frequency and magnitude with which multiple occurrence landslide events occur, requires a different set of parameters in addition to those used for individual landslides. The 'magnitude' of this type of event can be represented by measures of severity of landsliding within the affected the area. This first requires definition of 'affected area'. Theoretically, this measure is defined as the landsurface area within an envelope enclosing the cluster of landslides that occurred during the event. In practice, this is difficult to define because of the presence of landslides lying outside the main cluster of occurrences. Nevertheless, the parameter, 'affected area' is seen as a useful measure of event magnitude. In the case of earthquake-triggered landslides, KEEFER (1984) has demonstrated that there is a close relationship between the magnitude of the forcing process (earthquake magnitude) and the magnitude of the affected area.

Initial definition of the 'affected area' then allows other associated measures of magnitude to be employed. These include: the percentages of the affected area subject to erosion, transport and deposition (respectively or collectively), density of landslides (number per unit area), and material displaced per unit area (sometimes inaccurately expressed as a lowering or denudation rate of the 'affected area'). Alternatively, area and volume measurements may be expressed in absolute terms. The use of some of these measures is illustrated in Table 1.

Table 1. Magnitude parameters for multiple occurrence landslide episodes derived for three regions, North Island, New Zealand (GLADE 1997).

Magnitude parameter	Hawke's Bay	Wairarapa	Wellington
Study area (km^2)	49.04	122.72	25.16
Total landslide scar area (km^2)	0.16	0.81	0.10
Scar area (%)	0.32	0.66	0.40
Number of landslides	19,189	6,080	1,027
Landslide density (no/ha)	3.90	0.50	0.40
Mean scar area (m^2)	8.30	132.70	97.90
Mean volume (m^3/m^2)	2.20	0.73	0.78
Total volume displaced (m^3)	337,726	588,981	78,430
Mean scar volume (m^3)	4.08	96.90	76.4
Study area denudation (mm)[1]	6.80	4.8	3.1

[1] Denudation refers to specific landslide episodes.

2.3 *Frequency-magnitude relationships: rivers versus hillslopes*

Early frequency-magnitude research was driven by the debate over *uniformitarianism* versus *catastrophism* with respect to the geomorphic work done in the landscape. WOLMAN & MILLER (1960) thus designed their work to isolate the 'dominant discharge' which, measured by the product of its frequency and magnitude, was the class of event that achieved the largest amount of work in the period of record. Theoretically, the dominant landslide magnitude is also obtainable for landsliding by obtaining the product of frequency and magnitude of landslides in the record on either a spatial or temporal basis.

In practice, however, there are number of differences between the event data that is obtained from a given channel reach in a fluvial system and the event record of landslides. First, landslide events are much less frequent than streamflow events and, in order to obtain a comparable database, the landslide observational period must be longer. Consequently, as the required observational period increases, there is a greater chance that the evidence of smaller events will be obliterated or overlooked thus affecting the overall frequency distribution. Second, unlike river gauging stations, the location of landslide sites generally varies with each event thus introducing variation in the suite of controlling factors, including those which influence landslide magnitude. Finally, the concept of geomorphic work, which in the discharge event is a function of magnitude and frequency, may not be adequately represented by these factors in the case of landsliding. The reason for this is that the magnitude of a landslide actually represents the mass of material 'displaced' and not true work in terms of mass multiplied by distance moved. Some landslides, although large in volume, may be displaced only a few centimetres and remain in that position for hundreds of years thus playing only a small role in landform evolution.

A stable relationship between frequency and magnitude and the reliable designation of 'dominant' or 'effective' events can only be achieved if the system is in a state of equilibrium. While this may prevail for certain types of channel system over short periods, it is rarely the case in a landscape affected by landsliding. Landsliding is a self-annihilating process. Its very action tends to destroy the conditions which favour its initiation. Deep-seated landslides reduce the critical height or slope as a consequence of their action. In addition there is evidence (PRESTON, this volume) that regolith landslides events successfully exhaust the available material and alter terrain conditions with each event, changing the triggering-thresholds with time. Thus the scale of mass movement initiated by a given magnitude of triggering rainstorm or seismic shock may change with time. The rate at which thresholds vary with time is not well understood.

A study of south Taranaki, North Island, New Zealand provides an indication of how the 'effective' landslide event may change in the course of landform evolution (CROZIER & PILLANS 1991). In this area, continual uplift and exposure of marine sediments, at a rate of up to 1 mm per year for over two million years, has produced a landscape of young marine terraces at the coast and deeply dissected, older terraces and hill country inland. By comparing terraces of different ages it is possible to substitute space for time. This shows that, in the first half million years of slope development, modal valley side slope angles (approximately 30 degrees) are produced by shallow rainstorm-triggered landslides continually removing the regolith as it forms. Landslide activity of this type occurs somewhere in the region about every five or six years. However, for terrain that has experienced longer periods of uplift and fluvial dissection, sufficient relief is formed to produce stress within the bedrock itself. This allows seismic activity to trigger deep-seated landslides. These landslides occur at a frequency of hundreds to thousands of years; they are high magnitude-low frequency events compared with the effective events operative in the earlier stages of landform evolution. The deep-seated landslides destroy the shallow landslide equilibrium slopes and produce an entirely different topography. As the 'residence time' of landslide topography produced in this way (relaxation time) is greater than the return period for deep-seated landslide events (transient form ratio – BRUNSDEN & THORNES 1979), the topography is evidently undergoing irreversible change.

3 *Methods of investigation*

There are two fundamental approaches to determining the temporal frequency-magnitude behaviour of landslide activity, empirical and deterministic. The deterministic approach is still in its infancy and relies on predetermined relationships between the external triggering mechanism and the balance of stresses within the slope. The empirical approach, on the other hand, has been developed over a longer period and relies on the establishment of a record of landslide occurrence.

Methods for establishing the required record of landslide activity can be classed into two main groups, 'on-site' and 'off-site'. The on-site group includes:

- direct monitoring of hillslopes and existing landslides. This method is generally used to record reactivations rather than first-time failures and involves the use of sophisticated sensors such as strain gauges, inclinometers, and precise survey of surface and subsurface positional indicators (GILLON & HANCOX 1991).
- survey and direct field dating of actual landslide remnants. Many standard relative and absolute dating techniques have been successfully applied to landslides (LANG et al., in press, MCSAVENEY et al. 1991, BULL 1996, CROZIER et al. 1995, WEISS 1988).
- the use of documentary sources, including sequential air and ground photo coverage (BRUNSDEN & IBSEN 1994, GLADE 1996, GUZZETTI et al. 1994, THOMAS & TRUSTRUM 1984).
- Surrogate histories of landslide activity built up indirectly from records of associated infrastructural damage (SORRISO-VALVO 1997).
- Techniques which provide indirect physical evidence of movement, eg. Dendrogeomorphological studies of debris impact on trees (VAN STEIJN 1996, FANTUCCI & MCCORD 1995).

'Off-site' records rely mainly on stratigraphic analysis of sediment in depositional sites (PAGE et al. 1994, PAGE & TRUSTRUM 1997). Because of the obliteration of evidence in the active geomorphic environment of hillslopes, off-site records have the potential to extend over longer periods than on-site records. On the other hand, the signature of landslide activity recorded off-site may be obscured by complex transportational behaviour within the source catchment.

Whether established by on-site or off-site methods, a record of landslide activity can be used either directly or indirectly to obtain a measure of frequency. The direct approach is simpler and is achieved by dividing the number of landslide events or classes of landslide event by the length of the observational record. This provides the 'historical frequency' of events.

The indirect approach, in contrast, is more complex but has greater potential for forecasting landslide activity and determining the mass movement response to climatic change and other triggering factors. The indirect approach couples the forcing-process with a process-response. For rainfall-triggered landslides this involves establishing an initiating threshold between rainfall parameters and landslide occurrence (CROZIER 1989, JULIAN & ANTHONY 1994, GLADE et al., in press). For earthquake-triggered landslides, the threshold between non-occurrence and occurrence is usually either a value of shaking intensity or earthquake magnitude (KEEFER 1984).

Thresholds established in this way simply measure the susceptibility of the terrain under study to the landslide-triggering process. Clearly, inherent stability conditions and consequently thresholds will vary from place to place. A reliable regional threshold, however, may be used to determine the probability of occurrence (statistical frequency) of landslide activity by reference to the frequency-magnitude distribution of events for the triggering agent.

In situations where there is only a limited record of previous landslide activity, the empirical method is of little value. However, with the development of combined hydrology, slope stability, computer simulation models (ANDERSON 1996), a deterministic approach offers a promising alternative. This can be used to establish initiating thresholds for landslides in terms of rainfall input, and porewater pressures (either positive or negative). Essentially, in a validated model with the appropriate parameters for slope hydrology and strength, rainfall can be 'applied' incrementally, until on-going iterative stability analysis indicates failure. The rainfall total at failure thus represents a landslide-triggering rainfall threshold and this can be applied to frequency-magnitude distribution of the climate record to obtain probability of landslide occurrence. As with all modelling of this type, the set of parameters employed represent only a sample of what exists in the field. The representativeness of this approach is limited and it is likely to produce its most reliable results for individual slopes rather than large areas of terrain.

4 *Frequency and magnitude findings*

A few examples are given here to indicate the type of frequency and magnitude information that has been obtained by using the research methods discussed above. In general, the most detailed information relates to small areas and short periods of time.

4.1 *Regional frequency*

GLADE (1997) has employed a range of methods to obtain regional frequency data for New Zealand. Much of the work involved the initial establishment of a comprehensive database (GLADE 1996) which, although extensive, was unable to yield sufficient information for magnitude analysis. Using this database, it was possible, however, to produce historical frequencies for all New Zealand regions. In addition, by using the indirect threshold approach, probabilities of occurrence were calculated for three regions studied in detail.

Three different threshold models were applied to all three regions: the daily rainfall model, which simply compares rainfall magnitude and landslide occurrence, the antecedent daily rainfall model, which includes also information on antecedent rainfall over a defined period, and the antecedent soil water status model, additionally reflecting the soil moisture condition.

In the following case study, results obtained by applying the simple daily rainfall model to the Wellington region, North Island, New Zealand, are used to demonstrate the establishment of broad landslide-triggering rainfall thresholds. Subsequently, the probability of daily rainfall exceeding these thresholds is calculated. The results are intended to represent region-wide conditions only and do not represent landslide frequencies for individual sites.

The daily rainfall model expresses the relationship between landslide occurrence and actual daily rainfall. Intervals of 20 mm have been used to classify the daily precipitation values. These values were derived from climatic stations within each region. Each value has been stored within the respective class as one count. The 24h rainfall values associated with landslide occurrence were classed in the same 20 mm intervals. Fig. 3 shows the resulting frequency distribution of both variables for the Wellington region.

As Fig. 3 indicates, there is no record of landsliding associated with rainfalls in the precipitation class 0 to 20 mm. In contrast, rainfall values greater than 140 mm have always been associated with landslide occurrence. Thus it is possible to establish two empirically based rainfall thresholds, the 'minimum' and 'maximum' threshold respectively. These thresholds enclose a probability range, in which landslide probability values are assigned to the rainfall classes. This probability value reflects the likelihood of landslide occurrence with a given rainfall and is calculated simply as the ratio of the number of recorded landslide triggering daily precipitation counts to the actual measured total counts of daily precipitation within each class.

As one might expect, the probability of landslide occurrence increases with increasing rainfall magnitude. The minimum daily rainfall required to trigger landslides appears to be 20 mm, irrespective of antecedent conditions. This implies that combinations of antecedent soil moisture and rainfalls below 20 mm have never been sufficient for the slope to attain 'critical water content' (CROZIER 1997). In contrast, every rainstorm event greater than 140 mm has triggered landslides in the past. This boundary represents the maximum probability threshold for the Wellington region. As shown in other analyses, maximum thresholds vary from region to region, reflecting different terrain susceptibility to landsliding (GLADE 1996).

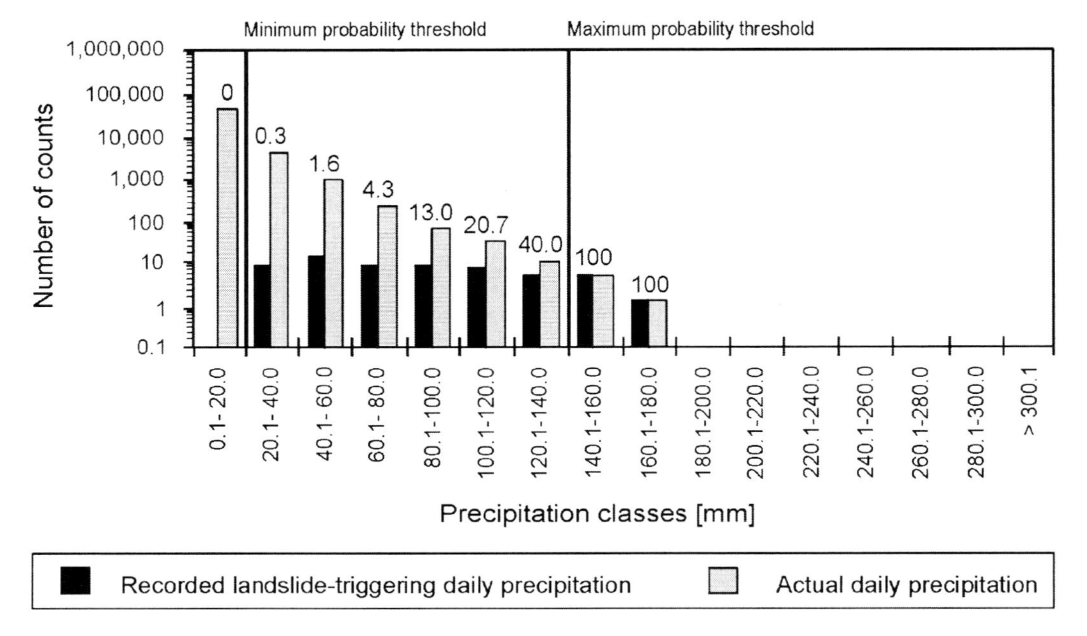

Fig. 3. Probabilities (%) of landslide occurrence associated with rainfall of a given magnitude in Wellington, New Zealand (note: a value of 50 means 50% of all measured daily rainfalls in a given category produced landslides in the past. The recording period is from 1862 to 1995).

The establishment of a maximum landslide-triggering rainfall threshold, in combination with the return period and the probability of occurrence of the respective rainfall magnitude, is a fundamental step in the evaluation of landslide hazard and risk analysis. Consequently, a frequency-magnitude analysis of daily rainfall has been carried out. Fig. 4 gives the return periods for the Wellington region for different rainfall magnitudes and shows the distinct relationship between the precipitation magnitude and return period.

To minimise the error of the trend line, the data set was separated into two parts. The first logarithmic function with an r^2 of 0.972 describes the relationship between these two variables for the rainfall magnitude of 0.1 to 100 mm with respective return periods of 0.004 to 2.5 years. Any rainfall magnitude above 100 mm and with a return period greater than 2.5 years is defined by the second logarithmic trend line with an r^2 of 0.983. The return period of the maximum probability threshold of 140 mm as taken from Fig. 3 is 20.1 years. Therefore, the Wellington region can expect at least every 20.1 years, a rainstorm which has a 100 % probability of triggering landslides. However, as indicated by the range between the maximum and minimum thresholds, there is also a chance (although of lower probability) that rainfalls with return periods less than 20.1 years will trigger landslides.

Another prognostic parameter, directly related to the return period, is the probability of occurrence within different periods. The maximum threshold of landslide-triggering rainfall, identified in Fig. 4 with a value of 140 mm, has a probability of occurrence on any day of 0.02%, within 30 days of 0.2% and within a year of 5% (Fig. 5).

Information on both return period and probability of occurrence are of particular interest for management and planning purposes. Established rainfall probabilities in combination with

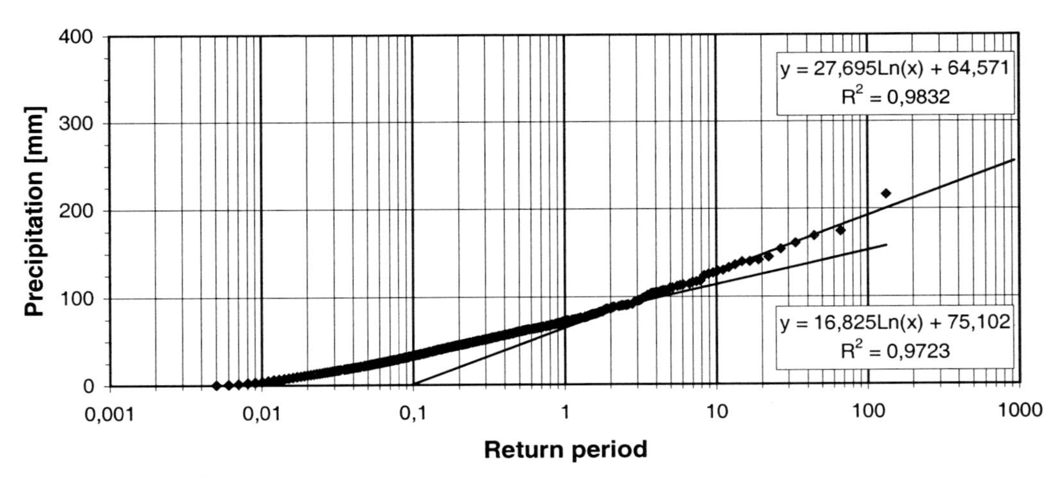

Fig. 4. Return periods of daily precipitation and fitted logarithmic trend lines in the Wellington region, New Zealand.

weather forecast can be used to indicate the likelihood of landslides triggered on the following day. Depending on the vulnerability of the region, the intensity of the following rainstorm event and past experience of landslide occurrence, suitable warning and mitigation measures can be taken by the appropriate organisations.

The major drawback of this analysis is, however, that neither landslide magnitude nor type of movement has been taken into account. This example shows frequency of landsliding only.

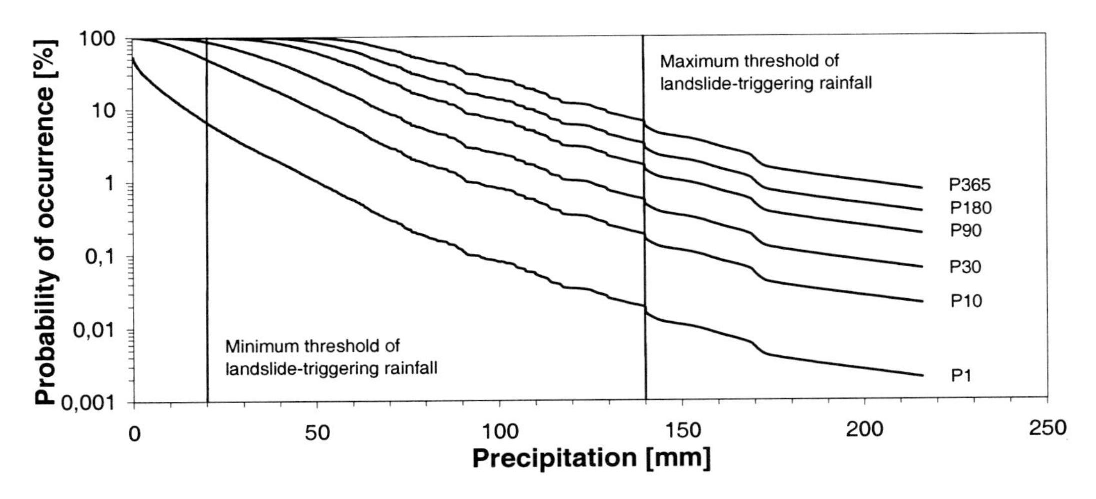

Fig. 5. Probability of occurrence of daily precipitation equalling or exceeding given values in Wellington, New Zealand. (Note: different lines refer to a probability of occurrence of a specific rainfall magnitude at each single day (p1), within a period of 10 days (p10), within a month (p30), etc. The empirically established minimum and maximum thresholds of landslide-triggering rainfall (140 mm) are shown by the thin vertical lines.).

Nevertheless, some qualitative consideration of the expected magnitude may be derived from the existing landslide data base. This database shows that each time the maximum threshold was passed, considerable damage occurred in the region. Correlation of landslide magnitude with triggering rainfalls below the maximum threshold, however, is not possible because of lack of standardisation in event records.

This case study provides an empirical method for assessing the frequency and magnitude of landsliding. Its further development as a means for answering geomorphic questions or as a tool for management depends on the establishment of comprehensive post-event recording systems.

4.2 *Frequency-magnitude and geomorphic work*

INNES (1985) carried out a spatial analysis of the frequency-magnitude relationships of 900 debris flows in a number of different sites in Scotland and Norway. His is one of the few empirical studies to examine geomorphic work accomplished in relationship to frequency and magnitude of landslides. Although this is essentially a spatial analysis, the debris flows studied had occurred as the result of a number of events taking place over a period of 500 years.

For the whole population as well as for individual multiple landslide events, the frequency of debris flows was found to decline with magnitude, in an apparently exponential fashion.

Work achieved by a given magnitude of landslide was measured as the percentage of the total amount of work done collectively by all landslides present. In this way INNES was able to test some of the conclusions put forward by WOLMAN & MILLER (1960). However, he discovered that the regular relationship between frequency and magnitude was not reflected in the work done by debris flows of a given magnitude. Work done by small and large debris flows was fairly similar in most sites although there was considerable variability. At one site all the work was done by debris flows of less than 20 m^3 in volume while at another, 30% of the work was completed by two flows of approximately 130 m^3. At yet another site, the work done by large flows was similar to that of small flows, indicating that despite the decreased frequency of larger flows, they were geomorphically significant owing to their magnitude.

Other studies have shown that when a forcing-process exceeds the threshold of initiation there is usually some form of positive relationship between the magnitude of the forcing agent and the magnitude of landslide response. As forcing processes such as rainfall, seismic shaking, and earthquake magnitude all exhibit clear negative relationships between event frequency and magnitude, it implies that, as indicated by INNES (1985), there will also be some form of negative relationship between the frequency and magnitude of landsliding. Indeed in spatial analysis of 8,000 landslide scars from the western Southern Alps of New Zealand, HOVIUS et al. (1997) demonstrate this negative relationship with a robust power-law magnitude frequency distribution. In this study, magnitude was represented by landslide surface area rather than volume.

Similar relationships between frequency and magnitude of landsliding have been found by temporal analysis of deposits in Lake Tutira, New Zealand (TRUSTRUM et al., this volume). In this study landslide magnitude is measured off-site by the thickness of sediment deposited during an event. As was the case in certain sites studied by INNES (1985), a disproportionate amount of lake sediment can be attributed to the low-frequency high-magnitude events.

From spatial analyses of separate multiple landslide events, close relationships have been found between the amount of rain falling on a slope and the percentage of the area eroded (EYLES & EYLES 1982, OMURA & HICKS 1991, GALLART 1995). GALLART presents a probabilistic model which describes the increase of landsliding as a function of event rainfall depth. He then calculates the product between a log-normal probability density function for rainstorm depth and a log-normal cumulative frequency distribution representing recurrence intervals. From the resultant peak value (maximum work event) in the distribution of products, the recurrence interval of the event carrying out the most work can be determined, in much the same way as was done by WOLMAN & MILLER (1960). Not surprisingly, regional results differ widely, with the Wairoa region of New Zealand yielding a recurrence interval for the maximum work event of 20 years compared to 1,000 years for the Llobregat valley in Spain. KEEFER (1984) has also found positive relationships between the Richter magnitude of measured earthquakes and the size of the area affected by landslides.

5 *Conclusion*

The frequency and magnitude of landslide activity varies both in time and space according to the susceptibility of the terrain and the potency of the triggering regime. Identifying the relationship between magnitude of the triggering agent and its response, by threshold analysis, provides a useful tool in identifying rates of geomorphic work, susceptibility of the terrain, and hazard. Considerable progress has been demonstrated in determining the probability of occurrence of landsliding on a regional basis. However, the success of this approach, whether achieved by deterministic or empirical modelling, depends on the quality of the data base. At present, it appears that the development of methodology has outpaced the development of suitable databases. Significant effort is still required to standardise parameters used and to establish consistent procedures for measurement and recording.

There is still insufficient data to test whether 'work' and 'effective event' theories derived from early fluvial studies have relevance for landslides. While smaller events are clearly more frequent, it appears that, in some areas, the biggest events on record do as much work with more geomorphic effect as the smaller events, despite their frequency.

References

ANDERSON, M.J. (1996): Developments in slope hydrology – stability models for tropical soils. – In: ANDERSON, M.J. & S.M. BROOKS (eds.): Advances in hillslope processes. – Wiley, 799–821

BULL, W.B. (1996): Prehistorical earthquakes on the Alpine fault, New Zealand. – Journ. Geophys. Res. **101** (B3): 6037–6050.

BRUNSDEN, D. & M.L. IBSEN (1994): The nature of the European archive of historical landslide data, with specific reference to the United Kingdom. – In: CASALE, R., R. FANTECHI & J.C. FLAGEOLLET (eds.): Temporal occurrence and forecasting of landslides in the European Community. – European Community: 21–70.

BRUNSDEN, D. & J.B. THORNES (1979): Landscape sensitivity and change. – Transact. Inst. Brit. Geogr. **4**: 463–484.

CROZIER, M.J. (1989): Landslides, Causes, consequences, and environment. – Routledge.

CROZIER, M.J. (1995): Landslide hazard assessment: a review of papers presented to theme G4. – In: BELL, D.H. (ed.): Landslides. – Proceedings of the sixth international symposium on landslides, Christchurch, February 1992, **3**: 1843–1848. Balkema, Rotterdam.

CROZIER, M.J. (1996): Magnitude-frequency issues in landslide hazard assessment. – In: MÄUSBACHER, R. & A. SCHULTE (eds.): Beiträge zur Physiogeographie. – Barsch Festschr., Heidelberger Geogr. Arb. **104**: 221–236.

CROZIER, M.J. (1997): The climate landslide couple: a Southern Hemisphere perspective. – Paleoclim. Res. **2**: 329–350.

CROZIER, M.J., R.J. EYLES, S.L. MARX, J.A. MCCONCHIE & R.C. OWEN (1980): Distribution of landslips in the Wairarapa hill country. – N.Z. Journ. Geol. Geophys. **23**: 575–586.

CROZIER, M J., M.S. DEIMEL & J.S. SIMON (1995): Investigation of earthquake triggering for deep-seated landslides, Taranaki, New Zealand. – Quatern. Internat. **25**: 65–73.

CROZIER, M.J. & B.J. PILLANS (1991): Geomorphic events and landform response in south eastern Taranaki, New Zealand. – Catena **18**(5): 471–478.

CROZIER, M.J. & N.J. PRESTON (1998): Modelling changes in terrain resistance as a component of landform evolution in unstable hill country – In: HERGARTEN, S. & H.J. NEUGEBAUER (eds.): Lecture Notes in Earth Sciences. **78**: 267–284, SpringerVerlag.

EYLES, R.J. & G.O.EYLES (1982): Recognition of storm damage events. – Proceedings of the eleventh New Zealand geography conference, Wellington 1981: 118–123.

FANTUCCI, R. & A. MCCORD (1995): Reconstruction of landslide dynamic with dendrochronological methods. – Dendrochronology **13**: 43–58.

GALLART, F. (1995): The relative geomorphic work effected by four processes in rainstorms: a conceptual approach to magnitude and frequency. – Catena **25**: 353–364.

GILLON, M.D. & G.T. HANCOX (1991): Cromwell Gorge landslides – a general overview. – In: BELL, D.H. (ed.) Landslides – Proceedings of the sixth international symposium on landslides, Christchurch, February 1992, **2**: 1451–1456, Balkema, Rotterdam.

GLADE, T. (1996): The temporal and spatial occurrence of landslide-triggering rainstorms in New Zealand. – In: R. MÄUSBACHER & A. SCHULTE (eds.): Beiträge zur Physiogeographie – Festschrift für Dietrich Barsch. – Heidelberger Geogr. Arb. **104**: 237–250.

GLADE, T. (1997): The temporal and spatial occurrence of rainstorm-triggered landslide events in New Zealand. – PhD Thesis, Department of Geography, Victoria University of Wellington, 380p.

GLADE, T. & M J. CROZIER (1996): Towards a national landslide information base for New Zealand. – N.Z. Geograph. **52**(1): 29–40.

GLADE, T., M.J. CROZIER & P. SMITH (in press): Establishing landslide-triggering rainfall thresholds using an empirical antecedent daily rainfall model. – Pure Appl. Geophys.

GUZZETTI, F., M. CARDINALI & P. REICHENBACH (1994): The AVI Project: A bibliographical and archive inventory of landslides and floods in Italy. – Environmental Management **18**(4): 623–633.

HOVIUS, N., C.P. STARK & P.A. ALLEN (1997): Sediment flux from a mountain belt, derived by landslide mapping. – Geology **25**(3): 231–234.

INNES, J. L. (1985): Magnitude-frequency relations of debris flows in northwest Europe. – Geograf. Ann. **67A**: 23–32.

JULIAN, M. & E.J. ANTHONY (1994): landslides and climatic variables with specific reference to the Maritime Alps of southeastern France. – In: CASALE, R., R. FANTECHI & J. C. FLAGEOLLET (eds.): Temporal occurrence and forecasting of landslides in the European Community. – European Community: 697–721.

KEEFER, D.K. (1984): Landslides caused by earthquakes. – Geol. Soc. Amer. Bull. **95**(4): 406–421.

LANG, A., J. COROMINAS, J. MOYA, L. SCHROTT & R. DIKAU (in press): Classic and new dating methods for assessing the temporal occurrence of mass movement. – Geomorphology.

MCSAVENEY, M.J., R. THOMSON & I.M. TURNBULL (1991): Timing of relief and landslides in Central Otago, New Zealand. – In: BELL, D.H. (ed.): Landslides. – Proceedings of the sixth international symposium on landslides, Christchurch, February 1992, **2**: 1451–1456, Balkema, Rotterdam.

OMURA, H. & D. HICKS (1991): Probability of landslides in hill country. – In: BELL, D.H. (ed.): Landslides. – Proceedings of the sixth international symposium on landslides, Christchurch, February 1992, **2**: 1045–1049, Balkema, Rotterdam.

OKUNUSHI, K., M. SONODA & K. YOKOYAMA (in press): Geomorphic and environmental controls of earthquake-induced landslides. – Physics and Chemistry of the Earth.

PAGE, M.J., N.A. TRUSTRUM & R.C. DEROSE (1994): A high resolution record of storm induced erosion from lake sediments. – N.Z. Journ. Paleolimnol. **11**: 333–348.

PAGE, M.J. & N.A. TRUSTRUM (1997): A late Holocene lake sediment record of erosion response to land use change in a steepland catchment, New Zealand. – Z. Geomorph. N.F. **41**(3): 369–362.

REIMER, W. (1995): Keynote paper: landslides and reservoirs. – In: BELL, D.H. (ed.): Landslides – Proceed-

ings of the sixth international symposium on landslides, Christchurch, February 1992, **3**: 1973–2004, Balkema, Rotterdam.
SORRISO-VALVO, M. (1997) Landsliding during the Holocene in Calabria, Italy. – Palaeoclim. Res. **2**: 97–108.
THOMAS, V.J. & N.A. TRUSTRUM (1984). A simulation model of soil slip erosion. Symposium on the effects of forest land use on erosion and slope stability. – Environment Policy Institute, East-West Center University of Hawaii, Honolulu: 83–89.
VAN STEIJN, H. (1996): Debris flow magnitude-frequency relationships for mountainous regions of Central and Northwest Europe. – Geomorphology **15**: 259–273.
VARNES, D.J. (1984): Landslide hazard zonation: a review of principles and practice. – UNESCO, Paris.
WEISS, E.E.J. (1988): Tree-ring patterns and frequency and intensity of mass movements. – In: BONNARD, C. (ed.): Proceedings of the fifth international conference on landslides: 481–483.
WOLMAN, M.G. & J.P. MILLER (1960): Magnitude and frequency of forces in geomorphic processes. – Journ. Geol. **68**: 54–74.

Address of the authors: M. J. CROZIER, School of Earth Sciences, Victoria University, P.O. Box 600, Wellington, New Zealand. T. GLADE, Geographisches Institut, Universität Bonn, Meckenheimer Allee 166, D-53115, Bonn, Germany.

Z. Geomorph. N.F.	Suppl.-Bd. 115	157–173	Berlin · Stuttgart	Juli 1999

Event-induced changes in landsurface condition – implications for subsequent slope stability

N.J. Preston, Bonn

with 4 figures and 7 tables

Summary. Erosional response to a triggering agent such as rainfall is commonly modelled on the assumption that thresholds for a given response are temporally invariant. In an active system this is unlikely to be the case. In the actively unstable New Zealand hill country, the process of erosion itself influences subsequent stability behaviour. The contemporary condition of the landsurface with respect to the relative distributions of erosional and depositional surfaces determines the status of factors which control slope stability and susceptibility to landsliding. An implication of this is that response to a given level of a triggering agent (e.g. rainfall) changes and thresholds must be considered unstable . This paper describes such a change in the relationship between rainfall input and erosional response. Changes in the distribution of landsurface condition over a 23 year period have been mapped. On the basis of these, the distribution of susceptibility to landsliding has been modelled, and shown to change. It is concluded that the relation of erosional response to rainfall input is not stable over time and that thresholds are unstable.

Zusammenfassung. Erosion als Folge von auslösenden Kräften (z.B. Niederschlag) wird allgemein nach der Annahme modelliert, daß die Grenzwerte für eine gegebene Reaktion zeitlich unveränderlich sind. In einem aktiven System ist das aber unwahrscheinlich. Im aktiv instabilen neuseeländischen Gebirgsland beeinflußt der Prozeß der Erosion das nachfolgende Stabilitätsverhalten. Der derzeitige Zustand des Reliefs in bezug auf die relative Verbreitung von Erosions- und Ablagerungsflächen bestimmt den Status der Faktoren, die die Hangstabilität und die Anfälligkeit für Erdrutsche beeinflussen. Eine Auswirkung davon ist, daß sich ein Ereignis bei einem gegebenem Niveau eines Auslösers (Niederschlag) ändert und Grenzwerte als variabel angesehen werden müssen. Diese Arbeit beschreibt eine solche Veränderung im Verhältnis zwischen Niederschlag und Erosionsereignis. Veränderungen in der räumlichen Verbreitung von Geländeeigenschaften wurden über einen Zeitraum von 23 Jahren kartiert. Auf dieser Basis wurde die Verbreitung der Disposition für Erdrutsche modelliert und Veränderungen gezeigt. Daraus wird geschlossen, daß das Verhältnis von Erosionsereignissen und Niederschlag über längere Zeit nicht stabil ist und daß Grenzwerte veränderlich sind.

1 *Introduction*

Erosion is widely recognised as a major constraint on productive land use in New Zealand's hill country. A young, geologically active terrain and a high frequency of erosion-triggering events (e.g. high intensity rainfall, earthquakes) predispose the landscape to high levels of natural erosion. Modification of the landscape through human agency has increased rates of erosion to the extent that depletion of the soil resource far exceeds its replenishment. The dominant erosional process in this terrain is shallow translational regolith landsliding, especially in areas where former forest cover has been removed. Although more accurately termed earth slides or, given the wet or saturated conditions under which they typically occur, earth

0044–2798/99/0115–0157 $ 4.25

flows (CRUDEN & VARNES 1996), the broad term *landslide* is used here to describe the characteristic form of mass movement involved in regolith removal.

Landslide occurrence is a function of both the potency of exogenic triggering agents and the inherent susceptibility to failure of a given terrain unit. Inherent susceptibility can be defined as the balance of stresses within a regolith unit, as determined by its geomechanical properties. Importantly however, inherent susceptibility is also subject to modification by mechanisms that act to filter the impact of exogenic triggers (CROZIER & PRESTON 1998). Notwithstanding this, in assessing the probability of landslide occurrence in a given region, it is customary to focus on the behaviour of exogenic triggering agents. In the New Zealand hill country this typically involves reference to the frequency/magnitude rainfall record with respect to historically established thresholds for landslide occurrence. Models expressing erosional response as a function of rainfall inputs have been established for many parts of New Zealand (e.g. PAGE et al. 1994).

The use and validity of these models has been questioned (CROZIER 1996). Such models employ the concept of a threshold rainfall input above which a given degree of erosional response is expected. Thresholds are established by empirically discriminating between those climatic conditions associated with landsliding and those which produce no landslide response (e.g. EYLES et al. 1978, GLADE 1998). Instable systems thresholds may be considered to be empirical constants, invariant over time. However, within recently perturbed, actively eroding systems, because of the dynamic behaviour of slope conditions and filtering phenomena within the terrain, it is thought that the relationship between energy input (rainfall) and erosional response is not simple, and unlikely that thresholds are in fact temporally or spatially stable (CROZIER & PRESTON 1998).

There is empirical evidence that the relationship between rainfall input and erosional response is not necessarily constant through time. Cyclone Bola in 1988, with 753 mm rainfall over four days, was the largest recorded rainfall event in the catchment of Lake Tutira in northern Hawke's Bay. In 1938 an event of similar magnitude – 692 mm over four days – produced nearly twice as much erosion as recorded by lake bottom sedimentation (PAGE et al. 1994). An anomaly such as this is explicable when viewed in the context of supply constraint, complex response and cyclic erosional behaviour (BUTLER 1959, CROZIER 1989, TONKIN 1994), or by invoking filtering phenomena within the terrain (CROZIER 1996). Use of rainfall magnitude to model erosional response assumes that erosion is not supply constrained, that the only factor determining the magnitude of erosional response is magnitude of a trigger. Realistically, this is unlikely ever to be the case, and certainly not in New Zealand where the frequency of high energy landslide-triggering rainfall events far exceeds the rate of regolith development. The implication is that threshold values of exogenic triggers for a given terrain are not constant through time. Further, if erosional activity and its effects on the landsurface influence thresholds, then thresholds will not be spatially constant either.

An example of variation in threshold behaviour is given here. Landsurface condition changes as a result of ongoing erosional processes, and especially in response to extensive landsliding episodes resulting from large rainfall events. Landsurfaces are classified here on the basis of recent erosional history, and hence with respect to the geomechanical resistance of materials. The landsurface condition of a small catchment in the east coast hill country of New Zealand's North Island has been mapped for two historical occasions. Site-specific susceptibilities to failure for each of these occasions are determined as a function of endogenic factors associated with these landsurface condition classes, and for a given level of energy input (rainfall magnitude is represented in stability equations using an index of porewater pressure). A change in susceptibility indicates a change in the relationship between energy input and erosional response, and a change in the triggering threshold can thus be inferred.

2 *The influence of erosion on geomechanical properties*

Endogenic factors that determine the balance of shearing and resistant stresses within the regolith include regolith depth and density, slope angle, strength parameters (cohesion and internal friction) and structural properties which control the development of porewater pressure (e.g. porosity, permeability, stratigraphy). Change in landsurface condition represents a change in the state of many of these stability-determining geomechanical properties of the regolith. Perhaps the most obvious influence of erosional activity can be seen in its effect on regolith depth. This is effectively reduced to zero with the occurrence of a landslide and increased downslope with the consequent deposition of debris. Similarly, local slope angle may change as a result of removal or deposition of material. Indeed, landsliding can be seen as a mechanism by which terrain adjusts to a distribution of regolith depths and slope angles in equilibrium with the contemporary energy regime.

Density of regolith material might also be expected to vary as a result of erosional processes. Both closely packed colluvial regolith and the unweathered surface materials of freshly exposed landslide scars are expected to be denser than undisturbed, weathered *in situ* soils (WILLIS 1995, PRESTON 1996). This has implications for stability in terms of the significance of density in strength/stress equations. Beyond this, however, changes in material density also have implications for soil moisture. Denser regolith is by definition less porous, permitting less infiltration and producing greater surface runoff. At the same time, because lower porosity is often linked to lower permeability, dense soils are also likely to be less well drained. Effects on soil moisture are likely to propagate off-site also. Impeded drainage at a given point may cause upslope ponding or build-up of high pore-water pressures, potentially inducing upslope failure. Conversely, enhanced runoff can lead to ponding and increases in pore-water pressures where flow lines converge.

The strength parameters cohesion and friction are related to soil structure, which is altered considerably by landsliding. Frictional strength is closely related to normal stress, which is itself dependent on depth, density and slope angle. Changes in landsurface condition consequent upon ongoing erosion are thus expected to have geomechanical implications.

3 *Contemporary landsurface condition classes*

Given the influence of landslide erosion on geomechanical properties within the regolith, five broad classes of Contemporary Landsurface Condition (CLC) can be defined.

1. *Undisturbed Material.* This is a surface that shows no morphological evidence of shallow regolith landsliding. It is characterised by a well-developed equilibrium soil showing pronounced horizonation. Surfaces that exhibit evidence of other erosional processes, e.g. soil creep, rilling, tunnel gullies, subsidence, are included within this class; the salient point is that landsliding has not occurred. Diagnostic features demonstrating the longevity of the soil may include the presence of *in situ* dateable tephras or a deep well-developed organic surface horizon.

2. *Landslide Scar*. The exposed shear plane of a landslide. Scars of many ages fall within this class as long as they have not yet experienced any significant soil redevelopment, e.g. through bedrock weathering, organic inputs, etc. Soil accumulations within a unit that is recognisable as a landslide scar should be classed separately as Landslide Debris.

3. *Landslide Debris*. Residual accumulation within a landslide scar, or a downslope accumulation of landslide-sourced material that can clearly be attributed on morphological grounds to a recognisably recent landslide event.

4. *Colluvium*. Material that cannot be sourced from any one individual landslide, but which originates from slope processes. Diagnostically, soil profiles will be stratified by successive deposition rather than through horizonation, although palæosols may be present. Poorly sorted angular clasts may be present.

5. *Alluvium*. A surface that can be recognised as being alluvially derived. In general, this is morphologically distinct from Colluvium, although at a micro-scale diagnoses will include particle sorting and imbrication.

These five classes broadly represent the possible states of the regolith, and the depositional processes giving rise to them, within the context of earth slide/flow erosion. However, in applying such a classification to the study of the distribution of geomechanical properties it becomes apparent that further refinement is required in two respects.

Landslide scars of different ages are expected to have different geomechanical properties. While soil depth recovery on landslide scars is a long-term process, initial soil redevelopment is quite rapid (TRUSTRUM et al. 1984). Not only does the depth of regolith change, soil redevelopment implies structural change, with significance for both strength parameters and hydrological behaviour. A distinction is thus drawn between New Scars and Old Scars. New Scars are those which exhibit bare surfaces, while Old Scars have experienced some degree of revegetation. This is, however, a somewhat arbitrary distinction; greater precision may be obtained by subdividing this class still further where the relative ages of landslide scars can be determined, e.g. from sequential aerial photography.

The Landslide Debris class also requires further refinement. While the surface expression of recent landslide debris is generally clear, the geomechanical implications are less so. The principal effect of debris deposition is of increasing local regolith depth, and local slope angle may change also. The effect of debris deposition on structural properties of the soil (including density, friction and cohesion) is more complex. To an extent regolith columns with Landslide Debris surfaces retain the geomechanical characteristics of the former surface type so that, with respect to potential shear failure, the properties of both the whole regolith column and those operating at the shear plane itself are of interest. For a regolith unit with recent deposition at its surface, overall density will be proportional to the densities of both buried and deposited material. Insofar as friction varies with normal stress (PRESTON 1996) it will be influenced by the increased depth of regolith. Cohesion as measured at the shear plane, however, is not expected to be greatly influenced by increased regolith depth and will be that of the buried regolith. Sub-classes of Landslide Debris are therefore recognised, dependent on the nature of the buried regolith – either Undisturbed, Old Scar or Colluvial. Deposition on to fresh scars is of course possible, and in many instances incomplete evacuation of scars is common. However, for classification purposes, it is very difficult to detect this remotely with any accuracy (e.g.

from an aerial photograph). Thus eight classes of Contemporary Landsurface Condition are recognised as being geomechanically significant and readily distinguishable:

1. Undisturbed surfaces
2. New Scar surfaces
3. Landslide Debris deposition on to an otherwise Undisturbed surface
4. Landslide Debris deposition on to a Colluvial surface
5. Landslide Debris deposition on to an Old Scar surface
6. Old Scar surfaces
7. Colluvial surfaces
8. Alluvial surfaces

4 *Application of the contemporary landsurface condition classification*

The Contemporary Landsurface Condition (CLC) classification system has been applied to the terrain within the catchment of Lake Waikopiro in northern Hawke's Bay, New Zealand (Fig. 1). This catchment is representative of the erosion prone soft-rock hill country that is widespread on the east coast of the North Island. It is formed on Tertiary sandstones and silty mudstones interbedded with limestones and conglomerates. These sedimentary rocks form part of the East Coast Deformed Belt and have been uplifted at approximately 1mm/y over the last 10 ka (Williams 1991), and dissected by incising rivers and streams. Topography consists of moderately steep to steep dissected hillslopes, with slope angles between 10° and 55°. Local relief is on the order of 250 m.

Underlying Tertiary marine sediments are relatively unweathered, and the regolith is largely composed of reworked colluvial material and airfall deposits. Æolian materials – tephra and

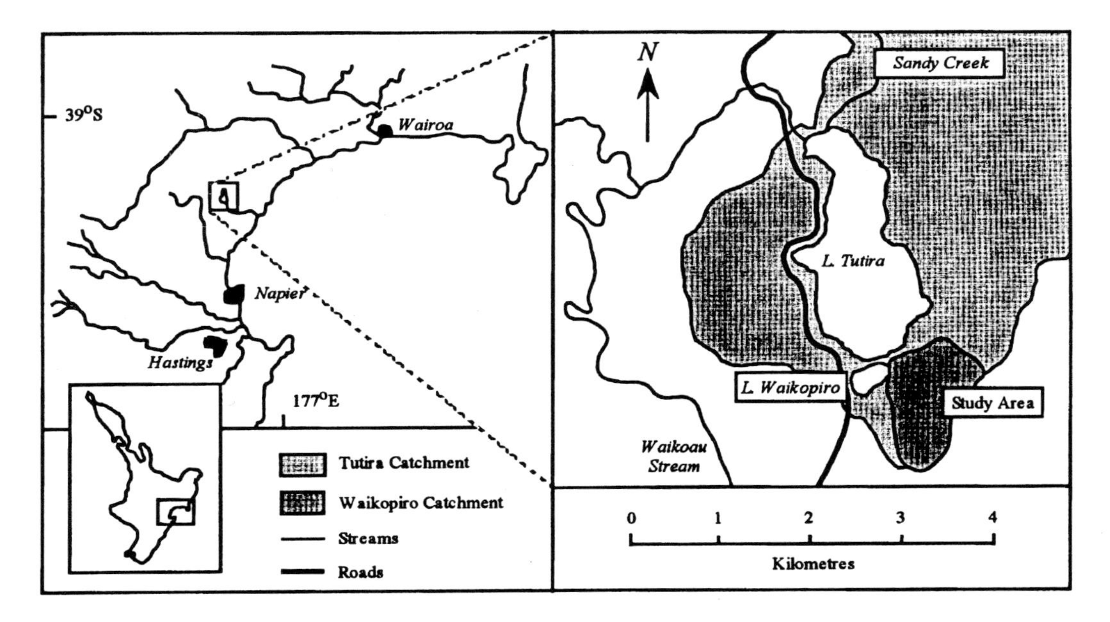

Fig. 1. Location of the study area in northern Hawke's Bay, New Zealand.

tephric loesses – are important as the region lies downwind of the Central Volcanic Zone, and contains some 14 identified Holocene tephras (EDEN et al. 1993), most notably the Taupo (1850 years B.P.) and Waimihia (3280 years B.P.). On much of the steep hillslopes, however, these tephras have been eroded, remaining only on the relatively undisturbed plateaux and interfluves.

This region is regularly subjected to erosion-inducing heavy rainfall. The recent erosional history of the catchment is reflected in its diachronous pattern of soil types. As the catchment is still very much within the evacuation stage of an erosion cycle (BUTLER 1959, TONKIN 1994), soil type distribution is controlled by sequential disturbance. Soil patterns on erosional surfaces are determined by lithology and bedrock stratigraphy, while on depositional surfaces it is location and the character of the depositional process (type and magnitude) that controls soil type. As a result, the regolith of the Waikopiro catchment does not conform to conventional soil catenas (WILLIS 1995). The various surfaces within the CLC classification have been characterised with respect to selected geomechanical properties of the regolith.

4.1 *Regolith depth*

Representative values of regolith depth have been obtained through field sampling. Non-parametric ranking techniques were used for comparison of samples as data are neither normally distributed nor taken from populations of equal variance. The Kruskal-Wallis ANOVA equivalent produces a p-value of <0.0005, while Kolmogorov-Smirnov individual comparisons produce p-values that are all <0.0005. Mean regolith depths for CLC classes are therefore considered to be significantly different (Table 1). Samples on Landslide Debris surfaces were of depositional material only, and do not include depth of substrate. Units with alluvial surfaces were deeper than the sampling probe (2.5 m), and have been characterised with a depth of 3 m.

4.2 *Density of regolith material*

Dry and saturated densities of regolith material, which can be used to model the density of regolith in a given section under a variety of soil moisture conditions, were calculated for all CLC classes. Initially, samples were taken from all horizons, i.e. a complete profile. Densities were weighted over the entire profile commensurate with relative thicknesses of horizons. However, little variation in density occurs within most of the profile. Subsequent samples were taken from only two horizons – the turfmat and the largely homogeneous horizon below this

Table 1. Mean regolith depths (m) for CLC classes.

	Colluvial	New Scar	Old Scar	Recent Debris	Undisturbed
Mean Depth	0.99	0.10	0.32	0.59	0.70
95% C.I.	±0.08	±0.01	±0.07	±0.07	±0.04
Range	2.25	0.22	1.44	1.47	1.92
Number of Samples	171	122	74	107	384

containing the postulated shear zone. Weighted estimates for whole profiles were derived from these samples, with the density of the shear zone material given a weighting of 0.85 and that of the turf mat 0.15, reflecting their proportions within the soil profile.

Colluvial and Undisturbed densities did not differ significantly. Only one sample of Old Scar material was taken. It was not significantly different in density from Colluvial material, and was treated as such. Densities of New Scar and Landslide Debris material were also not significantly different. Nor were either of these significantly different from the density of alluvially deposited material. There is thus only one distinction drawn for densities, reflecting in effect the recency of erosion or deposition – Undisturbed (including Colluvial and the Old Scar site) and Disturbed (New Scars, Landslide Debris and Alluvial). Differences in mean densities of these two classes – both dry and saturated – are significant using non-parametric statistical techniques. Mean values are presented in Table 2.

Table 2. Mean densities (kg/m^3) for CLC classes.

	Undisturbed	Disturbed
Dry Density (kg/m^3)	1210.4	1378.9
95% Conf. Interval	±59.0	±67.9
Range	554.8	730.1
Saturated Density (kg/m^3)	1694.8	1802.1
95% Conf. Interval	±39.8	±42.9
Range	358.0	498.4
Number of Samples	30	28

4.3 *Shear strength*

Strength parameters were derived from laboratory shear box testing of samples taken from the vicinity of the shear plane. These samples were found to be essentially cohesionless, with frictional strength varying as a function of normal stress (see PRESTON 1996). Equations relating frictional strength to normal stress are given in Table 3. Surfaces classed as Landslide Debris

Table 3. Equations relating friction to normal stress for CLC classes.

CLC Class	Friction Equation *	r^2	n
Undisturbed	$\phi' = 73.097\sigma^{-0.7724}$	0.96	36
Scars	$\phi' = 131.95\sigma^{-0.8708}$	0.99	38
Colluvial	$\phi' = 41.535\sigma^{-0.691}$	0.96	34
Old Scars	$\phi' = 29.018\sigma^{-0.6351}$	0.86	34
Alluvial	$\phi' = 0.5313\sigma^{-0.0368}$	0.00	9

*N.B. These equations were derived using normal stress (σ) expressed in units of kg/m^2, and produce friction values in tanform.

were assigned frictional strength values as appropriate to their substrates – either Undisturbed, Old Scar or Colluvial. For all classes there was very little difference in peak and residual strengths, partly a function of regolith texture, and partly because these materials have already undergone shear failure or have been otherwise exposed to ongoing slope processes, e.g. soil creep. Therefore the residual value has been used to characterise shear strength – with one exception. As discussed previously, Old Scars do not share the geomechanical attributes of other surfaces.

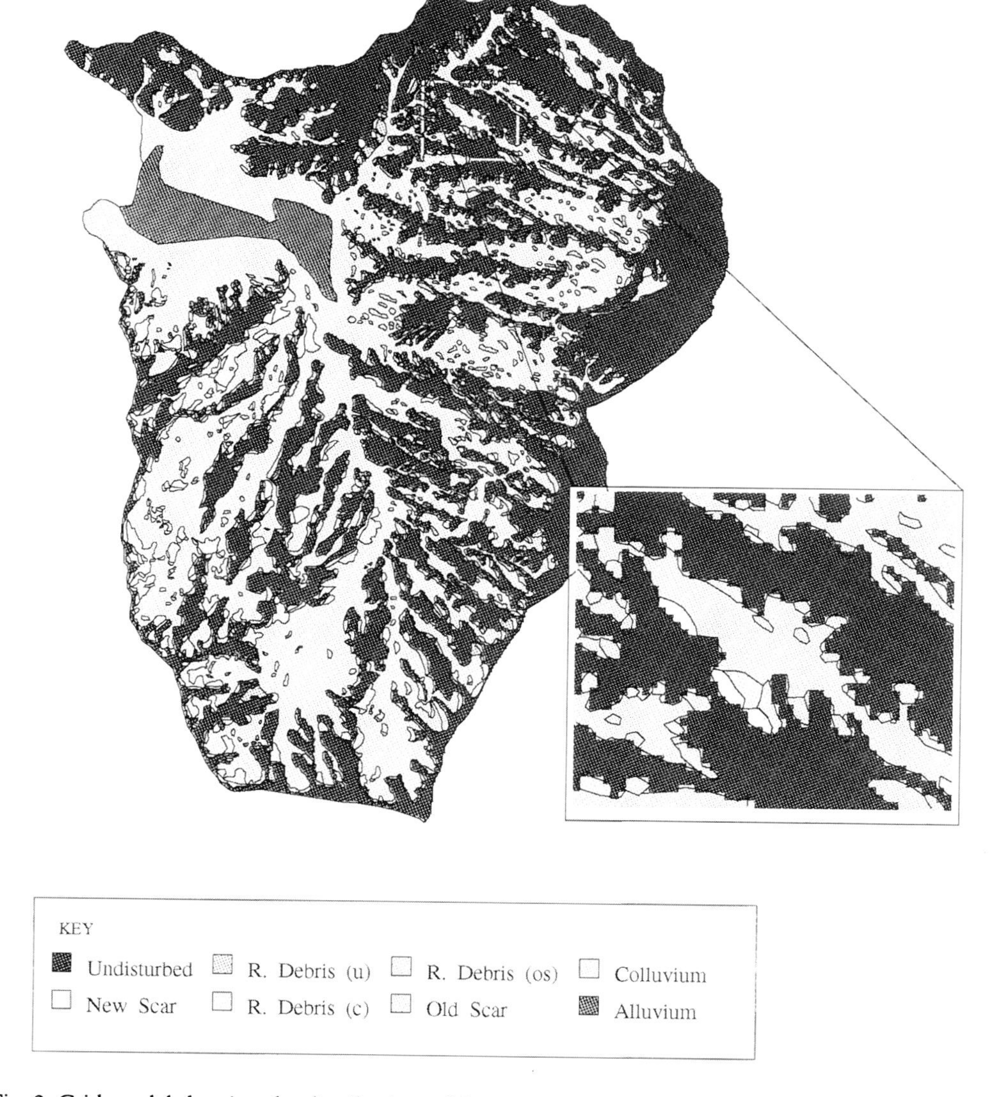

Fig. 2. Grid model showing the distribution of Contemporary Landsurface Condition in 1965. Grid cell classification closely matches mapped CLC polygons, as indicated by the overlaid line coverage in the enlargement.

With respect to strength, Old Scars were attributed with the peak frictional strength of colluvial samples, on the grounds that these soils are more likely to have developed in remnant colluvial material than from weathering of the scar surface, and on the assumption that they have not yet been sufficiently exposed to slope stresses to have been reduced to residual strength.

4.4 *Distribution of contemporary landsurface condition*

The distribution of CLC units in the Waikopiro catchment was derived through interpretation of aerial photography from two occasions (1965 and 1988). Over this 23 year period six episodes of extensive landsliding were recorded (GLADE 1996). Analysis of photographs from these two occasions allows the distribution of landsurface condition and the relative distribution of susceptibility to be compared after a series of geomorphically effective events. Analysis was done using Arc/Info GIS software, with an analytical framework constituting grids containing 26,589 cells, each representing a discrete 25m² area of terrain surface (Fig. 2). Separate grids were produced for each classification map. Overlaying these allows comparisons for individual cells. A little over six thousand cells (>20%) had undergone a change in CLC class over the period 1965 to 1988. Numerically this shift in landsurface condition is dominated by a decrease in Undisturbed and an increase in Colluvial surfaces (Table 4). Also of significance are increases in Landslide Debris (Colluvium) and Old Scars. Changes in frequency of Landslide Debris on both Undisturbed and Old Scar surfaces represent very minor changes within the context of the catchment as a whole. Changes in landsurface condition within the catchment are better illustrated with reference to Table 5, which lists numbers of cells from the various combinations of 1965 and 1988 CLC classes. Numbers in rows indicate the "destination" of cells with given CLC classes in 1965. Similarly, reading down columns, the "origin" of 1988 CLC class units is given. These changes are summarised below.

Of 1965's Undisturbed material, the greatest part (79%) remained Undisturbed in 1988, although significant numbers of cells were reclassified as scars and recent depositional surfaces. Overall, the distribution of Undisturbed material was reduced from 41% to 33% of the total

Table 4. Comparative distributions of CLC classes.

	1965		1988		
CLC Class	No. of Cells	%	No. of Cells	%	% Change
Undisturbed	10847	40.79	8798	33.09	−18.9
New Scar	641	2.41	671	2.52	+4.7
Debris (Undisturbed)	127	0.48	78	0.29	−38.6
Debris (Colluvium)	1099	4.13	1244	4.68	+13.2
Debris (Old Scar)	4	0.02	18	0.07	+350.0
Old Scar	1757	6.61	2062	7.76	+17.4
Colluvium	11600	43.63	13287	49.97	+14.5
Alluvium	514	1.93	429	1.61	−16.5
Totals	26589	100.0	* 26589	100.0	

*This total includes two cells representing surface water

Table 5. Changes in CLC classes from 1965–1988.

CLC Class 1965	Total	U	NS	D(U)	D(C)	D(OS)	OS	C	A
		CLC Class 1988							
Undisturbed (U) *	10845	8538	368	39	230	5	464	1201	0
New Scar (NS)	641	33†	46	10†	63	1†	362	126	0
Debris (Und.) (D(U))	127	8†	1†	1†	8	1†	13	95	0
Debris (Coll.) (D(C))	1099	13†	29	14	133	0†	43	867	0
Debris (O.S.) (D(OS))	4	0	0	1	0	0	2	1	0
Old Scar (OS)	1757	66†	54†	4†	91	7†	1040	495	0
Colluvium (C)	11600	140†	173	9†	718	4†	138	10398	20†
Alluvium (A)	514	0	0	0	1	0	0	104	409
Totals	26589	8798	671	78	1244	18	2062	13287	429

* Two cells that in 1965 were Undisturbed are now occupied by standing surface water
† These observations should be treated with caution; errors are greater than the value of the observation itself

catchment area. The origin of some of 1988's Undisturbed surfaces is paradoxical. Over the period concerned, development of Undisturbed soils is not anticipated. These observations have been interpreted as the result of error in photointerpretation. Quantification of this error is possible if it is assumed that it is random, and that it applies equally to the interpretation of other classes as well. Using these anomalous observations as a guide, margins of error for interpretation of landsurface condition in 1988 were calculated. Observations that become dubious in light of this are marked (†) in Table 5.

Most units that were New Scars in 1965 were recorded in 1988, by definition, as Old Scars. Some had failed again, while many had become depositional surfaces. New Scars in 1988 occurred principally in Undisturbed and Colluvial material, and in Debris or on existing scar surfaces to a lesser extent. There has been ongoing erosional activity on Landslide Debris surfaces, evidenced by the number of these cells subsequently recorded as scar surfaces, both New and Old. Numerically the most important change experienced by 1965 Debris surfaces has been a redesignation as Colluvial. 1988's Debris surfaces were for the most part Undisturbed or Colluvial surfaces in 1965. Most Old Scars were still just that. But many show evidence of ongoing erosional activity in that by 1988 they had become New Scars or depositional surfaces. Of 1965's Old Scars a greater frequency were recorded in 1988 as Debris on Colluvium rather than as Debris on Old Scars. This suggests that perhaps there has been one or more interim surface conditions applying to these units, i.e. the sequence of landsurface conditions for a given cell may have been: Old Scar (in 1965), Debris on Old Scar as a fresh failure occurred immediately upslope, Colluvium as material developed from fresh debris into a colluvial soil, and finally Debris on Colluvium in 1988. Such a sequence might be expected given the frequency of erosional activity in this area, and is consistent with an upslope migration of the erosion front. The numbers of 1988 Old Scars that were previously Undisturbed and Colluvial surfaces is a further indication of the dynamic erosional regime of the area. Many cells recorded in 1965 as Colluvial surfaces were classified in 1988 as Scars or Debris surfaces. The large increase in the extent of

Colluvium provides further evidence of continuing erosional activity. The least activity was associated with Alluvium, with only a small net encroachment of Colluvium onto Alluvial surfaces.

4.5 *Susceptibility*

Susceptibility was characterised with a Factor of Safety, defined in terms of HENKEL & SKEMPTON's (1954) infinite slope limiting equilibrium analysis technique, as the ratio of shear strength to shear stress. Both these parameters are dependent on geomechanical properties of the regolith. Factors of Safety were calculated on a cell-by-cell basis for both 1965 and 1988 landsurface condition distributions, using representative values of geomechanical properties as presented in Table 6. Values used in calculation of stresses for units with Landslide Debris were thus a composite of the respective values in Table 1 (e.g. a surface comprising Landslide Debris on an Old Scar was assigned a representative regolith depth of 0.91 m, comprising 0.32 m of Old Scar and 0.59 m of Landslide Debris). Although significant differences in density were found only on the basis of recency of disturbance, representative values as sampled were used for each class. As with depths, Landslide Debris surfaces were assigned composite density values, weighted according to the relative proportion of each type within the regolith column. Although shear strength of material at the shear plane is considered to be exclusively frictional, values of apparent cohesion were included in some instances. Surfaces with pasture cover (all except New Scar and Landslide Debris surfaces) were attributed with mechanical cohesion of 327.26 kg/m^2, derived from the turf mat membrane of pasture species (see PRESTON & CROZIER 1998). Similarly, the distribution of the regenerating tree species kanuka (*Kunzea ericoides*) was mapped and overlaid onto the grids; cells with this vegetative cover were attributed with a further increment of root induced mechanical cohesion of 330 kg/m^2 (O'LOUGHLIN et al. 1982, SIDLE et al. 1985, WATSON & O'LOUGHLIN 1985).

Shear strength and stresses vary with porewater pressure which, given its capacity for variability in real time compared to the more static geomechanical properties, is likely to be a key determinant of slope stability. Observations in the study area indicate that, at least in places, total regolith saturation does occur, as evidenced by the appearance of return overland flow (MERZ 1997). Conversely, determination of critical water table heights using back analysis on data taken

Table 6. Values used as attributes of CLC classes.

CLC Class	Depth (m)	Dry Density (kg/m^3)	Saturated Density (kg/m^3)	Friction Equation
Undisturbed	0.70	1180.2	1673.6	$\phi' = 73.097\sigma^{-0.7724}$
New Scar	0.10	1404.4	1821.4	$\phi' = 131.95\sigma^{-0.8708}$
Debris (Undisturbed)	1.29	1257.4	1722.3	$\phi' = 73.097\sigma^{-0.7724}$
Debris (Colluvium)	1.59	1301.0	1753.9	$\phi' = 41.535\sigma^{-0.691}$
Debris (Old Scar)	0.91	1322.8	1766.0	$\phi' = 29.018\sigma^{-0.6351}$
Old Scar	0.32	1270.8	1737.2	$\phi' = 29.018\sigma^{-0.6351}$
Colluvium	0.99	1270.8	1737.2	$\phi' = 41.535\sigma^{-0.691}$
Alluvium	3.00	1333.2	1782.6	$\phi' = 0.5313\sigma^{-0.0368}$

from surveyed landslide sites suggests that failure also occurs without the water table reaching the surface. There is clearly spatial variation in critical porewater pressures required for failure. However, for simplicity, discussion is confined here to results from modelling with a porewater pressure value represented by a water table depth to regolith depth ratio (m) of 0.75, which is thought to approximate the porewater pressure at which failure occurs. It is considered that adopting this value of $m = 0.75$ as an average over the catchment area allows for total regolith saturation in at least some sites, consistent with field observations. The important point with respect to comparisons over time is that they are made for a constant porewater pressure value.

Finally, the remaining variable required for calculation of stresses – slope angle – was derived directly from a digital elevation model generated from 5 m contours.

Although the Factor of Safety is an absolute measure, it can also serve as a measure of relative susceptibility. It is used in the latter respect here because calculated Factors of Safety in some cases have values less than unity. In theory this is impossible – a regolith unit with shear stresses greater than shear strength should have failed. At the other extreme, unrealistically high Factors of Safety have been calculated in some instances. These extremes are probably modelling artefacts. Average values for stability parameters have been used, so some unrealistic results might be expected. Further, it should be noted that the Factor of Safety is a measure of the susceptibility to failure of terrain that is subject to slope stresses. Analysis within a digital terrain model includes areas, such as plateaux and valley floors, which are not subject to slope stresses or failure, accounting for the presence of high Factors of Safety that would seem insensible for an inclined regolith unit.

Table 7 lists numbers of cells showing changes in Factors of Safety. Two scenarios are considered. In Scenario A, changes in Factors of Safety (ΔFS) are considered on the basis of observed absolute values, while Scenario B recognises that small shifts in susceptibility, either positive or negative, are probably not too significant and treats values within the range $-1.0 \leq \Delta FS \leq 1.0$ as representing no change (i.e. $\Delta FS = 0.0$). Under Scenario A, no change in susceptibility at all is recorded for only 1,443 cells (5.4%), and in fact a majority of cells became more susceptible to failure as a result of changes in landsurface condition between 1965 and 1988. However, the great majority of cells (21,153, 79.6%) recorded a change in Factor of Safety within the range $-1.0 \leq \Delta FS \leq 1.0$. If, as in Scenario B, these are considered to have experienced no change, a small net shift within the catchment towards greater stability can be inferred.

It is somewhat misleading, however, to consider changes in susceptibility without further qualification. A decrease in stability as recorded in Table 7 might, for example, simply represent a reduction in the Factor of Safety from 26 to 25. While becoming less stable, a cell experiencing such a change cannot be described as being particularly susceptible to failure. Fig. 3 compares Factor of Safety frequency distributions between 1965 and 1988. The upper tails of the distributions in Fig. 3(a) indicate that higher Factors of Safety were more common in 1988 than in 1965. Conversely, in 1988 there were fewer cells showing Factors of Safety at the low end of the range (Fig. 3(b)). The range of data plotted in Fig. 3(a) does not include the entire dataset; extremely

Table 7. Cells showing changes in Factors of Safety between 1965 and 1988 ($m = 0.75$).

	More Stable	%	Less Stable	%	No Change	%
Scenario A	11907	44.8	13239	49.8	1443	5.4
Scenario B	2739	10.3	2697	10.1	21153	79.6

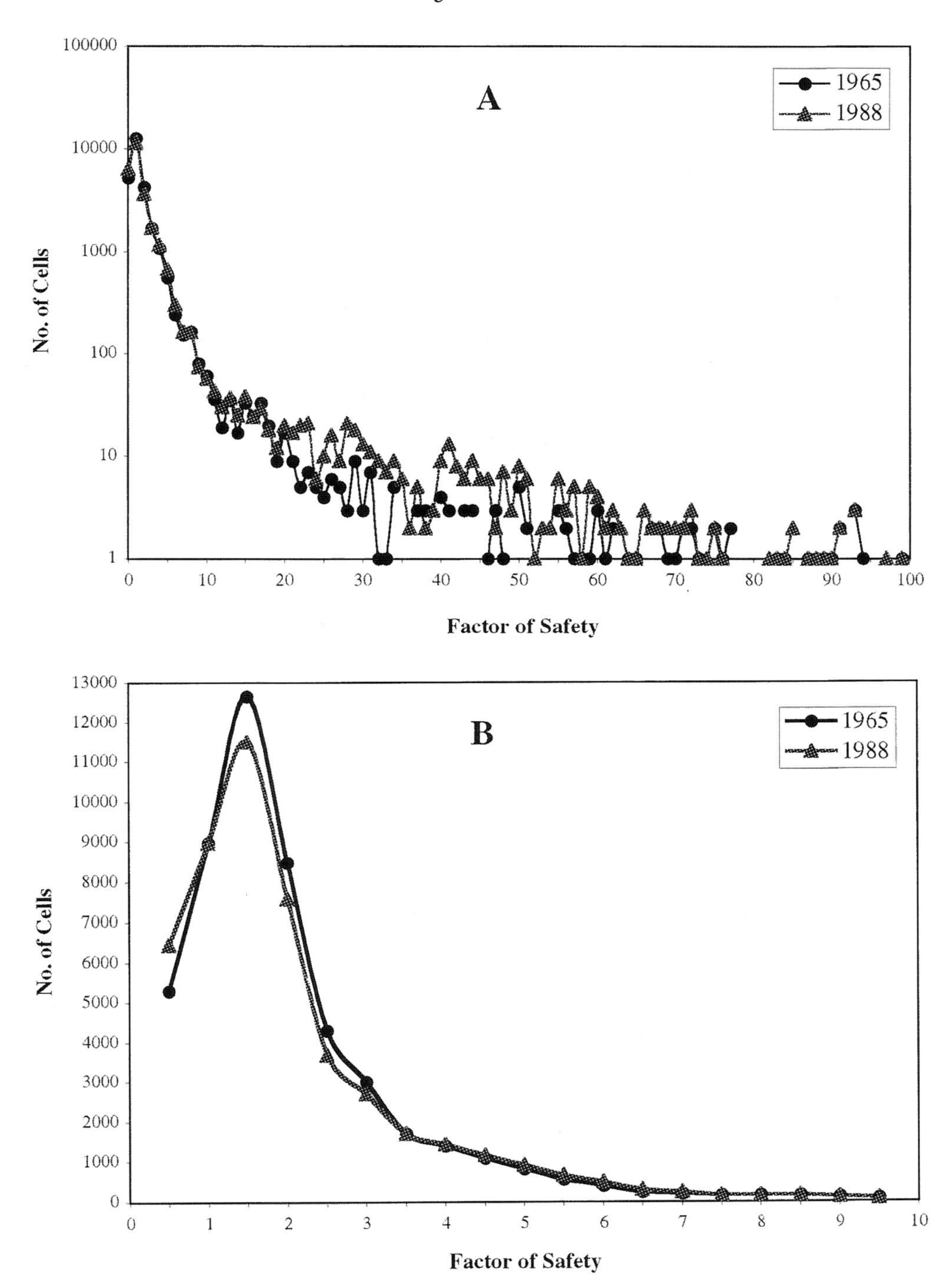

Fig. 3. Frequency distribution of Factors of Safety for (a) the catchment (b) lowest decile.

high values have been discarded. Nonetheless, the overwhelming majority of cells are included – 98.3% and 97.4% for 1965 and 1988 respectively. These last figures themselves suggest the direction of change, with fewer cells falling within this range in 1988. The overall susceptibility of the field area is indicated by the large proportion of cells with low Factors of Safety (Fig. 3(a)).

5 *Discussion*

Compared to the situation in 1965, in 1988 the terrain had greater resistance to failure for a given triggering event. In 1988 for a given porewater pressure (modelled as $m = 0.75$), a smaller portion of the catchment was as susceptible to failure as in 1965. This porewater pressure was not able to produce the same degree of landsliding as in 1965. Alternatively, greater average porewater pressures would have been required to generate an equivalent degree of landsliding. Accepting that the use of rainfall in modelling landslide occurrence is an acknowledgement of the significance of porewater pressure as a trigger of failure, it is reasonable to conclude that the threshold rainfall magnitude for a given erosional response has increased. Although not pursued in this study, further development of the current model may enable specific identification of critical porewater pressures.

While the catchment as a whole may have experienced a shift toward greater overall stability, with an anticipated reduced erosional response to a triggering event of given magnitude, two further points may be made. Firstly, note the increased frequency of very low Factors of Safety

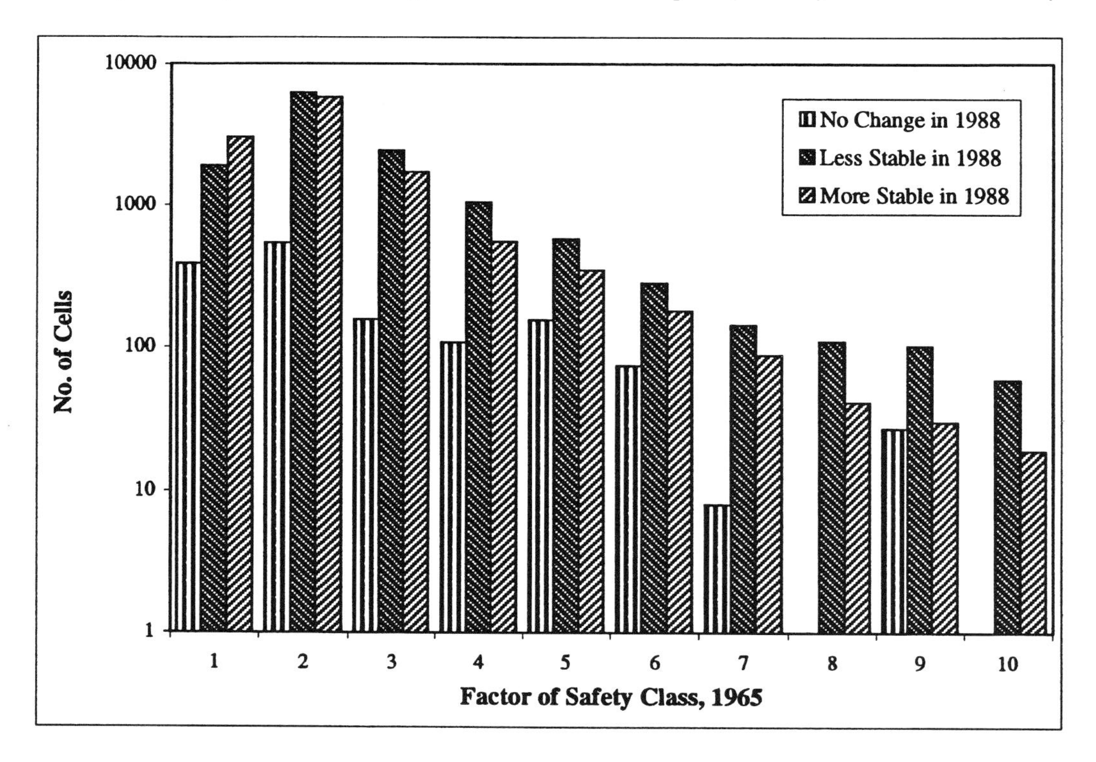

Fig. 4. Cell-specific changes in susceptibility, lowest decile.

in 1988 indicated in Fig. 3(b). This suggests that in a small rainfall event, capable of raising only these most highly susceptible cells above the threshold of instability, there may be an increased erosional response. However, for larger events with the capacity of substantially elevating porewater pressures in a greater number of cells, a lower response would have been anticipated in 1988 than in 1965. Secondly, it is interesting to consider the behaviour of the more susceptible terrain units within the catchment. SELBY (1993) cites Factors of Safety in the range 1.5 to 2.5 as being critical for landslide occurrence. Given that the Factor of Safety is treated here as a relative measure, and in keeping with the range of data presented in Figure 3(b), the change in susceptibility behaviour for the lowest decile has been analysed (Fig. 4). Of those cells that were most susceptible to failure in 1965 (FS Class 1 in Fig. 4), a majority had become more stable by 1988, presumably as a result of having experienced failure in the intervening period. Of the cells that were in the remaining nine most susceptible classes in 1965, however, there is a consistent trend of increased susceptibility by 1988, perhaps suggesting a cyclic element in susceptibility behaviour, with individual sites gradually becoming more susceptible. If this is indeed occurring, it is nevertheless against a background of decreasing overall catchment susceptibility, as indicated in Fig. 3. However, the behaviour of individual cells cannot be properly considered without some means of identification, and is probably not as simple as suggested here.

6 *Summary*

A Contemporary Landsurface Condition (CLC) classification system, specific to terrain experiencing shallow regolith landsliding, has been developed. Landsurface condition was used as a means of stratifying the catchment area for data collection, and has been found to reflect significant differences in geomechanical properties. The Contemporary Landsurface Condition of a small hill country catchment has been mapped for two occasions, 23 years apart, showing the change in landsurface condition resulting from successive erosional episodes. Not surprisingly, the spatial and frequency distributions of CLC units has changed over this period as ongoing erosional processes accomplish geomorphic work.

Selected geomechanical properties, pertinent to stresses occurring within the regolith, have been attributed to CLC classes. On the basis of these properties, the relative distributions of susceptibility to failure, defined with reference to the balance of stresses occurring within the regolith, have been determined for each of these occasions. Changes in susceptibility are not unidirectional; some areas within the catchment were more stable in 1988, others less so. Specifically, there is evidence that the most susceptible sites, as identified from the earlier occasion, tended to become more stable, presumably as a result of failure, while the remaining more susceptible sites were more likely to have been weakened. The distribution of relative susceptibilities is thus evolving over time. The catchment as a whole, however, exhibits a greater degree of stability, i.e. there was a larger area with higher Factors of Safety in 1988 than there had been in 1965. These susceptibilities were calculated using a constant porewater pressure – in effect the same energy (rainfall) input. The decrease in susceptibility is interpreted as an increase in the triggering threshold required to produce a given catchment-wide erosional response. This has occurred because of the change in the catchment's landsurface condition. The ongoing occurrence of erosional processes influences threshold behaviour, inducing an element of temporal variation. Erosional response within actively eroding terrain cannot therefore be modelled solely on the basis of rainfall frequency/magnitude relationships.

Acknowledgements

This work was funded in part by the Physical Geography Board of Studies and through a grant from the Internal Grants Committee, both of Victoria University of Wellington. My thanks to Mike Crozier and Thomas Glade for their constructive criticism of the manuscript.

References

Butler, B.E. (1959): Periodic phenomena in landscapes as a basis for soil studies. – CSIRO Australia Soil Publication No. **14**.

Crozier, M.J. (1989): Landslides: Causes, Consequences and Environment. – 2nd Ed., 252 pp., Routledge, London.

Crozier, M.J. (1996): Magnitude/frequency issues in landslide hazard assessment. – In: Mäusbacher, R. & A. Schulte (eds.): Beiträge zur Physiogeographie. Festschrift für Dietrich Barsch. – Heidelberger Geogr. Arb. **104**: 221–236.

Crozier, M.J. & N.J. Preston (1998): Modelling changes in terrain resistance as a component of landform evolution in unstable hill country. – In: Hergarten, S. & H.J. Neugebauer (eds.): Process Modelling and Landform Evolution. – Lecture Notes in Earth Sciences **78**, Springer, Heidelberg.

Cruden, D.M. & D.J. Varnes (1996): Landslide Types and Processes. – In: Turner, A.K. & R.L. Schuster (eds.): Landslides: Investigation and Mitigation. – Special Report **247**, Transportation Research Board, National Research Council, National Academy Press, Washington D.C.

Eden, D.N., P.C. Froggatt, N.A. Trustrum & M.J. Page (1993): A multiple-source Holocene tephra sequence from Lake Tutira, Hawke's Bay, New Zealand. – N.Z. Journ. Geol. Geophys. **36**: 233–242.

Eyles, R.J., M.J. Crozier & R.H. Wheeler (1978): Landslips in Wellington City. – N.Z. Geograph. **34(2)**: 58–74.

Glade, T.W. (1996): The temporal and spatial occurrence of landslide-triggering rainstorms in New Zealand. – In: Mäusbacher, R. & A. Schulte (eds.): Beiträge zur Physiogeographie. Festschrift für Dietrich Barsch. – Heidelberger Geogr. Arb. **104**: 237–250.

Glade, T.W. (1998): Establishing the frequency and magnitude of landslide-triggering rainstorm events in New Zealand. – Environmental Geology **35(2–3)**: 160–174.

Henkel, D.J. & A.W. Skempton (1954): A landslide at Jackfield, Shropshire, in a heavily over-consolidated clay. – Géotechnique **5**: 131–137.

Merz, J. (1997): Hydrological Investigations of a Hillside Affected by Landslides, Lake Tutira, New Zealand – unpubl. Diplomarbeit, Universität Bern, Switzerland.

O'Loughlin, C.L., L.K. Rowe & A.J. Pearce (1982): Exceptional storm influences on storm erosion and sediment yields in small forest catchments, North Westland, New Zealand. – National Conference Publication **82/6**: 84–91, Institute of Engineers, ACT, Australia.

Page, M.J., N.A. Trustrum & R.C. DeRose (1994): A high resolution record of storm-induced erosion from lake sediments, New Zealand. – Journ. Paleolimnol. **11**: 333–348.

Preston, N.J. (1996): Spatial and Temporal Changes in Terrain Resistance to Shallow Translational Regolith Landsliding. – unpubl. MSc(Hons) thesis, Victoria University of Wellington, New Zealand.

Preston, N.J. & M.J. Crozier (1999): Resistance to shallow landslide failure through rootderived cohesion in East Coast hill country soils, North Island, New Zealand. – Earth Surface Processes and Landforms **24**.

Selby, M.J. (1993): Hillslope Materials and Processes. – 2nd Edition, Oxford University Press.

Sidle, R.C., A.J. Pearce & C.L. O'Loughlin (1985): Hillslope Stability and Land Use. – American Geophysical Union.

Tonkin, P.J. (1994): Principles of soil-landscape modelling and their application in the study of soil-landform relationships within drainage basins. – In: Webb, T.H. (ed.): Soil-Landscape Modelling in New Zealand. – Landcare Res. Sci. Ser. **5**, Manaaki Whenua Press, Lincoln, New Zealand.

Trustrum, N.A., V.J. Thomas & M.G. Lambert (1984): Soil slip erosion as a constraint to hill country pasture production. – Proc. N.Z. Grassland Assoc. **45**: 66–76.

WATSON, A. & C.L. O'LOUGHLIN (1985): Morphology, strength and biomass of manuka roots and their influence on slope stability. – N. Z. Journ. Forestry Sci. **15(3)**: 337–348.
WILLIAMS, P.W. (1991): Tectonic geomorphology, uplift rates and geomorphic response in New Zealand. – In: CROZIER, M.J. (ed.): Geomorphology in Unstable Regions – Catena **18**: 439–452.
WILLIS, P.R. (1995): Soil/Landscape Modelling. Modelling the Spatial Distribution of Soil Characteristics using Landscape Features in Eroding Hill Country, Tutira, Hawke's Bay – unpubl. BSc(Hons) Research Report, Victoria University of Wellington, New Zealand.

Address of the author: N.J. PRESTON, School of Earth Science, Victoria University of Wellington, P.O. Box 600, Wellington, New Zealand.
Present address: Geographisches Institut, Universität Bonn, Meckenheimer Allee 166, D-53115 Bonn, Germany.

Z. Geomorph. N.F.	Suppl.-Bd. 115	175–189	Berlin · Stuttgart	Juli 1999

The frequency and magnitude concept in relation to rock weathering

A.S. Goudie and H.A. Viles, Oxford

with 5 figures

Summary. In their consideration of frequency and magnitude concepts in geomorphology, Wolman & Miller (1960) gave little attention to weathering. This paper investigates the concept of frequency and magnitude firstly in the context of a range of important, frequently studied weathering processes (e.g. solution, fire, thermoclasty, haloclasty, frost, and biological weathering) and their effectiveness in terms of rock breakdown, and secondly in terms of the totality of weathering and its effectiveness in terms of larger-scale geomorphological change. Finally, the importance of the ideas of frequency and magnitude as applied to weathering in today's geomorphology is examined.

Zusammenfassung. Wolman & Miller (1960) haben in ihren Überlegungen zu Häufigkeits- und Magnitudenkonzepten die Verwitterung kaum berücksichtigt. Die vorliegende Arbeit untersucht das Konzept von Häufigkeit und Größe zum einen im Kontext einiger wichtiger, oft studierter Verwitterungsprozesse (z.B. Lösung, Feuer, Hitzesprengung, Salzsprengung, Frostverwitterung und biogene Verwitterung) und ihrer Effektivität bezüglich des Gesteinszerfalls, zum anderen in Bezug auf die Gesamtheit der Verwitterungsprozesse und deren Bedeutung für geomorphologische Veränderungen in kleinen Maßstäben. Schließlich wird auch die Bedeutung der Idee von Häufigkeit und Größe in Zusammenhang mit Verwitterung in der heutigen Geomorphologie untersucht.

Introduction

In their analysis of the importance of the magnitude and frequency of forces in geomorphological processes Wolman & Miller (1960) neither mentioned the word "weathering" nor used any examples directly related to weathering processes. They concerned themselves primarily with palpably dynamic processes including the transport of sediment by wind and water, the shaping of river channels, the form and orientation of sand dunes, and the moulding of beach profiles. None the less, they concluded that "Analyses of the transport of sediment by various media indicate that a large portion of the 'work' is performed by events of moderate magnitude which recur relatively frequently rather than by rare events of unusual magnitude". If this is true for dynamic sediment transport processes, how more likely it is to be true for weathering processes, for these tend to come into the category of work that Wolman & Miller (p. 73) likened to that of a dwarf rather than to that of a man or a huge giant. "A dwarf", they averred, "works steadily and is rarely seen to rest."

In a later work Leopold, Wolman & Miller (1964) did give more explicit attention to weathering as a geomorphological process, stating on p. 97 that 'As a necessary prelude to erosion and mass wasting, weathering is a fundamental geomorphic process.' However, they went on to add that 'The importance of weathering in the preparation of land surfaces for action by agents of landscape sculpture is recognized; however, the detailed knowledge of

0044–2798/99/0115–0175 $ 3.75

weathering is meager and restricted to certain aspects.' In the 1960s detailed quantitative weathering studies were in their infancy, with progress being slow until the advent of suitable field (e.g. micro-erosion meter) and laboratory (e.g. scanning electron microscopy) techniques which permitted observations of the relatively slow and subtle weathering processes.

Weathering and its geomorphological role gives an ideal opportunity to investigate the notion that geomorphological processes (and/or events) of varying magnitudes and frequencies have different effects upon the landscape, in terms of work done. Weathering is, in comparison with most other earth surface processes, a slow but nevertheless highly important process, which often provides a constraint upon geomorphological change. In recent years laboratory-based, field-based and modelling studies of weathering have grown hugely (although numerical modelling techniques have not, as yet, been applied as widely within weathering studies as within other parts of geomorphology), and we now have much more data to utilise than in 1960 when WOLMAN & MILLER were writing. However, there are several aspects of the problem of conceptualising the frequency and magnitude of weathering which require initial clarification.

Firstly, we must identify what weathering processes (or groups of processes) are under consideration, as it is becoming increasingly clear that weathering consists of a huge number of individual, often highly inter-related mechanisms of rock breakdown, which are still imperfectly understood. Many studies have investigated the action of a limited range of weathering processes (e.g. by designing experiments in the laboratory or field to look at salt, frost or biological weathering), whilst others have concentrated on the broad results of a combination of weathering processes, as, for example, in studies of solutional denudation within drainage basins. Secondly, we must clarify how the effectiveness of individual weathering processes (or groups of weathering processes) is to be measured, or, to put it another way, how we assess the 'work done' by weathering. Are measurements of rates of weathering the object of interest? If so, there is much debate about the meaning and comparability of weathering rate measurements which may be derived in very different ways. Furthermore, there is much argument over the variation of weathering rates over different timespans (see, for example, the paper by COLMAN in 1981) which raises the problem of temporal scales in relation to the question of magnitude and frequency of weathering processes/events. One could also argue that the geomorphological effectiveness of weathering is best measured in terms of the amount of landform change occurring, rather than in the rate of weathering itself. In terms of landslides, for example, the geomorphological effectiveness of weathering may be a non-linear function of weathering rate. Depending upon the exact situation (which itself reflects he contigent nature of geomorphological systems) weathering may trigger a mass movement (e.g. FAN et al. 1996) by reducing catastrophically the strength of part of the profile below a critical threshold, or conversely, the same rate of weathering within a different context may deactivate a slip-prone slope by producing debris accumulating at the toe at a faster rate than erosion can remove it (e.g. coastal landslide examples). Perhaps this issue relates to spatial scale, combined with issues of location, of the geomorphological system of interest. At the heart of many of the problems discussed above is the difficulty of separating weathering from erosion in any meaningful sense; the two are inexorably intertwined in real geomorphological situations. Thus, it could be said that the whole task of examining the concept of frequency and magnitude in relation to weathering as a discrete entity is flawed, but on the other hand it is vital not to ignore weathering as a component of geomorphological change.

In the rest of this paper we investigate the concept of frequency and magnitude firstly in the context of a range of important, frequently studied weathering processes and their effectiveness in terms of rock breakdown, and secondly in terms of the totality of weathering and its effectiveness in terms of larger-scale geomorphological change. Finally, we examine the importance of the ideas of frequency and magnitude as applied to weathering in today's geomorphology.

Dissolved load transport and solutional weathering

For many geomorphologists, chemical weathering is intimately bound up with solutional erosion and the removal of material from drainage basins in solution. In 1964 LEOPOLD et al. further developed the concept of frequency and magnitude and analysed rates of removal of material in solution in streams (p. 77).

> "The data indicate that the greater the percentage of the total load carried in solution, the more significant from an erosional standpoint will be the flows of smaller magnitude."

This was a theme that was returned to by GUNN (1982) who looked at the frequency and magnitude characteristics of dissolved flows in 24 basins, most of which were underlain dominantly by carbonates and which were mostly located in cool temperate-humid environments. GUNN found that his conclusions were in some respects at variance with those of WOLMAN & MILLER, asserting (p. 509) "a very large part of the work done in the transport of dissolved material is by flows *greater than* the mean.". In non-carbonate basins from his sample he found "high magnitude, low frequency flows operational for only 5 per cent of the time transport more material than the flows less than or equal to the median". He attributed the differences between his data and those of WOLMAN & MILLER to the fact that their data drew heavily upon semi-arid catchments in the USA "where the solute concentrations associated with high flows may be up to an order of magnitude lower than those of baseflow. As a result, high flows transport a relatively small proportion of the total load." However, some further support for WOLMAN & MILLER's viewpoint comes from WEBB & WALLING (1982) who investigated the frequency and magnitude characteristics of solute transport in some drainage basins in Devon (S.W. England) and found (p. 21) 'that solute removal is mainly accomplished by flows at less than half-bankfull stage.' As WALLING & WEBB (1983) pointed out, on the basis of a global study of solute loads, concentrations of solutes in peak flows will tend towards those found in the storm rainfall (i.e. very low) due to the short residence time of much of the storm runoff.

In simple terms, dissolution of limestone and other soluble rocks and minerals (a weathering process) only occurs when water is present, and solutional denudation (a combination of the weathering process of dissolution and the removal of the dissolved products) only occurs where water flows significantly. Solution is one of the key examples of the intertwining of weathering and erosion (and most of the measurements of solution are measurements of the overall denudation and removal of dissolved load from catchments), as the movement of water facilitates further dissolution as well as removing the products of dissolution. The thermodynamics and kinetics of the dissolution reaction control the varying importance of high magnitude/low frequency flows versus low magnitude/high frequency flows in any one circumstance, thus it is very difficult to generalise. Where reactions are transport limited (i.e. the reaction occurs so fast that it is the supply of reactants and removal of products that is the key control) high magnitude flow events will produce more dissolution than low magnitude events. Where reactions are reaction rate limited (i.e. where the reactions are slow relative to the rate of flushing) a high magnitude flow event should have little impact on dissolution rates. According to BERNER (1978) mineral solubility is a major control on whether transport or reaction rate is the major control on dissolution rate, with higher mineral solubilities favouring transport control (e.g for gypsum, halite and other commonly occurring salts). If BERNER's model is correct, then mineral solubility should control the importance of flows of differing magnitude and frequency to dissolution rates. Recent work by TRUDGILL et al. (1996) explores the importance of mineral solubility in controlling the spatial and temporal characteristics of mineral dissolution within hillslope profiles and shows how, additionally, particle size distributions and soil moisture conditions play a role.

Recent work on karst hydrological systems indicates that dissolution rates can vary seasonally. For example, KIEFER (1994) studied temporal cycles in karst denudation in N W Georgia, U.S.A., and found solute transport (and thus karst denudation) in springs and streamwaters to be highly seasonal, relating primarily to seasonal variations in discharge. At the annual timescale, then, clearly notions of magnitude and frequency are important to explaining variations in dissolution rate. Furthermore, KIEFER found additional stochastic variation in dissolution rate related to shorter-term variations in discharge. As he puts it 'The resulting behaviour of TDI (the total denudation index) is one of 12 month 'long wave' cycles with embedded short-term stochastic variation.' (KIEFER 1994: p. 231). Similar findings also come from a consideration of speleothem luminescence microbanding (SHOPOV et al. 1994) which appears to show climatic control over a range of timescales, from storm-induced groundwater recharge events to seasonal and longer-term variations in calcite deposition. These examples illustrate a very important point about the relevance of WOLMAN & MILLER'S ideas to weathering studies today; that is, it is difficult to separate out fluctuations in process intensity occurring over different timescales. To put this more simply, short term 'events' are part of a whole hierarchy of geomorphological events and processes, and cannot easily be treated in isolation.

However, even high magnitude low frequency flows achieve relatively little in terms of landform development on short and medium time scales. For the most part, particularly in groundwater fed systems (including Karst), solutional processes are slow and not subject to dynamic thresholds. As DRAKE remarked (1984: p. 211), "The amount of solution over a period long enough to be geomorphologically significant is, therefore, the integral of a continuous function of time".

Weathering by fire

Turning to physical or mechanical weathering processes, it is apparent that one aspect of frequency and magnitude is particularly important, i.e. cyclical behaviour of climatic parameters such as temperature and moisture. However, in some cases less regularly distributed events may have great significance. One physical weathering process that may occur infrequently, but have a great influence on rock disintegration, is that associated with large fires. There are many observations that spalling of rock outcrops is caused by severe fires, the recurrence interval of which varies from an almost annual occurrence in some African savannas to tens or even hundreds of decades in some other environments. Because of the build up of combustible materials, infrequent fires may be able to achieve higher temperatures and therefore greater potential to cause rock disintegration. WEIN & MACLEAN (1983) provide a review of fire frequencies while GOUDIE et al. (1992) and ALLISON & GOUDIE (1994), from laboratory simulation, indicate how rock strength changes during a fire relate to different temperature and moisture conditions. They noted that substantial changes can occur in rock properties at temperatures as low as 200 °C, although the presence of moisture within the rock will tend to raise the temperature at which significant effects can be recorded. They also found that the duration of fire that was required to affect materials was not great, and in fact is sometimes less than one minute.

Further studies are required on the temporal distribution of geomorphologically-effective fire events and whether, indeed, bigger fires are always more effective than less dramatic events. The location of a fire event within the burning history of an area may also be crucial in explaining its geomorphological role. If, for example, a fire occurs not long after a previous large burn then the already spalled rock may respond differently to a rock which has not been affected by fire for many years. In most cases, fire primarily affects vegetation and the nature of the vegetation is crucial in determining the nature of the fire at ground level, and thus its geomorphological effectiveness.

Thermoclasty and haloclasty

Insolation weathering (thermoclasty), if indeed it is effective at all, which is still debated, can operate relentlessly on a daily basis in response to the daily temperature cycle or to more frequent perturbations caused by the presence of clouds etc. (JENKINS & SMITH 1990). Rapid drenching (and associated cooling) by rain may on a less regular basis creates a short, sharp shock for rock surfaces. Rock properties (such as albedo and thermal conductivity) will determine how different rocks respond to particular temperature regimes (WARKE & SMITH 1994) (Fig. 1). Thus, considerations of the effectiveness of different magnitude and frequency of events cannot be considered in isolation from the nature of the rock, or weathering surface, itself. The debate over the effectiveness of thermoclasty and wetting and drying as weathering processes illustrates the difficulty of identifying which weathering processes are operating in any field situation, and also the difficulty of relating laboratory simulations to the real world. These problems, in turn, make it quite difficult to come to any firm conclusions over importance of weathering-producing events of varying magnitude and frequency. HALL (1993) has attempted to monitor rock temperatures and moisture levels on Livingston Island, in the South Shetland islands, Antarctica using rock blocks. He found that wetting and drying events (but not freezing events) were common on open and north facing sites, whereas south facing sites showed continually high rock moisture levels favouring chemical weathering. Thus, this study suggests that the magnitude and frequency distribution of weathering-effective events varies hugely over small areas and may be crucial in determining which weathering process is most active, as well as the efficiency of that process.

Salt weathering (haloclasty) is another weathering process which may occur in response to frequent environmental cycles. For example, in hot deserts ground surface temperatures may frequently cycle across the approximately 32 °C temperature threshold that determines whether or not both sodium sulphate and sodium carbonate occur in a hydrated or non-hydrated form. If one makes the assumption that an air temperature of c. 17 °C translates into a day time rock surface temperature of c. 32 °C (the transition temperature for sodium carbonate and sodium sulphate) then that value is crossed daily upon between 5 and 9 months of the year depending on the desert station selected. In other words, there may typically be around 150 to 270 days in the year in which rock temperature conditions are favourable to the salt hydration mechanism of rock decay occurring.

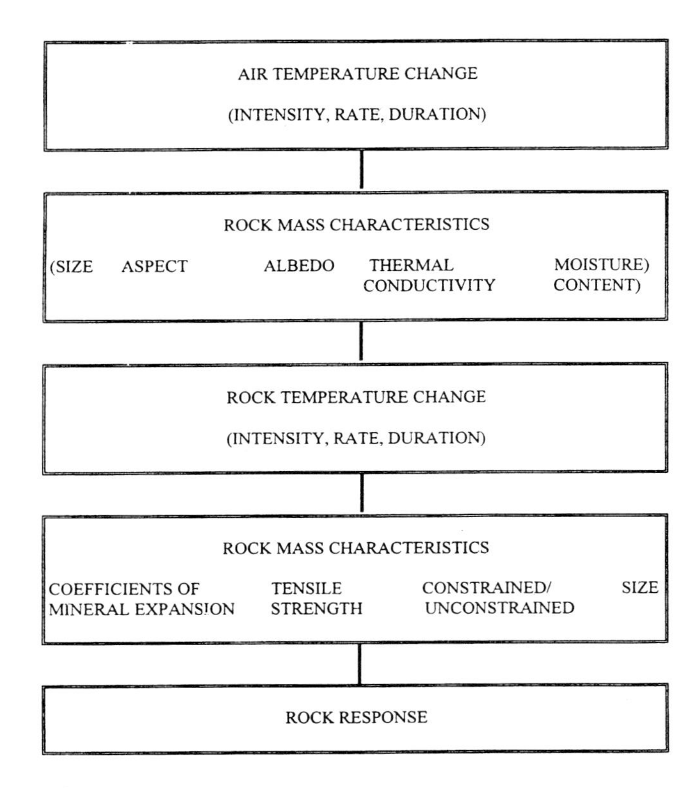

Fig. 1. Insolation weathering (thermoclasty).

Even salts, such as sodium chloride, which do not hydrate at above freezing ambient temperatures, may because of their deliquescence, alternately dissolve and re-crystallise during the course of a normal daily cycle of temperature and humidity. Other salts (e.g. sodium nitrate or magnesium sulphate) may respond to rather more extreme and possibly rarer thresholds (GOUDIE 1993) , but even in such cases suitable conditions may persist for several months in the year for daily cycles to operate. Likewise in a coastal spray zone, salt weathering is likely to occur on many days in the year, particularly under tropical conditions when strong evaporite concentration takes place.

Fig. 2 illustrates some of the main types of cycle that may influence the frequency of salt weathering activity. This figure illustrates the complexity of trying to make statements about the relative effectiveness (in terms of weathering) of events of different magnitudes and frequencies, as a range of event cycles operate together to produce salt weathering. As also seen in our discussion of weathering by fire, the effectiveness of salt weathering events will be partially controlled by the effectiveness of previous events, which may have altered the rock surface conditions significantly.

In some cases the past can have a large control on present weathering. For example, on calcareous building surfaces affected by polluted air, a major form of weathering results from the conversion of calcite to gypsum, producing a gypsum crust. Over time, this crust (because of differences in properties between it and the underlying stone) tends to blister and exfoliate,

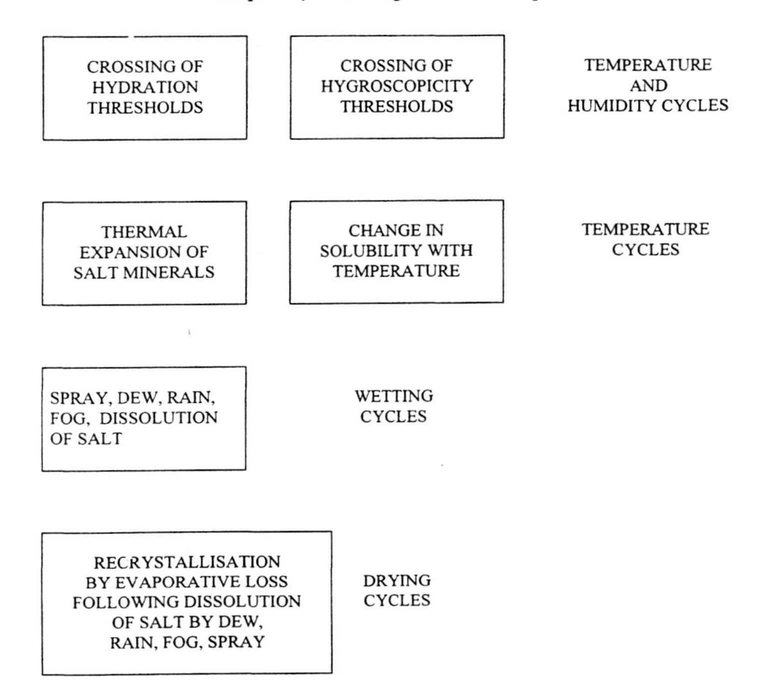

Fig. 2. Cycles in salt weathering (haloclasty).

revealing a highly weathered, friable layer underneath. A highly complex series of weathering processes operating at the micro and meso-scales are involved in this crust formation, blistering and exfoliation. Untangling the influence of differing events (e.g. air pollution episodes, temperature and humidity cycles) is hugely difficult.

Frost weathering

The relationships between frost weathering and the frequency and magnitude of changes in temperature and moisture are fraught with controversy and uncertainty. As THORN (1979: p. 211) pointed out, "Despite numerous assertions of the efficacy of freeze-thaw cycle weathering the process lacks definitive verification .... the paucity of field process studies means that the presence of the requisite conditions had not been established in most alpine and arctic regions." His own field study in the Colorado Front Range, USA, "casts serious doubt on the real effectiveness of freeze-thaw weathering in this classic nival environment." (p. 226).

Experimental simulations in the laboratory show no very consistent pattern of results. As MCGREEVY (1981: p. 57) pointed out, "Because of the number of factors to be considered (rock type, amount of moisture, intensity of freezing, amplitude of freeze-thaw cycles, number of freeze-thaw cycles), the results of the experiments are not strictly comparable in most cases and it is not easy, therefore, to make valid generalisations. Conflicting reports have thus ensued." So, for example, TRICART's (1956) pioneering simulations, using "Siberian" and "Icelandic" cycles indicated that the former were more effective in promoting rock breakdown than the latter, suggesting that the number or frequency of freeze-thaw cycles was less important than the intensity of freezing. JERWOOD (1987) in her study of combined salt and frost weathering also found that, in general, intense regimes are more effective in promoting breakdown than mild regimes. On the other hand, some other experimenters obtained results which were at stark variance with this view (e.g. WIMAN 1963 and POTTS 1970), suggesting that freezing intensity was less important than the number of temperature oscillations above and below freezing.

Some of the experimental work on frost weathering can also be criticised, in part because the experimental conditions may be based on the use of air temperature cycles which have no clear relationship to the temperature conditions existing in a particular rock mass, but also because there has been some emphasis on creating experimental conditions which favour fast rates of breakdown which may have little basis in reality. Moisture regimes, for example, have often been used which exceed those likely to be attained naturally (MCGREEVY & WHALLEY 1985).

More theoretical work on frost weathering throws some light on the critical conditions necessary for certain frost weathering mechanisms to operate. For example, WALDER & HALLET (1985) suggested that crack propagation was effective in a temperature range between -4 and -15 °C with very slow cooling rates (0.1–0.5 °C per hour). Fast cooling rates and very low temperatures will tend to preclude migration of water to freezing centres, particularly in rocks with large pores, thereby preventing mechanisms associated with ice segregation and crack propagation from operating. They argued that crack growth does not require the temperature to oscillate around 0 °C and that it is prolonged freezing periods which favour crack growth.

There are unfortunately relatively few studies that relate field records of freeze-thaw weathering to field records of actual temperature and moisture regimes. A notable exception here is the work of FAHEY & LEFEBVRE (1988). They found in Ontario that rockfall debris liberation, an indicator of weathering, if there was an adequate moisture supply, was favoured by long periods (three to five days) of comparatively intense freezing (with temperatures at -8 to 10 °C at 1 cm depth) followed by temperatures well over 0 °C. They found that shorter more frequent events (one to two days) were less productive, suggesting that the duration of freeze is more important than intensity.

It is also important to recognise that rock characteristics (especially pore size) may determine the temperature at which freezing takes place in the rock body, that small laboratory samples will undergo very different temperature cycles than in situ bedrock masses, and that solute content of pore water may also affect freezing behaviour (HALL 1988).

An interesting predictive model of frost weathering has been developed by MATSUOKA (1990) and tested in Japan, Svalbard and Antarctica (MATSUOKA 1991). The model relates frost shattering rate to the effective freeze-thaw frequency, the degree of saturation and the rock mass tensile strength. Based on laboratory-derived criteria, and calibrated using field data the model states:

$$R_s = 1.5 \times 10^{-3} S_r^{16} N_e S_{tf}^{-2}$$

Where: R_s = Rate of frost shattering
S_r = Degree of saturation
N_e = Number of effective freeze-thaw cycles
S_{tf} = Rock mass tensile strength

This equation shows the huge importance of degree of saturation to the rate of frost shattering. Even where there are many effective freeze-thaw cycles, if rock saturation is too low there will be minimal frost damage.

Fig. 3 illustrates some of the factors involved in determining the frequency and effectiveness of frost weathering activity, and indicates how it is not simply a question of the magnitude and frequency of climatic events, but how there is also a strong influence of rock properties. Furthermore, Fig. 3 shows how the response of the rock can influence the future effectiveness of frost weathering events through its control of rock strength and rock surface microclimate (where debris is produced which 'blankets' the underlying fresh rock surface).

Biological weathering

In the 1960s, when WOLMAN & MILLER produced their major ideas on magnitude and frequency in a geomorphological context (WOLMAN & MILLER 1960, LEOPOLD et al. 1964), the impor-

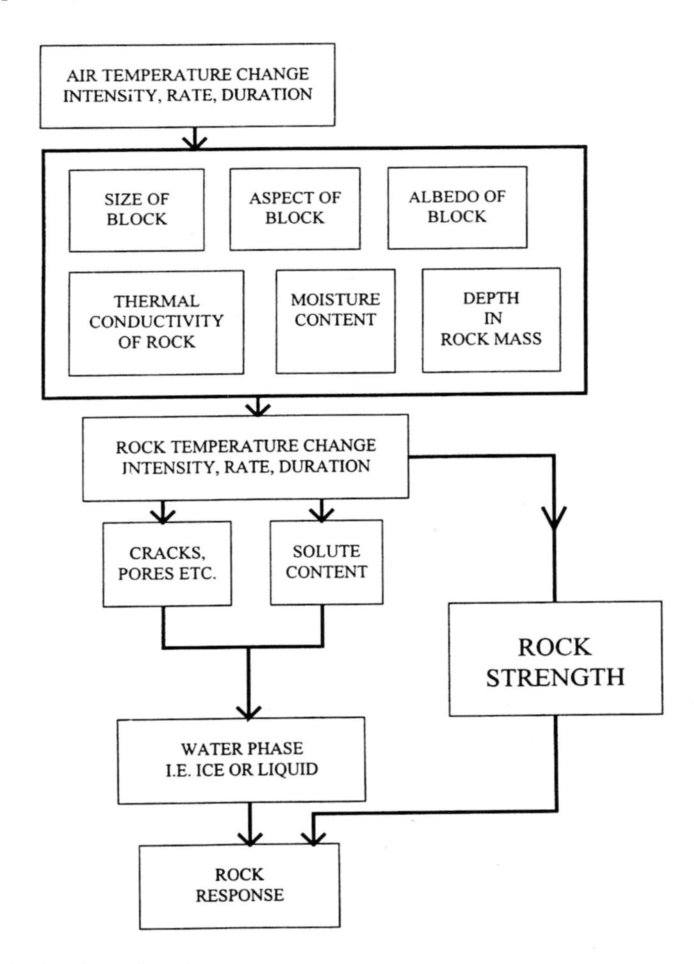

Fig. 3. Frost weathering (cryoclasty).

tance of biological agents of weathering had received very little serious attention. Since then there has been an explosion of interest in how organisms influence weathering, both directly and indirectly, through a range of chemical and mechanical processes. Work from a range of environments, as reviewed below, shows how organisms sometimes respond to climatic events, further reinforcing their effects; often mediate the impact of climatic events of varying magnitude and frequency on rock surfaces; and also introduce their own events operating over different magnitude and frequency distributions.

Lichens growing on bare rock surfaces have been shown to have a range of weathering effects. The early investigations of FRY (1927) showed the huge potential for mechanical weathering by lichens, especially on wetting and drying. MOSES & SMITH (1993) provide additional experimental evidence for the effectiveness of such action on limestone surfaces, and thus it seems that lichens may magnify the effects of wetting and drying on some rock surfaces. Further work needs to be done to see whether the magnitude and frequency of wetting and drying cycles has any major control over the effectiveness of the biomechanical action produced.

In other circumstances, organisms may act to reduce the effectiveness of climatic cycles on physical weathering processes and dull the impact of major rainfall events on dissolution. In terms of physical weathering, for example, biofilms and other rock surface organic coatings may set up a more consistent microenvironment reducing the impact of temperature and humidity fluctuations. Thus, in the central Namib desert, coloured lichens growing on boulders may alter the thermal response of the rock, as well as keeping the surface relatively moist, reducing the reactive surface area, and protecting the underlying rock from airborne salts. However, in this area lichens grow preferentially on less exposed boulder surfaces, and above the zone exposed to maximum attack from saline ground- and surface water, thus showing a spatially patchy distribution. Furthermore, the protective role of lichens and biofilms will only last while they are growing, and extreme climatic events (e.g. periods of severe drought, or high winds) may disturb the cover and thus reinstigate physical weathering activity.

In terms of dissolution within drainage basins, forest ecosystems have been found to play various roles. Work by VELBEL (1995), for example, illustrates how in many catchments biological uptake of mineral nutrients makes an important contribution to overall dissolution rates especially where forests are aggrading.

Organisms also bring to weathering systems an important dynamic of their own, setting up a whole new range of events of varying magnitude and frequency. For example, organism processes vary over the diurnal cycle, such as the contribution of biologically-produced carbon dioxide to the dissolution of limestone on walls in arid areas and in coastal pools which varies clearly between night and day (DANIN 1983, SCHNEIDER 1976). On a slightly longer timescale, but also within coastal and arid settings, grazing by molluscs and other animals on rock-dwelling algae and bacteria produces an important erosion of the weathered rock surface, thereby paving the way for future weathering activity. Several studies have been made of such combined bioerosive processes, but few workers have attempted to quantify the spatial and temporal patterning of grazing events and their impacts on weathering rates (SHACHAK et al. 1987, CHAZOTTES et al. 1995). Finally, biological effects on weathering can vary hugely over seasons. Fig. 4 shows some of the influences on biological weathering, and the different magnitude and frequency distributions of events and processes involved.

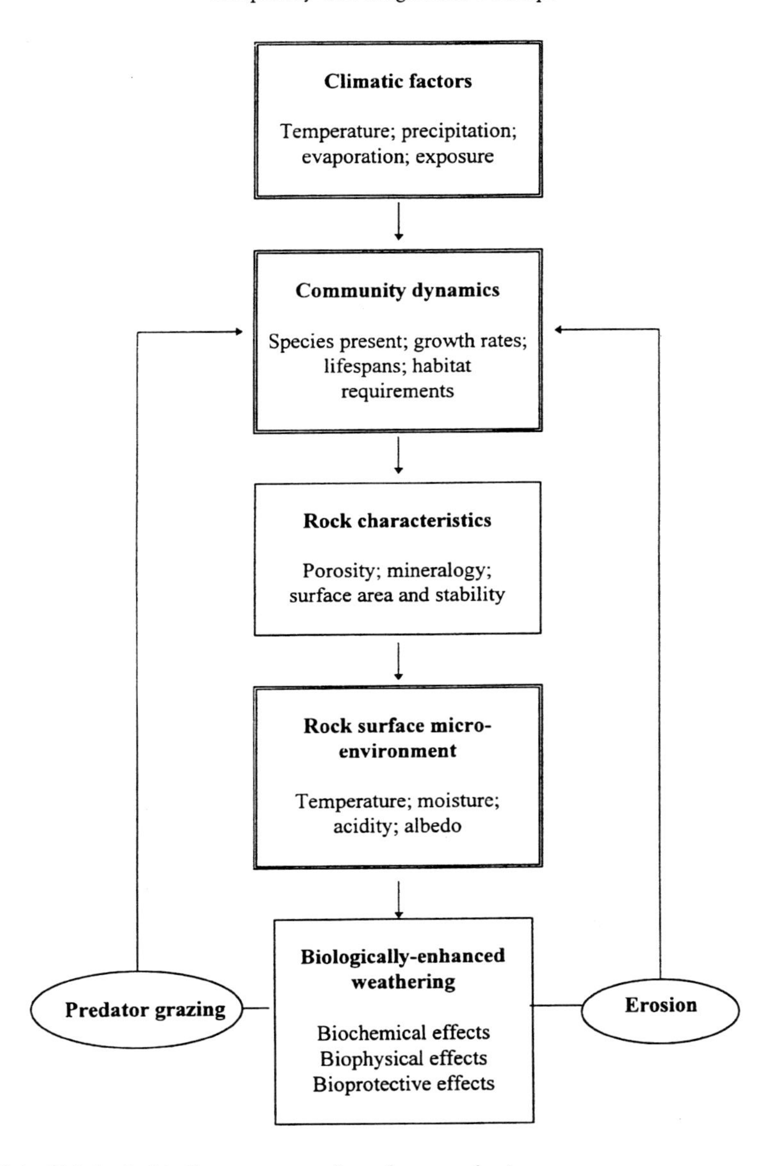

Fig. 4. A model of biological influences on rock surface weathering.

N.B. All boxes with double lines around have magnitude and frequency distributions which affect the progress of weathering.

Weathering and geomorphological change

The discussion above has identified a range of ways in which the notion of magnitude and frequency has relevance to various individual weathering processes. However, in most real world environments, a host of weathering processes act together to produce an overall weathering rate, or jointly make a contribution to landform development. How relevant are ideas of magnitude and frequency to investigating overall weathering rates and their contribution to geomorphological change? In most environments weathering rates are very slow, usually much less than 1 mm per year, and thus the contribution of shorter term events (e.g. those lasting one or two days) is difficult to measure using even the most accurate techniques. Even if almost all one year's weathering was concentrated into one event it would be very difficult to

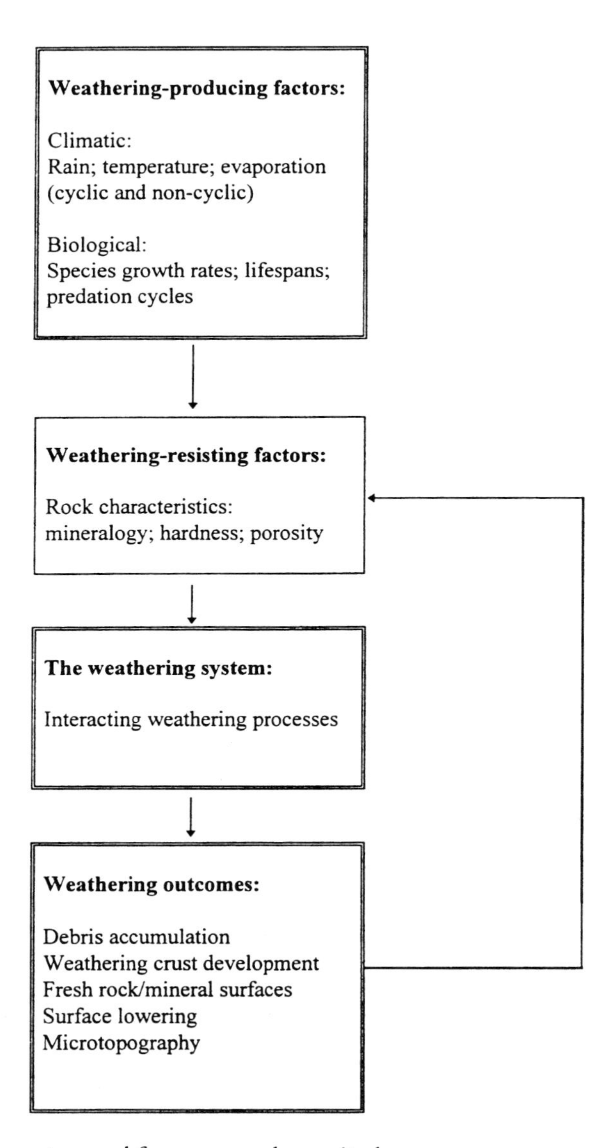

Fig. 5. The weathering system and frequency and magnitude.

N.B. Each of the three boxes with double lines around have important magnitude and frequency distributions which affect the progress of weathering.

measure this in the field in terms of surface lowering or topographical change. The sluggish and variable response of many surfaces also requires comment here. Rocks which weather through granular disintegration, for example, (such as the calcarenites studied by SPENCER 1981 on Grand Cayman Island) may show sporadic weathering as measured by surface lowering, which is probably unrelated to weathering events, but may be highly related to climatic events (such as storms or high winds) which remove the weathered debris from the surface. Indeed in many weathering systems, removal of debris provides a clear control on future weathering rates and thus the magnitude and frequency distribution of events which remove debris is also of great importance.

Conclusions

As shown in Fig. 5, there are three main interrelated components of weathering systems to which ideas of magnitude and frequency of events are relevant. Firstly, as illustrated in several examples the magnitude and frequency distribution of events producing weathering is of importance, and is largely controlled by climate. However, several of our examples have shown that even for one weathering process a range of different climatic events, each possessing different magnitude and frequency distributions, may be involved including cyclical and non-cyclical events. Furthermore, these event distributions are embedded in longer-term variations in climatic parameters (e.g. seasonal, decadal and century-scale) which also exert a control on the behaviour of weathering over short timespans.

Secondly, the actual weathering processes themselves have a distinctive magnitude and frequency distribution, related to the rock or mineral properties of the surfaces undergoing weathering as well as to the events discussed above. In the sort of geomorphological circumstances investigated by WOLMAN & MILLER (1960) the resistance to geomorphological change was low relative to the forces. However, in the case of most weathering systems, resistance can be high relative to the forces exerted, introducing a further element of complexity into the system. Thus, for example, several temperature and humidity cycles may be needed before any actual rock breakdown occurs in a salt-impregnated rock, because of the high resistance to change. Thirdly, the future action of weathering-inducing events is highly influenced by the nature of removal of debris in all but congruent dissolution situations, thus introducing some feedback into the simple notion of magnitude and frequency. Finally, many of the examples presented above illustrate that ideas of magnitude and frequency applied to weathering need also to consider thresholds, that size of events may be very important in that some small events may do nothing because a threshold has not been reached.

References

ALLISON, R.J. & A.S. GOUDIE (1994): The effects of fire on rock weathering : an experimental study. – In: ROBINSON, D.A. & R.B.G. WILLIAMS: (eds.): Rock Weathering and Landform Evolution. – 41–56, Wiley, Chichester.

BERNER, R.A. (1978): Rate control of mineral dissolution under earth surface conditions. – Amer. Journ. Sci. **278**: 1235–1252.

CHAZOTTES, V., T. LE CAMPION-ALSUMARD & M. PEYROT-CLAUSADE (1995): Bioerosion rates on coral reefs: interactions between macroborers, microborers and grazers (Moorea, French Polynesia). – Palaeogeogr., Palaeoclimat., Palaeoecol. **113**: 189–198.

COLMAN, S.M. (1981): Rock-weathering rates as functions of time. – Quatern. Res. **15**: 250–264.

DANIN, A. (1983): Weathering of limestone in Jerusalem by cyanobacteria. – Z. Geomorph. N.F. **27**: 413–421.

DRAKE, J.J. (1984): Theory and model for global carbonate solution by groundwater. – In: LAFLEURE, R.G. (ed): Groundwater as a geomorphic agent. – 210–226, Allen and Unwin Inc., Boston.

FAN, C.H., R.J. ALISON & M.E. JONES (1996): Weathering effects on the geotechnical properties of argillaceous sediments in tropical environments and their geomorphological implications. – Earth Surf. Proc. Landforms **21**: 49–66.

FAHEY, B.D. & T.H. LEFEBVRE (1988): The freeze-thaw weathering regime at a section of the Niagara escarpment on the Bruce Peninsula, Southern Ontario, Canada. – Earth Surf. Proc. Landforms **13**: 293–304.

FRY, E.J. (1927): The mechanical action of crustaceous lichens on substrata of shale, schist, gneiss, limestone and obsidian. – Ann. Botany **XLI**: 437–460.

GOUDIE, A.S. (1993): Salt weathering simulation using a single-immersion technique. – Earth Surf. Proc. Landforms **18**: 369–376.

GOUDIE, A.S., R.J. ALLISON & S.J. MCCLAREN (1992): The relations between Modulus of Elasticity and temperature in the context of the experimental simulation of rock weathering by fire. – Earth Surf. Proc. Landforms **17**: 605–615.

GUNN, J. (1982): Magnitude and frequency properties of dissolved solids transport. – Z. Geomorph. N.F. **26**: 505–511.

HALL, K. (1988): A laboratory simulation of rock breakdown due to freeze-thaw in a maritime Antarctic environment. – Earth Surf. Proc. Landforms **13**: 369–382.

HALL, K. (1993): Rock moisture data from Livingston Island (Maritime Antarctic) and implications for weathering processes. – Permafrost and Periglacial Processes **4**: 245–253.

JENKINS, K.A. & B.J. SMITH (1990): Daytime rock surface temperature variability and its implications for mechanical rock weathering : Tenerife, Canary Islands. – Catena **17**: 449–59.

JERWOOD, L.C. (1987): Laboratory simulations of frost and salt weathering, with particular reference to chalk. – D.Phil. thesis (unpubl.), University of Sussex.

KIEFER, R.H. (1994): Temporal cycles of karst denudation in northwest Georgia, U.S.A. – Earth Surf. Proc. Landforms **19**: 213–232.

LEOPOLD, L.B., M.G. WOLMAN & J.P. MILLER (1964): Fluvial processes in Geomorphology. – W.H. Freeman, San Francisco and London.

MCGREEVY, J.P. (1981): Some perspectives on frost shattering. – Progr. Phys. Geogr. **5**: 56–75.

MCGREEVY, J.P. & W.B. WHALLEY (1985): Rock moisture content and frost weathering under natural and experimental conditions : a comparative discussion. – Arct. Alpine Res. **17**(3): 337–46.

MATSUOKA, N. (1990): The rate of bedrock weathering by frost action: Field measurements and a predictive model. – Earth Surf. Proc. Landforms **15**: 73–90.

MATSUOKA, N. (1991): A model of the rate of frost shattering: Application to field data from Japan, Svalbard and Antarctica. – Permafrost Periglac. Proc. **2**: 271–281.

MOSES, C.A. & B.J. SMITH (1993): A note on the role of *Collema auriforma* in solution basin development on a Carboniferous limestone substrate. – Earth Surf. Proc. Landforms **18**: 363–368.

POTTS, A.S. (1970): Frost action in rocks : some experimental data. – Transact. Inst. Brit. Geograph. **49**: 109–24.

SCHNEIDER, J. (1976): Biological and inorganic factors in the destruction of limestone coasts. – Contrib. Sedimentol. **6**: 1–112.

SHACHAK, M., C.G. JONES & Y. GRANOT (1987): Herbivory in rocks and the weathering of a desert. – Science **236**: 1098–1099.

SHOPOV, Y.Y., D.C. FORD & H.P. SCHWARCZ (1994): Luminescent microbanding in speleothems: High-resolution chronology and paleoclimate. – Geology **22**: 407–410.

SPENCER, T. (1981): Microtopographic change on calcarenites, Grand Cayman Island, West Indies. – Earth Surf. Proc. Landforms **6**: 85–94.

THORN, C.E. (1979): Bedrock freeze-thaw weathering regime in an alpine environment, Colorado Front Range. – Earth Surf. Proc. **4**: 211–228.

TRICART, J. (1956): Etude expérimentale du probléme de la gélivation. – Biult. Periglacj. **4**: 285–318.

TRUDGILL, S.T., J. BALL & B. RAWLINS (1996): Modelling the solute uptake component of hillslope hydro-

chemistry: Are flow rates and path lengths important during mineral dissolution? – In: ANDERSON, M.J. & S.M. BROOKS (eds): Advances in hillslope processes – Vol 1: 295–324, John Wiley, Chichester.
VELBEL, M.A. (1995): Interactions of ecosystem processes and weathering processes. – In: TRUDGILL, S.T. (ed.): Solute modelling in catchment systems. – 193–209, John Wiley, Chichester.
WALDER, J. & B. HALLET (1985): A theoretical model of the fracture of rock during freezing. – Bull. Geol. Soc. Amer. **96**: 336–346.
WALLING, D.E. & B.W. WEBB (1983): The dissolved load of rivers: a global overview. – IAHS Publ. **141**: 3–20.
WARKE, P.A. & B.J. SMITH (1994): Short-term rock temperature fluctuations under simulated hot desert conditions: some preliminary data. – In: ROBINSON, D.A. & R.B.G. WILLIAMS (eds.): Rock Weathering and Landform Evolution. – 57–90, Wiley,Chichester.
WEBB, B.W. & D.E.WALLING (1982): The magnitude and frequency characteristics of fluvial transport in a Devon drainage basin and some geomorphological implications. – Catena **9**: 9–23.
WEIN, R.W. & D.A. MACLEAN (eds.) (1983): The role of fire in northern circumpolar ecosystems. – Wiley, Chichester.
WIMAN, S. (1963): A preliminary study of experimental frost weathering. – Geograf. Ann. **45A**: 113–21.
WOLMAN, M.G. & J.P. MILLER (1960): Magnitude and frequency of forces in geomorphic processes. – Journ. Geol. **68(1)**: 54–74.

Address of the authors: Professor A.S. GOUDIE and Dr. H.A. VILES, University of Oxford, School of Geography, Mansfield Road, Oxford, OX1 3TB, UK.

Z. Geomorph. N.F.	Suppl.-Bd. 115	191–217	Berlin · Stuttgart	Juli 1999

Geomorphology of faulting: The Wairarapa Fault, New Zealand

Rodney Grapes, Wellington, New Zealand

with 14 figures and 2 tables

Summary. The Wairarapa Fault is the most active fault of the North Island Dextral Fault belt in southern North Island, New Zealand, with a late Quaternary slip rate of between c. 6.0 ± 0.5 – 11.5 + 0.8/–1.1 m/ka. This accounts for a almost one third to half the 21 mm/yr strike-slip component of the Pacific – Australian plate convergence at 37 mm/yr. During the M8 + 1855 earthquake surface rupturing occurred along the Wairarapa Fault for a distance of at least 148 km. Geomorphic evidence indicates that faulting was accompanied by dextral slip of 12 ± 1 m, maximum uplift of 6.5 m near the SW end of the fault (Turakirae Head) that decreased inland NE along the fault to about 0.5 m, and regional uplift and north-west tilting of some 5000 km^2 of the southern part of the North Island. Geomorphic evidence of Holocene displacements are indicated from deformation of the Last Glacial aggradation surface, progressive displacement of a flight of younger degradation terraces (Waiohine River), uplifted beach ridges (Turakirae Head) and regional uplift pattern of the 6.5 ka shoreline across the Wellington peninsula. The pattern of Holocene vertical displacement is similar to that which occurred in 1855, i.e. an increase in the dextral/vertical ratio NE along the fault and NW decrease in the amount of regional uplift. Individual dextral displacements on the Wairarapa Fault can be described as seven seismic events with average slip values of between 10 and 12 m although at least three, possibly four, displacements are considered to be missing from the Holocene paleoseismic record. Maximum cumulative dextral displacement of 127 ± 3 m for the Last Glacial aggradation surface could have been the result of 10 or 11 large earthquakes with magnitudes similar to that in 1855. Ages of three seismic events prior to 1855 at 1.85, 4.1, and 5.7 ka have been estimated from the uplifted beach ridges at Turakirae Head and conform to a slip predictable pattern based on a 6.5 ka age for the highest stranded beach ridge, uniform uplift rate of 3.8 mm/yr and the date of the last uplift in 1855. Correlation of beach ridge uplifts with the four youngest dextral displacements on the Wairarapa Fault indicates a 6 ± 0.5 m/ka slip rate. Except for the beach ridges preserved at Turakirae Head, only a minor part of late Quaternary uplift of the Rimutaka Range west of the Wairarapa Fault can be attributed to vertical movement on the fault. The main cause can probably be attributed to crustal shortening and folding in combination with crustal flexuring to form the Wanganui Basin west of the Dextral Fault Belt in southern North Island.

Zusammenfassung. Die Wairarapa Verwerfung ist mit einer spätquartären Bewegungsrate von ca. 6,0 ± 0,5 bis 11,5 + 0,8/–1,1 m/ka die aktivste Verwerfung des Dextralen Verwerfungsgürtels im Süden der neuseeländischen Nordinsel. Diese Bewegung steht für annähernd ein Drittel bis zu einer Hälfte des 21 mm/Jahr Transversalverschiebungsanteils der Pazifisch-Australischen Plattenkonvergenz von 37 mm/Jahr. Während des M8 + Erdbebens von 1855 kam es zu Bruchbildung in der Erdoberfläche entlang der Wairarapa Verwerfung auf einer Länge von mindestens 148 km. Geomorphologische Indizien zeigen, daß die Bruchbildung mit einer dextralen Verwerfung von 12 ± 1 m, einer maximalen Erhebung von 6,5 m in der Nähe des südlichen Endes der Verwerfung (Turakirea Head), die nordöstlich ins Innere des Landes entlang der Verwerfung auf ca. 0,5 m abnimmt, und einer regionalen Hebung mit nordwestlich gerichteten Schrägstellung von ca. 5000 km^2 des nördlichen Teils der Nordinsel begleitet war. Geomorphologische Indizien von holozänen Verschiebungen sind die Deformation der letzten glazialen Ablagerungsoberfläche, die progressiven Verschiebungen jüngerer Denudationsterrassen (Waiohine River), gehobene Strandwälle (Turakirea Head) und das regionale Hebungsmuster der 6,5 ka Küstenlinie über der Wellington Halbinsel. Das Muster der holozänen vertikalen Verschiebungen ist ähnlich dem von 1855, d.h. es

0044-2798/99/0115-0191 $ 6.75

gibt eine Steigerung im dextral/vertikalen Verhältnis der Bewegungen entlang der Verwerfung und eine nordwestliche Abnahme des regionalen Hebungsbetrags. Einzelne dextrale Verschiebungen der Wairarapa Verwerfung können als sieben seismische Ereignisse mit durchschnittlichen Bewegungsraten von 10–12 m beschrieben werden, obwohl angenommen wird, daß mindestens drei, wahrscheinlich vier Ereignisse in den holozänen paläoseismischen Aufzeichnungen fehlen. Der maximale kumulative dextrale Verschiebungsbetrag von 127 ± 3 m für die letzte glaziale Ablagerungsoberfläche könnte das Ergebnis von 10 oder 11 großen Erdbeben mit einer Stärke vergleichbar mit dem von 1855 gewesen sein. Die Alter von drei seismischen Ereignissen vor 1855 wurden ausgehend von den gehobenen Strandwällen bei Turakirae Head auf 1,85; 4,1 und 5,7 ka geschätzt. Damit entsprechen sie einem Vorhersagemuster der Bewegungen, das auf einem Alter von 6,5 ka für den höchsten Strandwall, einer gleichförmigen Hebungsrate von 3,8 mm/Jahr und dem Jahr 1855 mit dem letztem Hebungsereignis basiert. Korrelationen von Strandwallhebungen mit den vier jüngsten dextralen Verschiebungen auf der Wairarapa Verwerfung lassen auf eine Bewegungsgeschwindigkeit von 6 ± 0,5 m/ka schließen. Außer bei den Strandwällen, die bei Turakirae Head erhalten sind, kann nur ein geringer Teil der spätquartären Hebung des Rimutaka Höhenzugs westlich der Wairarapa Verwerfung den vertikalen Bewegungen an der Verwerfung zugeschrieben werden. Die Hauptursache liegt wahrscheinlich in einer Krusteneinengung und -faltung zusammen mit Krustenverbiegung, die das Wanganui Becken westlich des Dextralverwerfungsgürtels im Süden der Nordinsel geformt haben.

Introduction

The age of a landscape often depends on how rapidly it evolves. In the case of faulting new landscape features are created almost instantaneously. In New Zealand, evidence from small scale geomorphic features typically provides information on the nature, timing and distribution of faulting and seismicity over a time scale of thousands to tens of thousands of years. Longer term tectonically controlled features such as the growth of a mountain range evolve over an order of magnitude longer and do not usually preserve evidence of short-term variations in uplift which although an integral part of the process, are superimposed on the long-term trend. In this paper, interaction between tectonic and geomorphic processes related to oblique plate convergence in the southern part of the North Island of New Zealand is examined with respect to (i). the nature, timing and magnitude of displacement on the Wairarapa Fault during a single earthquake in 1855 and the Holocene paleoseismic record as interpreted from small scale geomorphic features and regional uplift pattern, and (ii). the relationship between vertical displacement on the Wairarapa Fault and uplift of the Rimutaka Range west of the fault.

Tectonic setting

The southern North Island, New Zealand, is dominated by subduction of the oceanic Pacific Plate beneath the continental Australian Plate along the Hikurangi Trough. The morphotectonic elements of the convergent margin are shown in Fig. 1. Offshore the dip of the subduction zone is shallow but steepens westward so that beneath Wellington the subduction interface is about 25 km deep (Robinson 1986). The relative plate motion at the latitude of Wellington is about 37 mm/yr at 261° calculated using the NUVEL-1A Pacific/Australian plate motion and is oblique to the overall northeast trend of neotectonic structures in a fold-thrust belt developed at the eastern margin of the Australian Plate (Fig. 1). As a consequence of oblique subduction, the central and western parts of southern North Island are traversed by five NE

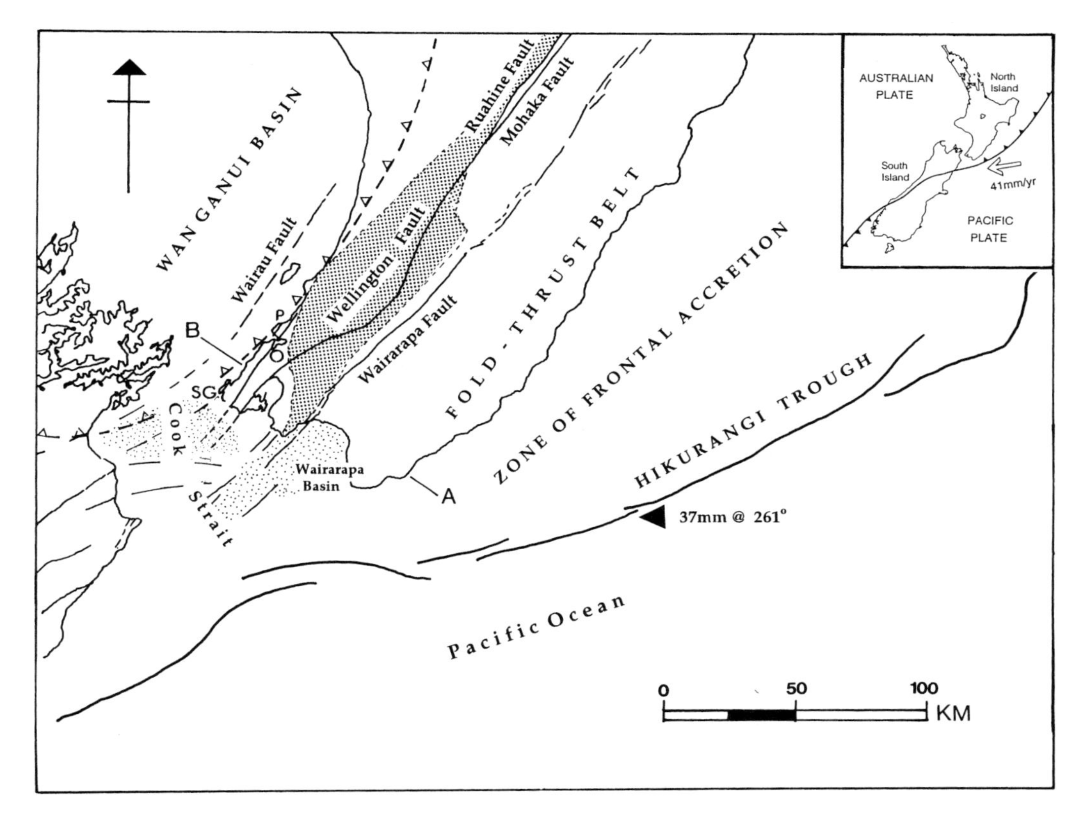

Fig. 1. Map showing tectonic setting of southern part of the North Island of New Zealand and major morphotectonic features associated with oblique convergence (37 mm/yr @ 261°) between the Australian continental plate and the Pacific oceanic plate along the Hikurangi Trough. Morphotectonic divisions after VAN DER LINGEN (1982), LEWIS & PETTINGA (1993). Grey toned area = Southern Axial Ranges (Rimutaka – Tararua ranges); **SGP** = Shepherds Gully and Pukerua faults; **O** = Ohariu Fault. Location of the Wairarapa Basin, Wairau Basin to west (light stippled areas) and associated faults from CARTER et al. (1989). Dashed line with arrows delineates eastern margin of the Wanganui Basin (STERN et al. 1992). Sections along Line A-B are shown in Fig. 14.

striking, active dextral faults termed the North Island Dextral Fault Belt (BEANLAND 1995) that extend from Cook Strait through the Southern Axial Ranges of basement greywacke. In Cook Strait the faults either die out or are associated with with fault-bounded basins of Cenozoic sediments (Fig. 1). In the Wellington region, the faults, from west to east, are the Wairau (offshore), Shepherds Gully – Pukerua (partly offshore), Ohariu, Wellington and Wairarapa Faults (Fig. 1) with estimated late Quarternary slip rates of 4–7 mm/yr (LENSEN 1976, GRAPES & WELLMAN 1986); 1.7 mm/yr (MIYOSHI et al. 1987); 1.5–2.5 mm/yr (MIYOSHI et al. 1987); 4.7–6.6 mm/yr (BERRYMAN 1990, GRAPES et al. 1984, VAN DISSEN et al. 1992, GRAPES 1993), and 11.5 mm/yr (GRAPES 1991) respectively. The faults show evidence of repeated movement over the past 125,000 years as determined from offset geomorphic features and the rates account for nearly all the margin-parallel 21 mm/yr strike slip motion proposed by BEANLAND (1995). West of the dextral fault belt a major lithospheric-scale downwarp forms the elliptical-shaped Wanganui Basin (Fig. 1) that contains about 4 km of shallow-water Plio-Pleistocene marine sediments (e.g. STERN et al. 1992).

The Wairapapa Fault

The Wairarapa Fault was first mapped as an active trace northeast from the coast at Palliser Bay for 124 km by ONGLEY (1943) who, according to the fashion of the time, recognised the vertical but not the dextral displacements (strike-slip movement on New Zealand's faults was not generally recognised until the 1950's). Later work along parts of the fault by WELLMAN (1955), LENSEN (1969) and LENSEN & VELLA (1972) established the dominant dextral nature of the faulting. GRAPES & WELLMAN (1988) presented detailed maps on a scale of 1:17,000 for the best defined 90 km length of the fault and listed 78 displacements.

A map showing the extent of the Wairarapa Fault as discussed in this paper is given in Fig. 2. Although the fault is mappable as a continuous line for about 75 km on scales of 1:250,000 or smaller, it is actually a series of straight sections separated by sinistral jogs of up to 500 m that have resulted in the development of prominent bulges (GRAPES 1991). Normal faults branch off from several of the mapped bulges and cross the Wairarapa Plain to the east (Fig. 2).

For most of its mapped length, the best defined part of the Wairarapa Fault forms the boundary between Late Jurassic greywacke of the Rimutaka Range to the west and Holocene fluvial gravels of the Wairarapa Plain to the east (Fig. 3). South of Lake Wairarapa the fault is difficult to trace but its continuation is indicated by a 100–200 m wide crush zone within the greywacke undermass. Where exposed near the coast at Palliser Bay the contact between crushed greywacke and Pleistocene-Holocene sediments is complicated by thrusting and possible gravity collapse (GRAPES & WELLMAN 1993). Offshore, seismic reflection profiles (CARTER et al. 1988) show that the Wairarapa Fault can be traced for at least 16 km southwest of Turakirae Head into Cook Strait (Fig. 1). From about 45 km to 70 km inland from the coast the fault cuts Miocene-Pliocene sediments and in places exhibits a reversal in throw, being upthrown on the southwest side rather than the northwest side as is usual to the south. Northeast of Mauriceville in northern Wairarapa (Fig. 2) the line of the fault is uncertain but its continuation is probably indicated by a series of subparallel traces within Late Miocene mudstone; the Dreyers Rock fault zone (ORBELL 1962), Alfredton and Pa Valley faults (LENSEN 1968, 1970, KELSEY et al. 1995) and further northeast to Hawkes Bay by way of the Makuri – Mangatoro faults (ONGLEY 1944, KELSEY et al. 1995) and the Waipukurau – Poukawa fault zone (KINGMA 1971, FROGGATT & HOWARTH 1980), (Inset in Fig. 2).

(i) *The 1855 earthquake*

On 23 January 1855 a large earthquake occurred that was felt over almost the whole land area of New Zealand from at least Auckland in the north to Dunedin in the south. The earthquake was severely damaging to the early settlements in the southern half of the North Island, particularly Wellington and Wanganui, and the northern part of the South Island, with a maximum intensity of MMIX-X. Surface rupturing occurred along the Wairarapa Fault and about 5,000 km^2 of land in the southern part of the North Island was uplifted and tilted with possible tectonic subsidence affecting a part of northeastern South Island. The earthquake resulted in ground damage over a wide area with expulsion of water and sand, lateral spreading, differen-

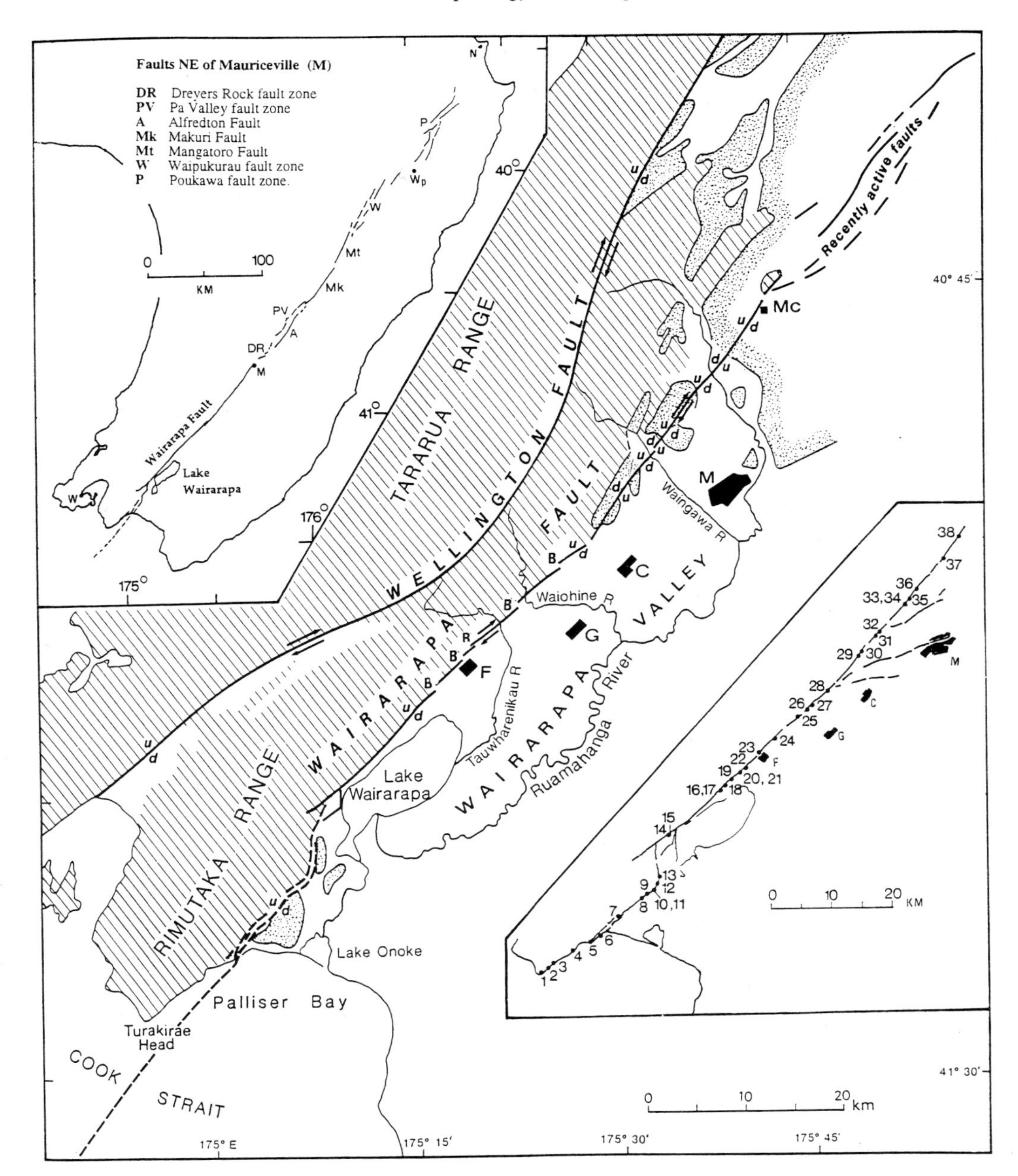

Fig. 2. Map showing extent of the Wairarapa Fault and associated geology. Cross hatching = greywacke of the Rimutaka and Tararua ranges; stippled area – Late Cenozoic sediments; B = bulge (see text); u, d = upthrown and downthrown side of the Wairarapa Fault; **F** = Featherston; **G** = Greytown; **C** = Carterton; **M** = Masterton; **Mc** = Mauriceville. **Upper inset** shows probable continuation of the Wairarapa Fault NE from Mauriceville to Hawkes Bay (see text); **W** = Wellington; **Wp** = Waipukurau; **N** = Napier. **Lower inset** shows numbered locations of 1855 displacement sites along the Wairarapa Fault (values listed in Table 1 and plotted in Fig. 6).

Fig. 3. Oblique aerial photo looking southwest across the trace of the Wairarapa Fault from Featherston (top right corner) to the Waiohine River (bottom). The Rimutaka Range lies to the west; the Wairarapa Plain to the east.

tial subsidence and landsliding. Seiches were observed to accompany the earthquake and a tsunami was generated in Cook Strait with waves up to 9–10 m high. The water level in Wellington Harbour oscillated for 8–12 hours in response to tsunami and seiche generated by the differential uplift across the harbour and probable horizontal displacement of the harbour floor caused by dextral slip along the Wairarapa Fault (GRAPES & DOWNES 1997). Numerous aftershocks were experienced, several damaging on the northeast coast of the South Island. An isoseismal map of the felt intensities of the 1855 earthquake together with areas of ground damage, and extent of faulting is shown in Fig. 4. Areas enclosed by MMIX, MMVIII and MMVII isoseismals of the earthquake greatly exceed those of the 1929 Buller and 1931 Napier earthquakes both with surface wave magnitudes of Ms 7.9. In particular, the major axis of the MMIX isoseismal of the 1855 earthquake is twice that of the 1929 earthquake, indicating a considerably higher magnitude estimated by DARBY & BEANLAND (1992) and GRAPES & DOWNES (1997) to be M8.2–8.3 making it New Zealand's largest historic earthquake.

Historic evidence of faulting : The best geological account by Sir CHARLES LYELL (1856, 1868) summarised information given to him in London by three eyewitnesses. LYELL stated that

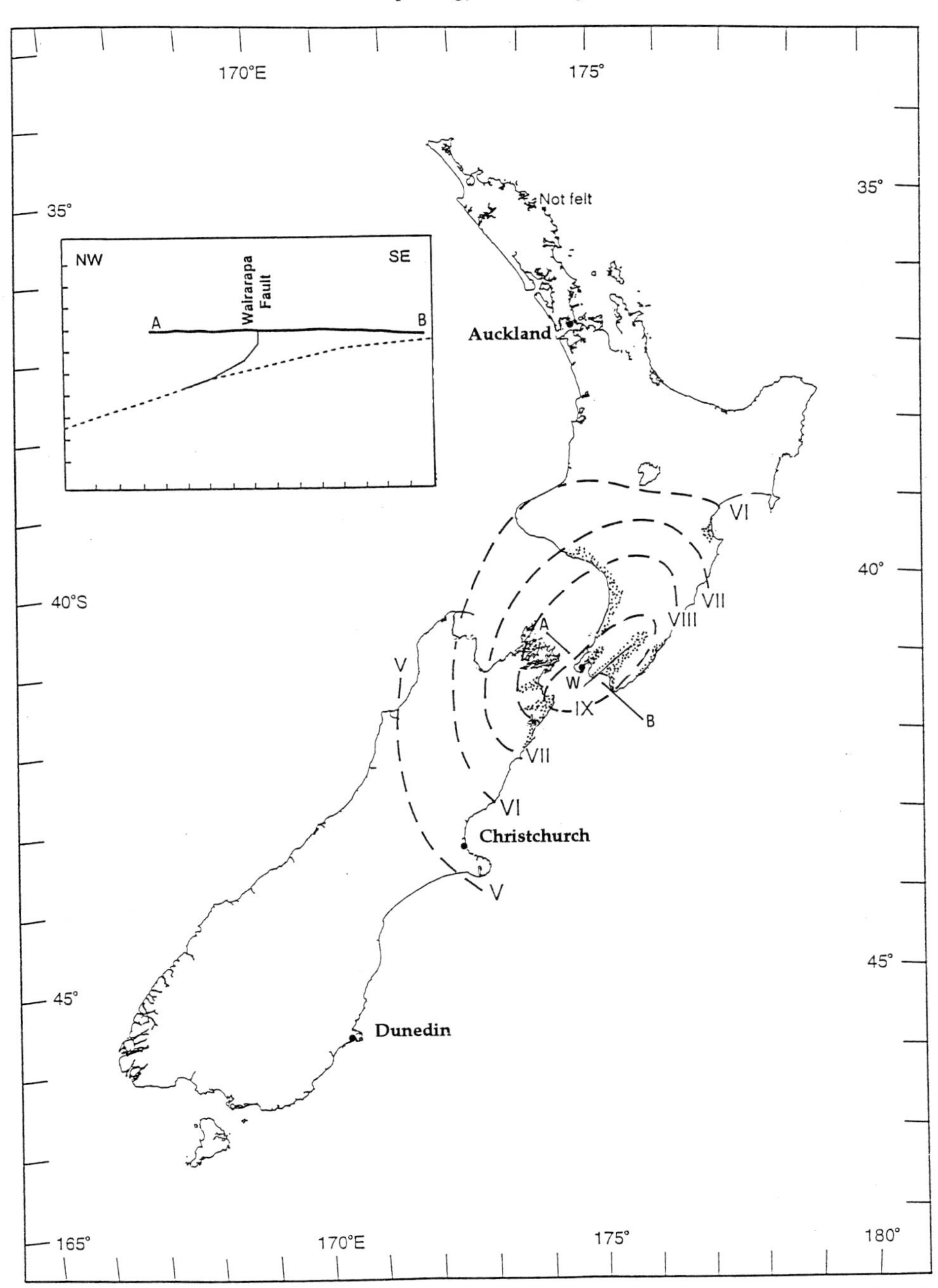

Fig. 4. Isoseismal map of 1855 earthquake including area of ground damage (stippled) and length of rupture along Wairarapa Fault (after GRAPES & DOWNES 1997) W = Wellington. **Inset** – cross section along line A-B showing inferred listric nature of the Wairarapa Fault (after DARBY & BEANLAND 1992). Dashed line = top of the Pacific oceanic plate. Each horizontal/vertical interval division = 10 km.

faulting during the earthquake produced a perpendicular cliff of fresh aspect 2.7 m high along the base of a continuous east-facing escarpment marking the junction between the greywacke of the Rimutaka Range and "Tertiary" gravels of the Wairarapa Plain and that the fault break could be followed inland from the coast at Palliser Bay for 90 miles (144 km). In several places the fault was marked by a fissure or fissures some 1.8–2.7 m broad that were either open or filled with mud and loose earth. Along the western side of Palliser Bay there was uplift of 2.7 m which gradually decreased westwards to almost nothing at the west coast of the southern part of the North Island and 30 km west of the projected line of the fault.

In March 1862 CRAWFORD, the Wellington Provincial geologist, noted the appearance of the fault on the south side of the Waiohine River: "Here the split or fissure may be observed which was caused by the earthquake of 1855, and the western side of which, or that nearest the mountains, stands at a height of several feet (CRAWFORD 1869, gives a value of 4ft) above the rest of the plain. The fissure may be observed all along the western side of the Wairarapa Valley for a distance of 60 miles, and was clearly produced by the rise of the main range, and not by the sinking of the plain" (CRAWFORD 1880), (Fig. 5).

The 1855 faulting was also recognised by MCKAY (1892) who depicted the fault as a perfectly straight active rent extending northeast from Palliser Bay for 35 miles (56 km), far short of the 90 miles stated by LYELL. MCKAY mentioned that historical (presumably 1855) movement could be traced from Featherston to the sea on the west side of Palliser Bay and he recognised that there had undoubtedly been earlier movements on the fault.

These accounts make it certain that conspicuous vertical movement took place in 1855 on what is now known as the Wairarapa Fault. However, in 1855 there were no fences or roads

Fig. 5. **a.** The inferred 1855 fault scarp (upthrown to the west) near Pigeon Bush, south of Featherston. The height of the fault scarp is 1.6 m. (near Locality 20; Fig. 2). **b.**View looking NE along the Wairarapa Fault showing the remains of a fissure developed along the line of the 1855 rupture. The apparent upthrow on the right side (east) of the fault is the result of the dextral displacement of the hill slope in the background (i.e. towards the observer) on the right side of the fault (between Featherston and Locality 24; Fig. 2).

crossing the fault except at Featherston (Fig. 2) where a bridle track from Wellington to the Wairarapa crossed the fault at an oblique angle and there was no mention of any horizontal displacement along the line of rupturing.

Present day evidence : Until European settlement in 1845 the trace of the Wairarapa Fault was almost entirely covered by forest. Now about two-thirds is cleared of which about one half is smoothed for farming. Clearing has exposed the main fault scarp so that it can be seen to advantage from the Wairarapa Valley side (e.g. CRAWFORD 1880, MCKAY 1892) but farming has generally destroyed the details of many 1855 earthquake features and more are likely to be destroyed unless they are recognised as being valuable. Although some of the open fissures mentioned by LYELL are now exposed (Fig. 5), most have been filled, either naturally or by man when the land was cleared. Nevertheless, superb examples remain of displaced streams and stream channels, protruding half scarps, sags and rents, particularly in areas that have been cleared for grazing within the last 50 years or so (Fig. 6), while near vertical soil scarps are preserved where forest cover still exists.

A detailed ground survey and air photo examination has been made of geomorphic features developed along a 90 km fault length from the coast at Palliser Bay to just north of Mauriceville. There are 38 localities where ***minimum*** displacements can be measured (dextral offsets ranging from 9 to 13 m at 23 localities, and vertical offsets across steep soil scarps ranging from 2.3 m to 0.5 m at 19 localities) that are considered to represent the 1855 movement. Localities are numbered in Fig. 2 and the amount of dextral and/or vertical displacement and the type of feature offset are listed in Table 1. As a well-defined line, the 1855 rupture appears to end at locality 38, near Mauriceville (Fig. 2). There are several fresh-looking, subparallel rents within Tertiary mudstone that vary between 4 km and 1 km in length and trend northeast for about 15 km from locality 32 (Fig. 2). The smallest displacements are between 5 m – 9 m dextral and 0.6 m – 0.3 m vertical (LENSEN 1970) and may indicate the NE continuation of the 1855 rupture.

The inferred 1855 displacement values listed in Table 1 are plotted against distance along the fault in Fig. 7. Dextral displacement varies between 9–13 m along the entire fault length examined with a mean of 11.6 ± 1 m. Vertical displacement decreases northeast along the fault from 2.7 m at the Palliser Bay coast to 0.5 m about 88 km inland. Thus the continuous uplift of 2.7 m for 144 km reported by LYELL (1856; 1868) is incorrect. From the position where the fault reaches the coast, the 1855 uplift continued southwest for another 11 km along the western side of Palliser Bay as indicated by a stranded beach ridge, increasing to a maximum of 6.5 m followed by a decrease to about 4.5 m at Turakirae Head (Figs. 6a; 7). Adding this distance to the 88 km of well-defined faulting along the Wairarapa Valley, the 15 km extension northeast of Mauriceville, and the 16 km submarine extension of the fault into Cook Strait, gives a minimum length of fault rupture during the 1855 earthquake of 148 km. Equations of BONILLA et al. (1984) that provide estimates of surface magnitude (Ms) from dextral displacement and minimum fault rupture length yield magnitudes of 7.8–7.9 and 7.7 respectively (standard deviation of ± 0.3 Ms units) for the 1855 event (cf. GRAPES & DOWNES 1997).

Surface rupturing and uplift that occurred on the Wairarapa Fault resulted in the sudden uplift and tilting of some 5000 km^2 of the south western part of the North Island and probable tectonic subsidence of up to 1.5 m in the lower Wairau Plain of Marlborough in the South Island. Localised maximum uplift of about 6.5 m occurred near Turakirae Head (WELLMAN 1967, MCSAVENEY & HULL 1995). Based on contemporary and subsequent measurements of

Table 1. Smallest displacements along the Wairarapa Fault from Turikirae Head to 1 km north of Mauriceville (localities shown on Fig. 2).

Locality No	Displacement (m) Vertical	Dextral	Offset Feature
1	4.9	–	Beach ridge
2	5.6	–	Beach ridge
3	6.5	–	Beach ridge
4	2.7	–	Rock platform
5	2.7	–	Rock s on marine platform[1]
6	3.0	–	Beach ridge[2]
7	2.3	–	Fault scarp
8	–	11.5	Stream
9	–	12.5	Stream
10	1.9	–	Fault scarp
11	2.0	–	Step in river flood plain
12	1.9	–	Fault scarp
13	–	9.5	Abandoned stream channel
14	1.8	–	Fault scarp
15	1.7	–	Fault scarp
16	–	12.3	Stream
17	–	12.0	Stream
18	–	11.5	Stream
19	1.5	11.0	Stream
20	1.0	12.0	Abandoned stream channel
21	–	12.0	Protruding half scarp
22	1.5	12.5	Stream
23	1.8	–	Fault scarp[3]
24	1.0	12.5	Stream
25	1.2	–	Fault scarp[4]
26	1.0	12.0	Abandoned stream channel
27	1.3	–	Fault scarp[5]
28	1.5	12.5	Stream
29	–	12.0	Stream
30	–	12.0	Stream
31	–	12.5	Stream channel
32	–	12.5	Stream channel
33	–	12.0	Stream
34	1.0	13.0	Stream
35	–	13.0	Stream
36	0.5	12.0	Terrace riser
37	0.7	12.0	Terrace riser
38	0.5	11.5	Terrace riser
		n = 23	
		12.0 ± 1.0	

* Dextral displacement ± 0.5 m or less
Vertical displacment ± 0.2 m or less

Localities 1–5 selected from a continuous profile of Beach Ridge C surveyed by Wellman (1967)
Localities 4 and 7–15 from Grapes (unpublished data)
Localities 15–38 from Grapes & Wellman (1988) except where otherwise indicated.

[1] Uplift of white line of nullipores coating rocks originally at and just below low tide level (Lyell 1856, 1868)
[2] McKay (1901)
[3] Crawford (1866)
[4] Jackson (1883)
[5] Grapes (unpublished data)

Fig. 6. Photos showing 1855 and older displacement features along the Wairarapa Fault. **a**. Oblique aerial view looking NE from Turakirae Head along the western Palliser Bay coast showing the succession of uplifted beach ridges. The Wairarapa Fault lies offshore. The lowest beach ridge is the present day forming storm beach ridge. The next highest ridge represents the 1855 uplift (about 5 m at Turakirae Head). **b**. View looking west at the Wairarapa Fault scarp which is about 5 m high. Two abandoned river channels end at the fault scarp. The two channels are dextrally displaced 12 and 22 m with respect to the vegetated incised stream gorge that cuts through the fault scarp to the right. The 12 m displacment represents the 1855 movement; the 22 m displacement represents the 1855 plus the previous earthquake displacement. (See text). (Locality 20; Fig. 2).

c. View looking SW along the trace of the Wairarapa Fault (dashed line) showing an incised stream that was dextrally displaced 12.5 m (a – b) during the 1855 earthquake (Locality 22; Fig. 2). **d**. Oblique aerial view looking SW across the trace of the Wairarapa Fault showing a prominant bulge developed on the upthrown side of the fault where there is a sinistral jog in the fault trace. The deformed surface is the youngest aggradation (Waiohine) surface in the Wairarapa Valley.

uplift along the south and west coasts of the Wellington peninsula and vertical displacement along the Wairarapa Fault (Fig. 7), a tentative uplift map for the 1855 earthquake is shown in Fig. 8. The anticlinal bulge indicated by the beach ridge uplift profile near Turakirae Head (Fig. 7) is centred over the axis of the Rimutaka Anticline where it intersects the coast and is inferred to indicate localised uplift of the anticline, at least to a point where there is a marked offset in the summit ridge line near Mount Matthews (Fig. 8).

(ii) *Older displacements on the Wairarapa Fault*

At many localities along the Wairarapa Fault alluvial terraces have been vertically and horizontally displaced by repeated movement on the fault. Dextral displacement of the highest, most extensive loess-free surface by the Wairarapa Fault is 127 ± 3 m. This value represents the mean of ten measurements ranging from 120–130 m (± 5 m error limit for each measurement) at nine localities. In contrast to dextral displacement, vertical displacement of the surface is irregular. For example, the maximum throw of the surface is up to 18 m on the northeast side of the Waiohine River whereas it is only 7 m on the southwest side of the river. This and other localities where rapid changes in vertical displacement of the surface occur appear to be related to growing bulges caused by sinistral sidesteps on the fault trace (Fig. 6d), (GRAPES 1991). For most of its length the surface is upthrown on the northwest, or Rimutaka Range side, but where the fault cuts Cenozoic cover strata northeast of the Waiohine River some places are upthrown to the southeast (Fig. 2). These reversals in throw may in part be due to vertical splay faulting at the crest of the upbulged greywacke undermass along the fault below the Cenozoic cover as shown by seismic profiling (CAPE et al. 1990).

An c. 85 km long profile showing variation in throw of the highest loess-free surface along the Wairarapa Fault is given in Fig. 9. and indicates that there is a general northeast decrease and in particular a marked decrease in vertical displacement to <10 m a few km northeast of the Waiohine River, that coincides with a 15° change in strike of the fault (Fig. 2). This pattern

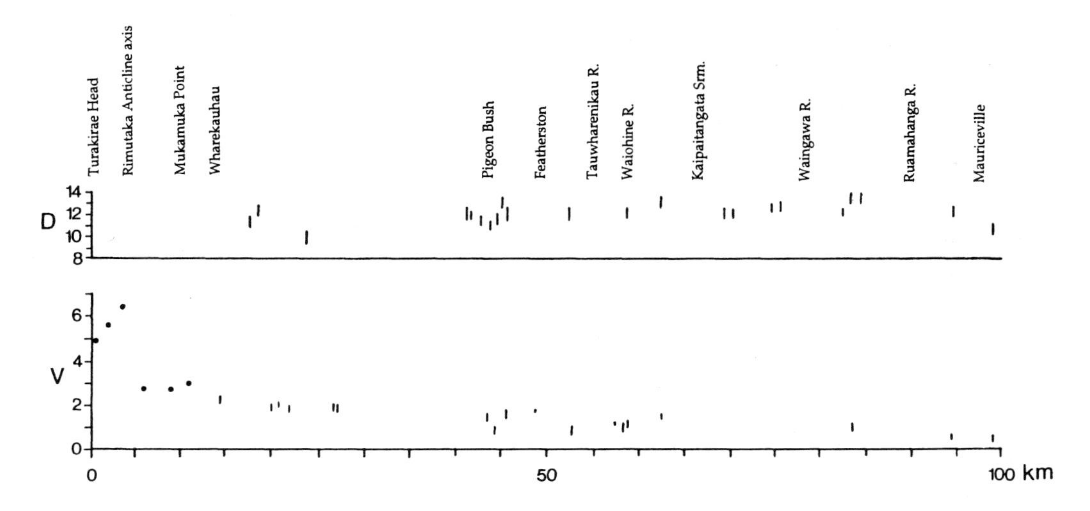

Fig. 7. SW-NE profiles of inferred 1855 dextral (D) and vertical (V) displacement (m) along the Wairarapa Fault (data from Table 1).

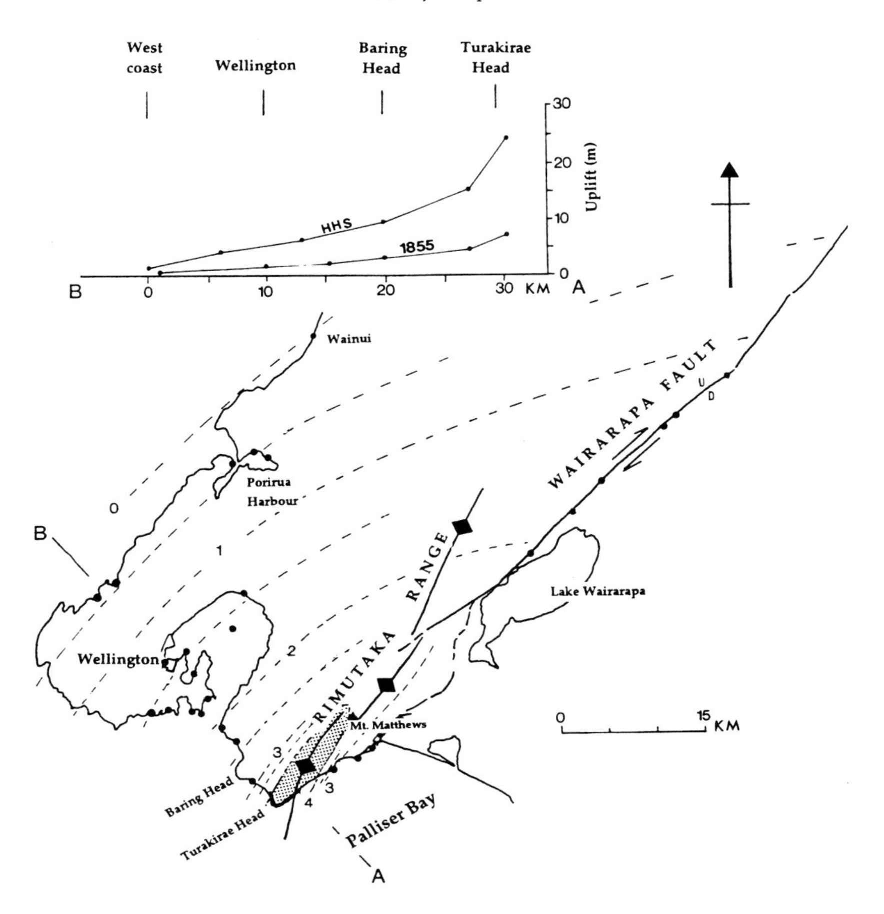

Fig. 8. Uplift map of 1855 earthquake (0.5 m contours) constructed from contemporary and subsequent coastal uplift data and selected vertical displacement values along the Wairarapa Fault (localities shown by dots; thick line around Turakirae Head denotes area of continuous survey of Beach Ridge C by WELLMAN (1967)). The subsidiary domed uplift of 4–6.5 m (shaded area) near Turakirae Head is constructed from the profile of Beach Ridge C made by WELLMAN (1967) and is constrained by location of the axis of the Rimutaka Anticline (line with outward pointing arrows) where it intersects the coast. The domed uplift is extended inland to the jog in the anticline axis near Mt. Matthews. **Inset** – Uplift profiles for **1855** beach and highest Holocene shoreline (**HHS**) across the Wellington peninsula projected onto a plane parallel to line **A** – **B**. Data points from LYELL (1868), CRAWFORD (1858), WELLMAN (1967), STEVENS (1973), MCFADGEN (1980), OTA et al. (19 81); this paper and GRAPES (unpubl.data).

of decreasing uplift northeast along the fault is therefore similar to the change in vertical displacement that occurred during the 1855 earthquake (Fig. 7). A similar pattern is evident in the highest Holocene shoreline profile northeast from Turakirae Head (Fig. 9) with respect to the 1855 uplift (Fig. 7). Also, the regional uplift and northwest tilting that occurred during the 1855 earthquake appears to be a continuation of the trend over the last 6.5 ka, i.e. when sea level rose to its present day level (GIBB 1986), (Fig. 8). Projection of the 1855 and highest

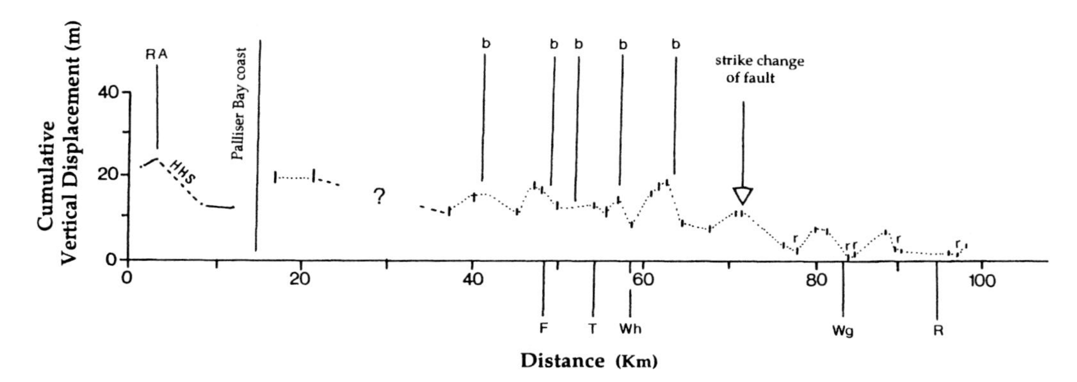

Fig. 9. Profile showing cumulative vertical displacement of the Last Glacial aggradation surface (**ag3** surface in Table 2) along the Wairarapa Fault (data from GRAPES & WELLMAN 1988, 1993, GRAPES 1993 and unpublished data) and cumulative uplift of the highest Holocene shoreline (**HHS**) along the western side of Palliser Bay (dashed where inferred), (data from WELLMAN 1967 – Beach Ridge F corrected for change in position of the 1855 beach ridge; GRAPES, unpubl. data). **RA** = position of Rimutaka Anticline axis; **b** = bulges developed along the Wairarapa Fault (Fig. 2); **r** = reversal in upthrown side of Wairarapa Fault, i.e. to the SE with respect to usual upthrown side to the NW; **F** = Featherston; **T** = Tauwharenikau River; **Wh** = Waiohine River; **Wg** = Waingawa River; **R** = Ruamahanga River

Holocene shoreline uplift profiles shows that they intersect approximately 3 km west of the west Wellington coast line (zero uplift contour), (see also WELLMAN 1967). This is almost coincident with the change to a negative uplift pattern predicted from the listric fault model of DARBY & BEANLAND (1992) and interestingly, with the area of crustal flexture that marks the beginning of subsidence to form the Wanganui Basin (Fig. 1) modelled by STERN et al. (1992).

On the true left bank of the Waiohine River, the Wairarapa Fault displaces a flight of river terraces (Fig. 10) providing a record of earthquake displacements at a single locality. The uppermost terrace (ag3) displaced by the fault is the youngest loess-free aggradation surface (GRAPES & WELLMAN 1988, GRAPES 1991), while the lower terraces are degradational. The faulted terraces at Waiohine River have been described in detail by LENSEN & VELLA (1971) who list displacements for six of the degradation terraces. The smallest dextral displacement was found to be 12 m and is considered to record the 1855 movement. Displacement values and terrace nomenclature used by GRAPES & WELLMAN (1988) are compared with those of LENSEN & VELLA (1971) in Table 2. There are slight differences in the dextral and vertical displacement values, and because one of LENSEN and VELLA's terrace treads is regarded as a channel, six degradation terrace treads are recognised below the highest aggradation surface with respect to the seven mapped by LENSEN & VELLA (Fig. 10; Table 2).

Additional to the displacements preserved at Waiohine River and to those inferred to be of 1855 origin, a total of 53 dextral displacements along the best exposed part of the Wairarapa Fault from the southwestern end of Lake Wairarapa to north of Mauriceville are listed by GRAPES & WELLMAN (1988). The combined data set is averaged as seven specific dextral displacements as shown in Fig. 11. It is clear that some groupings are more poorly constrained than others but compared with the total number of dextral displacements recorded along the whole length of the fault, there are two displacements missing at Waiohine River, i.e. 22.5 ±

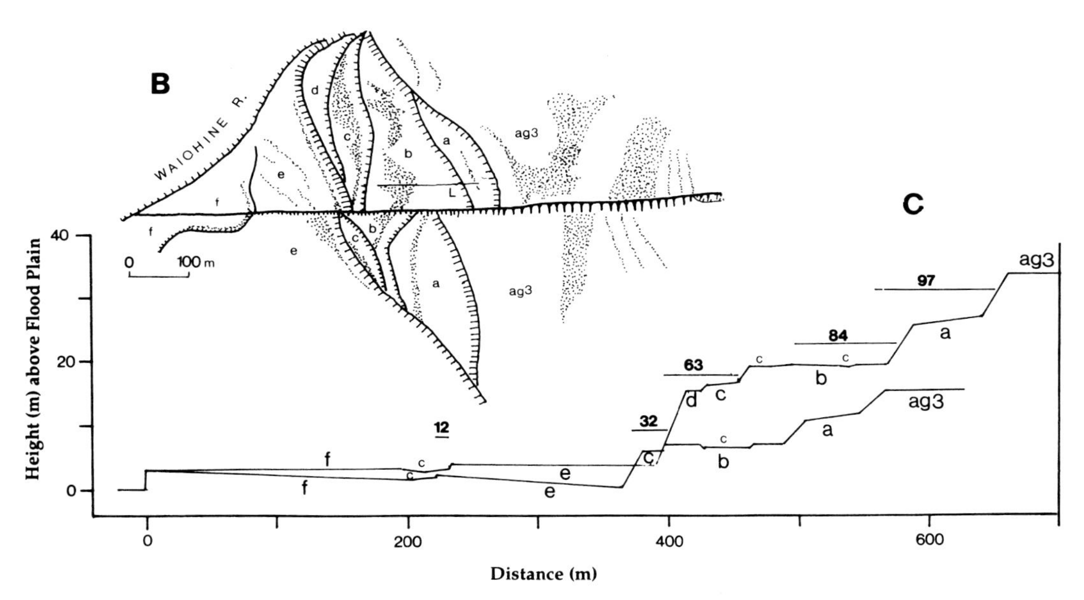

Fig. 10. The faulted terraces at Waiohine River. **A**. Oblique aerial photo looking west across the Wairarapa Fault (upthrown to the west) and showing the displaced terraces on the true left side of the Waiohine River. The most extensive and highest terrace is the Last Glacial (Waiohine) aggradation surface; the lower terraces are degradational and have been cut into Late Glacial gravels of the youngest aggradation phase in the Wairarapa Valley. **B**. Map of Waiohine faulted terraces (after GRAPES & WELLMAN 1988). Terrace nomenclature together with dextral and vertical displacement values are listed in Table 2. Dotted areas = channels; Fault scarp shown as thick tapered barbs; terrace risers shown as thin barbs. L = lineament. **C**. Cross sections of upthrown and downthrown sides of the Waiohine faulted terrace sequence. Dextral displacement values (± 1 m) given and measured from the mid points of corresponding risers on either side of the fault.

1.2 and 44.8 ± 4.2; the rest are in good agreement with the adopted displacement groupings. Of the total number of displacements preserved along the Wairarapa Fault, the last four have been 11 ± 1 m whereas the next four displacements range between 15 m and 29 m and almost certainly do not represent all earthquake events.

Aggradation surface age and estimate of dextral slip rate

Rivers flowing east from the Axial Ranges into the Wairarapa Valley are incised into alluvial gravels which aggraded during the Last Glacial advance between about 25–10 ka. The flights of terraces such as those preserved at Waiohine River provide a record of climatic variation and superimposed tectonic uplift (VELLA 1969, VELLA et al. 1988, GRAPES 1991) where alternating periods of colder drier and warmer wetter climate resulted in the formation of aggradation terraces and degradation terraces respectively.

In the Rangitikei area west of the Axial ranges and in northern Wairarapa, the Last Glacial aggradation is recognised as three surface subsets (MILNE 1973, MARDEN et al. 1986). The surfaces are characterised by little or no loess cover and the absence of a 22.5 ka rhyolitic tephra (Kawakawa tephra) which occurs within loess blown from the Last Glacial aggradation surface onto all older and higher aggradation terraces. The base of the loess is radiocarbon dated at 25 000 ± 800 yr B. P. and the top is dated at 9480 ± 100 yr B. P. (MILNE & SMALLEY 1979) consistent with the position of the tephra in the lower third of the loess section. In northern Wairarapa the Last Glacial surfaces are considered to range in age from c.18 ka B.P (highest surface) to about 10 ka B.P (MARDEN et al. 1986, MARDEN & NEALL 1990). In central and southern Wairarapa only one Last Glacial surface (the Waiohine Surface) has been recognised (VELLA 1963, PALMER 1984, GRAPES 1991). Based on available data from the southern North Island (summarised in GRAPES 1991), the surface is considered to be 11 ± 1 ka, i.e. the youngest Last Glacial surface, and gives a Holocene dextral slip rate for the Wairarapa Fault of 11.5 + 0.8/−1.1 m/ka.

Earthquake frequency and magnitude

The observed pattern of uplift that has affected the southwest part of the North Island appears to have been associated with movement on the Wairarapa Fault during large, possibly subduction-related, earthquakes, the most recent in 1855. An estimate of the number of large earthquakes since the age of the Last Glacial surface can be made from the number of dextral displacements along the Wairarapa Fault. There are few single event fault displacements worldwide that exceed the 12 m dextral displacement that occurred on the Wairarapa Fault in 1855 (e.g. the San Andreas Fault; SIEH et al. 1978), and this value must be considered close to the maximum for any coseismic fault displacement. It therefore seems likely that at least three displacement values (i.e. at about 54, 73 and 112 m) could be missing from the total number of displacements on the Wairarapa Fault over the time interval in question (Fig. 11) and that the maximum cumulative displacement of 127 ± 3 m over this period is the result of 10, possibly 11 earthquakes with magnitudes similar to that inferred for 1855.

Apart from the 1855 movement, the timing of previous displacements on the Wairarapa Fault is essentially unknown. However, the uplift along the western side of Palliser Bay and the

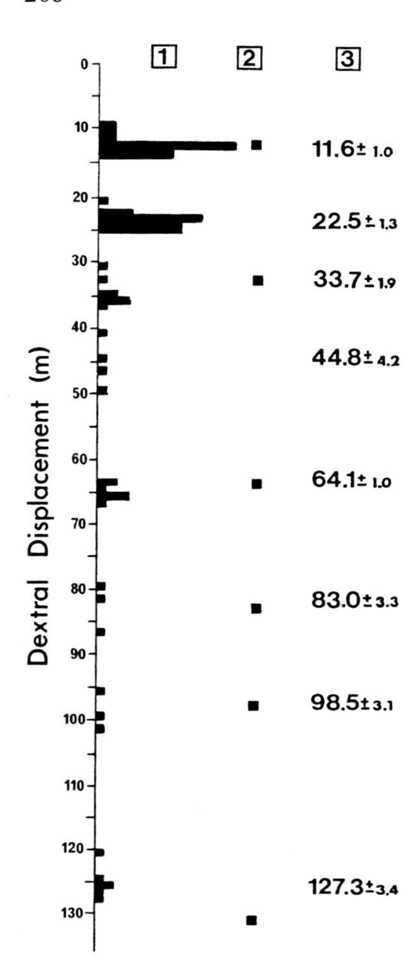

Fig. 11. Histogram showing dextral displacements along the Wairarapa Fault (data from GRAPES & WELLMAN 1988, GRAPES, unpubl. data). Column 1–81 dextral displacements from Palliser Bay to Mauriceville; Column 2 – Displacements at Waiohine River (see Table 2); Column 3 – Means of inferred displacement groupings ± one standard deviation.

displacement along the Wairarapa Fault that occurred in 1855 suggest this coupled relationship might also apply to previous earthquakes (WELLMAN 1972). Recent work by McSAVENEY & HULL (1995) have established that there are four uplifted beach ridges at Turakirae rather than five concluded by previous workers (e.g. WELLMAN 1967, STEVENS 1969, 1972), (Fig. 12). The discrepancy has resulted from misidentification of the present day forming beach storm ridge which is situated 2–3 m above the highest fetch of waves, a height similar to the amount of 1855 uplift stated by LYELL (1856, 1868). It was not realised by earlier workers that the well-preserved beach ridges at Turakirae Head are built by severe storms and lie inland from, and several meters above, the shoreline of the time.

Maximum uplift values (heights above present day forming storm beach ridge), (WELLMAN 1967) near the axis of the Rimutaka Anticline, modified by redefinition of the 1855 beach ridge (Table 2), are used to construct a time verses uplift diagram (Fig. 12A) based on a 6.5 ka age for the highest and oldest beach ridge at 24.5 m above the present day storm beach, a uniform uplift rate of 3.8 m/ka and the date of the last uplift in 1855. The construction best fits

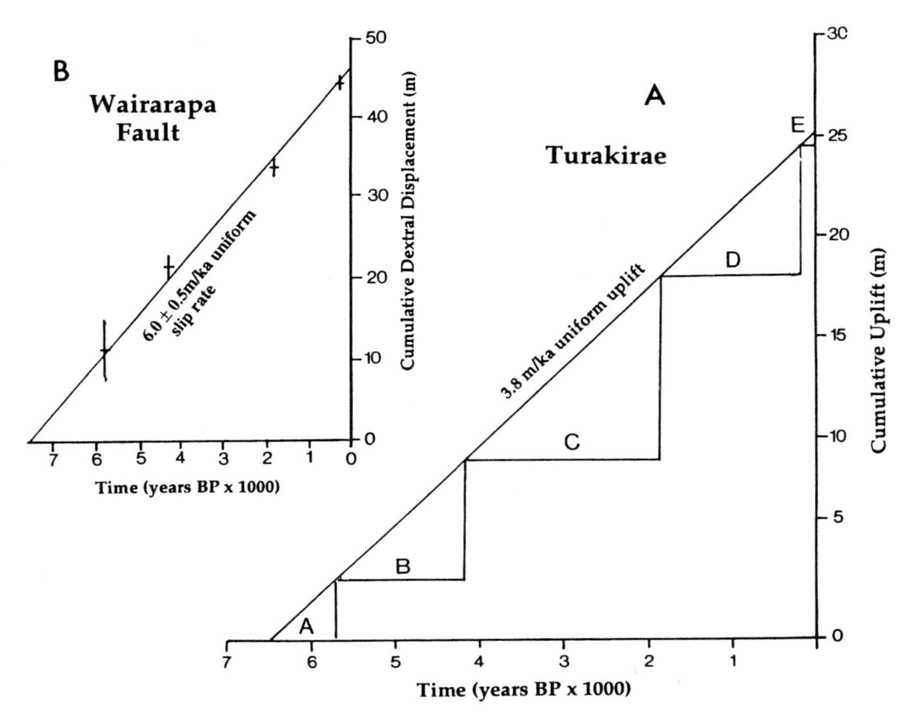

Fig. 12. **A.** Step function diagram of earthquake uplifts where the axis of the Rimutaka Anticline intersects the coast near Turakirae Head. Earthquake uplift times are contrained by uniform uplift rate of 3.8 m/ka over the last 6.5 ka and conform to a slip-predictable pattern (see text). **B.** Plot of last four dextral displacements on Wairarapa Fault verses time. Earthquake times from **A.** Dextral displacement values from Fig. 11.

a slip predictable pattern (e.g. SHIMAZAKI & NAKATA 1980), i.e. time to next earthquake is proportional to displacement during the next earthquake (uniform uplift rate is maintained by slip), and gives uplift ages of the three beach ridges older than 1855 of 1.8–1.9, 4.1 and 5.7 ka BP (cf. WELLMAN 1967, MOORE 1987, MCSAVENEY & HULL 1995; Table 2). The derived uplift ages are in accord with the ^{14}C ages of shells and wood from the 1855 and earlier beach ridges listed by MOORE (1987), MCSAVENEY & HULL (1995), and the presence of the 1.8 ka Taupo Pumice on the 1855 beach and its absence from the previously uplifted beach.

If earthquake uplifts previous to 1855 are caused by movement on the Wairarapa Fault, then by simple enumeration the four distinct uplifted beach ridges at Turakirae can be paired with the four youngest dextral displacements on the Wairarapa Fault (Table 2). A plot of the last four dextral displacements on the Wairarapa Fault verses earthquake time derived from the Turakirae data (Fig. 12B) shows that they approximate a 6 ± 0.5 m/ka dextral slip rate. This is about half that of the longer term average rate of 11.5 + 0.8/−1.1 m/ka and implies: (i). that some earthquake events are missing from the sequence of stranded beach ridges at Turakirae Head; (ii). the inferred correlation between uplift at Turakirae and dextral movement along the Wairarapa Fault is invalid; (iii). Holocene slip rate is non uniform and has decreased over the last 6 ka, or (iv). the Late Glacial aggradation surface is older than 11 ± 1 ka.

Table 2. Displacements along the Wairarapa Fault at Waiohine River and Turakirae Head (all values in metres) together with estimated uplift ages of beach ridges at Turikirae Head.

Waiohine River												Turakirae Head												
Lensen & Vella (1971)						Grapes & Wellman (1988)						Wellman (1967)			McSaveney & Hull (1995)			This Paper			Uplift Age +			
Tce	Tu	Td	V	D	D/V	Tc	Tu	Td	V	D	D/V	BR	U	D/V	BR	U	D/V	BR	U	D/V	Wellman (1967)	Moore (1987)	McSaveney & Hull (1995)	This Paper
a	33.0	14.3	18.3	nd	nd	ag3	33.0	14.5	18.5	130.0	7.0													
b	26.1	11.3	14.7	99.0	6.7	a	25.8	10.5	15.5	97.0	6.3													
c	18.9	6.6	12.3	85.5	6.9	b	18.7	6.5	12.0	84.0	7.0													
d	18.0	6.0	12.0	66.9	5.6	c	14.0	5.6	8.4	63.0	7.5													
e	16.2	6.0	10.2	nd	nd	------ not recognised ------						F	27.0	2.4	5	24.0	1.9	A	24.5	1.8	5.6	5.5–6.5	7.2	5.7
f	–	–	–	–	–	d	12.5	nd	nd	nd	nd	E	24.0	1.9	4	21.0	1.9	B	21.5	1.6	5.1	4.6–5.4	5.1–5.4	4.2
g	3.6	0.0	3.6	32.1	9.0	e	3.6	0.0	3.6	32.0	8.9	D	18.0	1.9	3	15.5	1.6	C	15.5	1.5	3.2	2.9–3.4	2.2	1.8–1.9
-------- missing --------						-------- missing --------						C	9.0	2.5	2	6.4	1.5	D	6.5	1.9	0.6	0.45	(1855)	(1855)
h	2.7	1.8	0.9	12.0	13.0	f	2.8	1.8	1.0	12.0	12.0	B	2.5	4.8	1	PD	1.8	E	PD		(1855)	(1855)	PD	PD
												A	PD	-							PD	PD		

Tce = terrace or surface tread/riser; ag3 = youngest of three main aggradation surfaces in Wairarapa Valley
a, b, c etc. = degradation terraces cut into gravels of ag3 with a being the oldest and highest;
missing terrace = 22.5 ± 1.3 m dextral displacement; nd = not determined.

Tu = height of tread on upthrown side (NW) of fault above present river flood plain.

Td = height of tread on downthrown side (SE) of fault above present river flood plain.

V = vertical fault displacement; D = Dextral fault displacement; D/V = ratio of dextral to vertical displacement.

BR = beach ridge designated F to B (Wellman 1967), 5 to 1 (Moore 1987; McSaveney & Hull 1995) or A to E (this paper. Nomenclature allows for future uplift) with F, 5 or A being the oldest; B, 2 and D being the 1855 uplifted beach ridge; PD = Present day forming beach ridge.

U = cumulative uplift height of beach ridge above present day forming beach (emergence height). Note : The 2.5 m high beach ridge (Beach Ridge B of Wellman 1967) is now recognised as the present day storm ridge rather than the 1855 uplifted beach (McSaveney & Hull 1995)

D/U = ratio of dextral displacement along Wairarapa Fault to amount of beach ridge uplift (dextral valves Fig. 10).

\+ uplift ages of beach ridges prior to 1855 in Kyrs BP.
uplift ages of Wellman (1967) assuming that the amount of each uplift is proportional to the total time from the previous to the following uplift.
uplift ages of Moore (1987) from ^{14}C dating and amount of each uplift being proportional to total time from previous to following uplift.
uplift ages of McSaveney & Hull (1995) from ^{14}C dating and fit to slip-predictable pattern.
uplift ages of this paper discussed in text and derived from Fig. 11.

The well preserved, distinctive nature and extent of the uplifted beach ridges at Turakirae Head (Fig. 6a) makes it unlikely that earthquakes uplifts are missing. The present storm beach became established within 50 years of the 1855 uplift (ASTON 1912) and previous beaches are likely to have been equally rapidly established following earthquake uplift. The correlation between uplift at Turakirae and displacement on the Wairarapa Fault initially suggested by WELLMAN (1971) has been criticised by SUGGATE & LENSEN (1972) who suggested that uplift at Turakirae may be more frequent than movement on the Wairarapa Fault due to displacement on other faults of the North Island Dextral Fault Belt. Although these faults show evidence of Holocene uplift on both their northwestern and southeastern sides by amounts <5 m (e.g. Wellington Fault; BERRYMAN 1990), longer term uplift is consistently to the northwest. As such, movement on these faults is unlikely to have affected Turakirae Head.

Independent evidence that strengthens the suggested coupled relationship between the age of uplift events at Turakiare Head and movement along the Wairarapa Fault comes from displaced stream channels at Pigeon Bush (Locality 20, Fig 2), (manuscript in preparation). At this locality there are two well defined displaced stream channels, each about 1.2 m deep that are beheaded against the 6 m high scarp of the Wairarapa Fault (Fig. 6b). The two channels have been displaced dextrally by 12 m and 11 m from a 7.5 m deep stream-cut gully on the upthrown side of the fault and represent the 1855 and penultimate earthquake displacement. The width of the 1855 displaced channel where it intersects the fan surface is 6.7 ± 0.3 m; that of the previous earthquake displaced channel is 7.7 ± 0.6 m. The surface width of the stream gully on the upthrown side of the fault is 13.6 ± 0.7 m, i.e. about the sum of the two displaced channel widths. The present-day forming channel has a surface width of 0.5 ± 0.05 m that represents the amount of erosion since the 1855 displacement (i.e 142 years at the time of survey). At this rate (other factors being equal), formation of the displaced channel surface widths of 6.7 m and 7.7 m could have taken 1.7–2.0 ka and 2.0–2.4 ka respectively. Both age ranges bracket the earthquake uplift times of 1850 yrs B. P. and 4.1 yrs B. P for Turakirae beach ridges D and C respectively (Table 2), where the older uplift date correlates with the sum of the two channel width formation age ranges of 3.7 and 4.4 ka (average 4.05 ka B. P).

The 6 ± 0.5 m/ka dextral slip rate on the Wairarapa Fault inferred from earthquake ages at Turakirae over the last 6 ka years may represent too small a time window of observation to reflect the overall Holocene average rate of 11.5 m/ka. It could also reflect part of an overall decelerating slip trend or possibly part of an active phase of episodic coseismic displacement along the fault. The history of the fault is not well enough known to indicate whether the simplifying assumption of constant slip rate is valid or not. Also, slip rate values depend on how well a particular reference surface, i.e. fluvial aggradation surface, is dated. The Last Glacial fan surface at Waiohine River is age trangressive in that it becomes younger away (eastwards) from the Rimutaka Range front and Wairarapa Fault and is lapped by Holocene sediments (VELLA 1963, TOMPKINS 1987) but lacks loess cover that could preserve wood for radiocarbon dating. If the surface displaced by the Wairarapa Fault is older than 11 ± 1 ka,(e.g. between 11 and 18 ka), the average slip rate would decrease accordingly, i.e. to 7 m/ka if 18 ka old. Clearly, application of alternative dating methods, e.g. weathering rinds; thermoluminescence, are needed to help contrain the age of alluvial surfaces displaced by the Wairarapa Fault.

Relationship between Wairarapa Fault and uplift of the Rimutaka Range

Asssuming that correlation between uplifts at Turakirae and dextral displacement on the Wairarapa Fault is valid, a plot of vertical displacement verses dextral displacement along the fault at two localities, where the axis of the Rimutaka Anticline intersects the coast near Turakirae Head and at Waiohine River, is shown in Fig. 13. The plot demonstrates a high ratio of dextral/vertical faulting at Waiohine River compared with Turakirae (Table 2) in accordance with uplift rates of 1.8 m/ka and 3.8 m/ka respectively over the last 6.5 ka, although in detail rates are variable. Overall there is an increase in the horizontal/vertical ratio of displacement northeast along the Wairarapa Fault (vertical rate decreasing), both long term and during the 1855 earthquake. Despite this, the uplift rate of the Rimutaka Range greywacke is high with late Quaternary rates up to 4.5 mm/yr (GHANI 1978). Only at Turakirae does uplift on the fault approach this rate because of close proximity to the axis of the Rimutaka Anticline. The decrease in uplift rate northeast along the fault is consistent with divergence of the fault from the anticline axis (Fig. 8). Therefore, as a first approximation, during each large earthquake: (i) vertical displacement along the Wairarapa Fault at any one locality will be controlled by the relative rate of uplift along the eastern flank of the Rimutaka Anticline, and (ii) the amount of dextral displacement during a single event will balance the late Quaternary slip rate of the North Island Dextral Shear Belt in the southern part of the North Island. Available geomorphic data indicate that the amount of single event dextral displacements appear to have been fairly constant since the Last Glacial aggradation.

An east-west Quaternary uplift profile shows the relationship between the position of the dextral faults and regional folding across southern North Island (Fig. 14). The faults are upthrown to the west, the topographic highs are anticlinal with faulted eastern margins and the topographic lows are synclinal. The overall structure can be characterised as a series of fault bounded northwest tilted blocks of basement greywacke with fault angle depressions formed

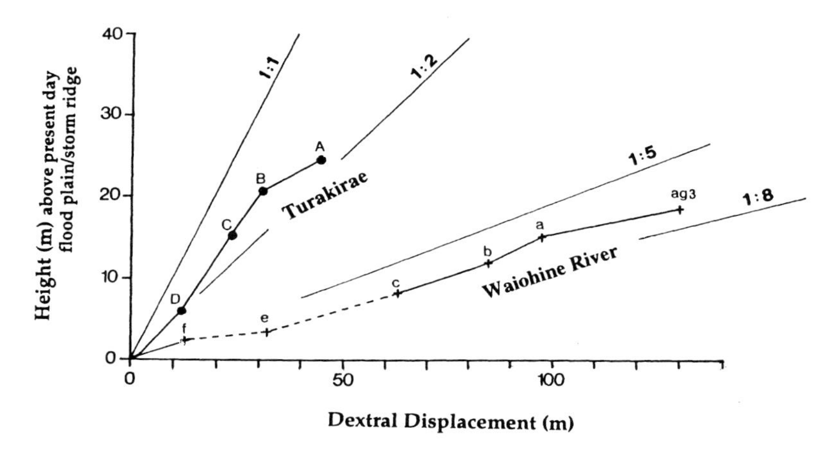

Fig. 13. Plot of dextral displacement on Wairarapa Fault verses height above present day flood plain level (Waiohine faulted terrace sequence) and present day storm beach ridge (uplifted beach ridges near Turakirae Head). Terrace and beach ridge nomenclature adopted in this paper are given in Table 2. Dashed lines in the Waiohine River profile indicate inferred missing displacement equivalents of beach ridge C and A uplifts (see text).

on the downthrown side of the Wairarapa, Wellington and Ohariu faults. Late Quaternary rock (= surface) uplift rates range up to 4.5 mm/yr along the Axial Range with the rate gradually decreasing westwards. Except for the southern part of the Wairarapa Valley east of the Wairarapa Fault, the entire coast is being uplifted and tilted northwest in the Wellington region (GHANI 1978, OTA et al. 1981).

Forward elastic dislocation modelling by DARBY & BEANLAND (1992) shows that the best-fit configuration that explains the 1855 earthquake rupture and associated pattern of uplift, tilting and subsidence is that of a listric Wairarapa Fault having a steep dip at the surface that decreases with depth to become tangential to the subduction interface of the westward dipping Pacific Plate at a depth of about 25 km (Inset in Fig. 4). The listric fault geometry allows successive seismic uplifts to be accommodated as (i) anticlinal folding adjacent to the shallow near-vertical part of the Wairarapa Fault and (ii) regional tilting west of the fault by crustal thickening and anticlockwise rotation due to updip ramping along the deeper, shallow dipping part of the fault in combination with crustal flexuring along the eastern margin of the Wanganui Basin, and the relatively "strong" Mesozoic basement greywacke of the Rimutaka Range uplifting along its eastern margin (Fig. 14B).

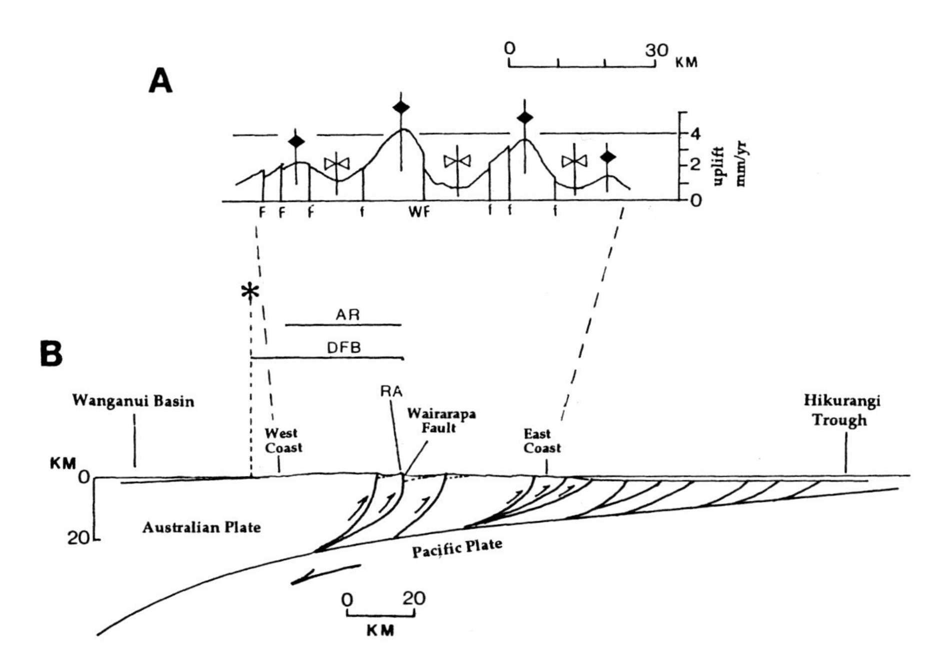

Fig. 14. A. Section (along line A-B in Fig. 1) showing late Quaternary uplift rates (mm/yr) (after GHANI 1978), in relation to faults of the North Island Dextral Fault Belt and growing folds. WF = Wairarapa Fault; F = major faults as shown in Fig. 1; f = minor faults not shown in Fig. 1. B. Crustal cross section showing relationship between tectonic elements of the Australian – Pacific plate convergent margin west of the Hikurangi Trough (see text). **AR** = Axial Ranges; **RA** = Rimutaka Anticline; **DFB** = Dextral Fault Belt; vertical line marked by asterist = intersection of 1855 and 6.5 ka (highest Holocene shoreline) uplift profiles.

Conclusions

1. The Wairarapa Fault has an average dextral slip rate between c. 6 ± 0.5–11.5 + 0.8/–1.1 m/ka over the last 10–20 ka that accounts for nearly one third to half the c. 21 mm/yr margin parallel slip rate associated with oblique subduction of the Pacific Plate along the Hikurangi Trough east of New Zealand.
2. The Wairarapa Fault shows evidence of repeated movement during the Holocene, the last occurring during the M8.2–8.3 1855 earthquake that resulted in a rupture distance of at least 148 km. Small scale geomorphic features indicate that 1855 dextral displacement ranged between 9–13 m (mean of 11.6 ± 1 m) while vertical displacement decreased NE along the fault from a maximum of 6.5 m near Turakirae Head to 0.5 m near Mauriceville in northern Wairarapa. Heights of the stranded 1855 beach and rock platforms indicates that uplift along the fault was accompanied by regional NW tilting of about 5000 km^2 of the southern part of the North Island.
3. The Last Glacial alluvial aggradation suface cut by the Wairarapa Fault has been dextrally displaced by 127 ± 3 m. This, in combination with progressively smaller dextral displacements preserved in a flight of degradation terraces at Waiohine River and elsewhere, suggest the possibility that 10 or 11 displacements similar to that in 1855 have occurred on the Wairarapa Fault since the Last Glacial aggradation surface formed between 10–20 ka. Vertical displacement of the Last Glacial surface is irregular, in part due to bulges developed where there is a sinistral sidestep in the fault trace, although there is a general decrease in vertical displacement of the surface NE along the fault similar to that in 1855. For the most part the fault is upthrown on the NW, or Rimutaka Range, side but where the fault cuts Cenozoic cover rocks it is sometimes upthrown on the SE side, possibly the result of short wave-length folding in the greywacke undermass.
4. Apart from the 1855 movement, the timing of previous Holocene seismic displacements on the Wairarapa Fault are unknown. However, four earthquake stranded beach ridges at Turakirae Head that are also considered to have been uplifted along the Wairarapa Fault as in 1855, suggests that they may correlate with the four youngest displacements along the inland part of the fault. Dating of the beach ridges based on published ^{14}C ages of shells and wood, a 6.5 ka age of the oldest and highest ridge at 24.5 m above the present day forming storm ridge, a uniform uplift rate of 3.8 m/ka, and uplifts approximating a slip predictable pattern, gives earthquake ages of 1.85, 4.1 and 5.7 yrs BP. These dates imply that: (i). four of the 10 or 11 coseismic displacements on the Wairarapa Fault occurred within the last 6 ka indicating an average slip rate of 6 ± 0.5 m/ka and, (ii). the possibility that slip rate has decreased and/or the Last Glacial surface dextrally displaced by 127 ± 3 m could be older than 11 ± 1 ka.
5. An increase in the dextral/vertical ratio NE along the Wairarapa Fault from between 1.1–1.2 near Turakirae Head to between 1.5–1.8 at Waiohine River 60 km to the NE, reflects a difference in uplift rates of 3.8 m/ka and 1.8 m/ka respectively. Only near Turakirae Head where the axis of the Rimutaka Range anticline intersects the coast, does the vertical faulting rate approach the maximum late Quaternary uplift rate of 4.5 m/ka. The decrease in uplift rate NE along the fault is consistent a N15°E divergence of the fault trace from the axis of the range. On the other hand, the amount of dextral displacement on the fault during individual earthquake events appears to have been fairly constant since the Last Glaciation and is probably controlled by the pattern of total Holocene slip rate partitioning over the Dextral Fault Belt in the southern part of the North Island.

6. A listric Wairarapa Fault geometry allows successive earthquake uplifts to be accommodated as: (i). anticlinal folding near the shallow near-vertical part of the fault and (ii). regional tilting of Mesozoic greywacke basement of the Rimutaka Range by a combination of crustal thickening and anticlockwise rotation along the deeper, shallow dipping part of the fault and crustal flexuring at the eastern margin of the Wanganui Basin some 35 km west of the Wairarapa Fault.

Acknowledgements

I would like to thank Professors Euan Smith and Dick Walcott and Dr. Tim Stern (School of Earth Sciences, Victoria University of Welligton) for constructive criticism, helpful comments and discussion of an early draft of the manuscript.

References

Aston, B.C. (1912): Raised beaches of Cape Turakirae. – Transact. N.Z. Inst. **44**: 208–213.
Beanland, S. (1995): The North Island Dextral Fault Belt, Hikurangi subduction margin, New Zealand. – Unpubl. PhD theses, Geology Department, Victoria University of Wellington.
Bonilla, M.G., R.K. Mark & J.J. Lienkaemper (1984): Statistical relations among earthquake magnitude, surface rupture length, and surface fault displacement. – Bull. Seismolog. Soc. Amer. **74**: 2379–2411.
Cape, L.D., S.H. Lamb, P. Vella, P.E. Wells & D.J. Woodward (1990): Geological structure at Wairarapa Valley, New Zealand, from seismic reflection profiling. – Journ. Roy. Soc. N.Z. **20**: 85–105.
Carter, L., K.B. Lewis & F. Davey (1988): Faults in Cook Strait and their bearing on the structure of central New Zealand. – N.Z. Journ. Geol. Geophys. **31**: 431–446.
Crawford, J. C. (1869): Essay on the geology of the North Island of New Zealand. – Transact. N.Z. Inst. **1**: 27 pp.
Crawford, J.C. 1880: Recollections of travel in New Zealand and Australia. – 468 pp., Trubner, London.
Darby, J.D & S. Beanland (1992): Possible source models for the 1855 earthquake, New Zealand. – Journ. Geophys. Res. **89**: 12375–12389.
Froggatt, P.C. & R. Howarth (1980): Uniformity of vertical faulting for the last 7000 years at Lake Poukawa, Hawkes Bay, New Zealand. – N.Z. Journ. Geol. Geophys. **23**: 493–497.
Ghani, M.A. (1978): Late Cenozoic vertical crustal movements in the southern North Island, New Zealand. – N.Z. Journ. Geol. Geophys. **21**: 117–125.
Gibb, J.G. (1986): A New Zealand regional Holocene eustatic seal-level curve and its implications to determination of vertical tectonic movements. – Roy. Soc. N.Z. Bull. **24**: 377–395.
Grapes, R.H. (1991): Aggradation surfaces and implications for displacements rates along the Wairarapa Fault, southern North Island, New Zealand. – Catena **18**: 453–469.
Grapes, R.H. (1993): Terrace correlation, dextral displacement, and slip rate along the Wellington Fault, North Island, New Zealand. – Comment. N.Z. Journ. Geol. Geophys. **36**: 131–133.
Grapes, R.H. & G. Downes (1997): The 1855 Wairarapa earthquake: analysis of historic data. – Bull. N.Z. Nation. Soc. Earthquake Engin. (in press).
Grapes, R.H., E.F. Hardy & H.W. Wellman (1984): The Wellington, Mohaka and Wairarapa Faults. – Geol.Dept. Publ. **28**: Victoria University of Wellington, 25pp.
Grapes, R.H. & H.W. Wellman (1986): The north-east end of the Wairau Fault, Marlborough, New Zealand. – Journ. Roy. Soc. N.Z. **16**: 245–250.
Grapes, R.H. & H.W. Wellman (1988): The Wairarapa Fault. – Res. School Earth Scis. Publ.**4**, 55pp.
Grapes, R.H. & H.W. Wellman (1993): Field guide to the Wharekauhau Thrust (Palliser Bay) and Wairarapa Fault (Pigeon Bush). – Geol. Soc. N.Z. Miscellaneous Publ. **79B**: 27–44.
Kelsey, H.M., S.M. Cashman, S. Beanland & K.R. Berryman (1995): Structural evolution along the inner forearc of the obliquely convergent Hikurangi margin, New Zealand. – Tectonics **14**: 1–18.
Kingma, J.T. (1962): Sheet 11 – Dannevirke (1st edition). – Geological Map of New Zealand, 1:250,000. Department of Scientific and Industrial Research, Wellington.

KINGMA, J.T. (1967): Sheet 12 – Wellington (1st edition). – Geological Map of New Zealand, 1:250,000. Department of Scientific and Industrial Research.

LENSEN, G.J. (1976): Hillersden and Renwick. – Late Quaternary Map of New Zealand 1:50,000. Department of Scientific and Industrial Research, Wellington.

LENSEN, G.J. (1968): Sheet N158 – Masterton. – Late Quaternary Map of New Zealand 1:63 360. New Zealand Department of Scientific and Industrial Research, Wellington, New Zealand.

LENSEN, G.J. (1970): Sheet N153 – Eketahuna. Late Quaternary Map of New Zealand 1:63 360. New Zealand Department of Scientific and Industrial Research, Wellington, New Zealand.

LENSEN, G.J. & P. VELLA (1971): The Waiohine faulted terrace sequence. – Roy. Soc. N.Z. Bull. **9**: 117–119.

LEWIS, K.B. & J.R. PETTINGA (1993): The emerging imbricate frontal wedge of the Hikurangi margin. – In: BALLANCE, P.F. (ed). South Pacific Sedimentary Basins. – Sedimentary Basins of the World **2**: 225–250. – Elsevier Science Publishers B. V., Amsterdam.

LYELL, Sir CHARLES (1856): Sur les Effects du Tremblement de Terre du 23 Janvier, 1855, a la Nouvelle Zelande. – Bull. Geol. Soc. France, ser. 2, No **13**: 661–667.

LYELL, Sir CHARLES (1868): Principles of Geology. – Vol. 2, 19th ed., 82–89, John Murray, London.

MCFADGEN, B.G. (1980): Age relationship between a Maori plaggen soil and Moa-hunter sites on the west Wellington coast. – N.Z. Journ. Geol. Geophys. **23**: 249–256.

MCKAY, A. (1892): On the geology of Marlborough and the Amuri district of Nelson (Part II). – N.Z. Geol. Surv.Rep. Geol. Explor. 1890–91, **21**: 1–28 (map).

MCKAY, W.A. (1901): Report on the geology of Cook Strait from Pencarrow Head to the Ruamahanga River, and the eastern slopes of the Ruahine Mountains between Tamaki and Makaretu Rivers. – Appendix to the Journal of the House of Representatives **C-10**; 28–34.

MCSAVENEY, M.J. & A.G. HULL (1995): A 7000-year record of great earthquakes at Turakirae Head, Wellington (Abstract). – Subduction systems and processes in New Zealand. New Zealand Geophysical Society Symposium, Sepember 1995, Victoria University of Wellington, Wellington, New Zealand.

MARDEN, M., I.K. PAINTIN, C.M. LEES & V.E. NEALL (1986): Woodville neotectonics and Quaternary stratigraphy field trip. – Geol. Soc. N.Z. Miscellaneous Publ. **35B**: B3 1–20.

MARDEN, M. & V.E. NEALL (1990): Dated Ohakean terraces offset by the Wellington Fault, near Woodville, New Zealand. – N.Z. Journ. Geol. Geophys. **33**: 449–453.

MILNE, J.D.G. (1973): River terraces in the Rangitikei Basin. New Zealand Soil Bureau maps 142/1, 142/2, 142/3, 142/4. – Wellington. New Zealand Department of Scientific and Industrial Research.

MILNE, J.D.G. & I.J. SMALLEY (1979): Loess deposits in the southern part of the North Islandf New Zealand: an outline of stratigraphy. – Acta Geol. Acad. Scient. Hungar. **22**: 197–204.

MIYOSHI, M., D. HERON & K.R. BERRYMAN (1987): Active faults and associated hazards in the Pauatahanui – Waikanae area, northwest Wellington. Unpubl. Report on investigations in sheet R26 to 1987. File 831/30. – Institute of Geological and Nuclear Sciences. (Fault slip-rate data quoted in Beanland 1995).

MOORE, P.R. (1987): Age of the raised beaches at Turakirae Head, Wellington: a reassessment based on radiocarbon dates – Journ. Roy. Soc. N.Z. **17**: 313–324.

ONGLEY, M. (1943): The trace of the 1855 earthquake. – Transact. Roy. Soc. N.Z. **85**: 205–212.

ORBELL, G.E. (1962): Geology of the Mauriceville district, New Zealand. – Transact. Roy. Soc. N.Z. **1**: 253–267.

OTA, Y., D.N. WILLIAMS & K.R. BERRYMAN (1981): Part sheets Q27, R27 & R28 – Wellington. 1st edition. – Late Quaternary tectonic map of New Zealand 1:50 000. With notes. Wellington, New Zealand. Department of Scientific and Industrial Research.

PALMER, A.S. (1984): Quaternary geology of Wairarapa. – Geol. Soc. N.Z., Miscellaneous Publ. **31B**: 20–59.

ROBINSON, R. (1986): Seismicity, structure and tectonics of the Wellington region. – Geophys. Journ. Roy. Astron. Soc. **87**: 379–409.

SHIMAZAKI, K. & T. NAKATA (1980): Time-predictable recurrence model for large earthquakes. – Geophys. Res. Letters **7**: 279–282.

SIEH, K.E. (1978): Slip along the San Andreas fault associated with the great 1857 earthquake. – Bull. Seismol. Soc. Amer. **68**: 1421–1448.

STERN, T.A., G.M. QUINLAN & W.E. HOLT (1992): Basin formation behind an active subduction zone: three-dimensional flextural modelling of Wanganui Basin, New Zealand. – Basin Res. **4**: 197–214.

STEVENS, G.L. (1969): Raised beaches at Turakirae Head, Wellington. New Zealand – Sci. Review **27**: 65–68.

STEVENS, G.L. (1973): Late Holocene marine features adjacent to Port Nicholson, Wellington, New Zealand. – N.Z. Journ. Geol. Geophys. **16**: 455–484.

SUGGATE, R.P. & G.J. LENSEN (1973): Rate of horizontal fault displacement in New Zealand. – Nature **242**: 518–519.

TOMPKINS, J. (1987): Late Quaternary pollen stratigraphy, geology and soils of an area south of Greytown, Wairarapa. – Geol. Soc. N.Z. Newsl. **77**: 5–6.

VAN DER LINGEN, G.J. 1982: Development of the North Island subduction system, New Zealand. – In: LEGGETT, J. K. (ed).: Trench-Forearc Geology. – Geol. Soc. London, Spec. Publ. **10**: 165–272, Blackwell Scientific Publications, Oxford.

VAN DISSEN, R J., K.R. BERRYMAN, J.R. PETTINGA & N.L. HILL (1992): Paleoseismicity of the Wellington-Hutt Valley segment of the Wellington Fault, North Island, New Zealand. – N.Z. Journ. Geol. Geophys. **35**: 165–176.

VELLA, P. (1963): Upper Pleistocene succession in the inland part of the Wairarapa Valley, New Zealand. – Transact. Roy. Soc. N.Z. **4**: 63–78.

VELLA, P., W.KAEWYANA & C.G. VUCETICH (1988): Late Quaternary terraces and their cover beds, northwestern Wairarapa, New Zealand, and provisional correlations with oxygen isotope stages. – Journ. Roy. Soc. N.Z. **18**: 309–324.

WELLMAN, H.W. (1955): New Zealand Quaternary tectonics. – Geol. Rdsch. **43**: 247–257.

WELLMAN, H.W. (1967): Tilted marine beach ridges at Cape Turakirae, New Zealand. – Journ. Geoscis., Osaka City Univers. **10**: 123–129.

WELLMAN, H.W. (1972): Rate of horizontal fault displacement in New Zealand. – Nature **237**: 275–277.

Address of the author: RODNEY GRAPES, Research School of Earth Sciences, Victoria University of Wellington, P.O. Box 600, Wellington, New Zealand.